The Kidney Today

Contributions to Nephrology

Vol. 100

KARGER

Basel · Freiburg · Paris · London · New York · New Delhi · Bangkok · Singapore · Tokyo · Sydney

The Kidney Today

Selected Topics in Renal Science

Volume Editor
G.M. Berlyne, Brooklyn, N.Y.

67 figures and 23 tables, 1992

KARGER

Basel · Freiburg · Paris · London · New York · New Delhi · Bangkok · Singapore · Tokyo · Sydney

Contributions to Nephrology

Library of Congress Cataloging-in-Publication Data
The Kidney today: selected topics in renal science / volume editor, G.M. Berlyne.
(Contributions to nephrology; vol. 100)
Includes bibliographical references and index. (alk. paper)
1. Kidneys – Pathophysiology. 2. Kidneys – Physiology.
I. Berlyne, Geoffrey Merton. II. Series: Contributions to nephrology; v. 100.
[DNLM: 1. Kidney. 2. Kidney Diseases. W1 CO778UN v. 100 / WJ 300 K4585]
RC903.9.K535 1992 616.6′1 – dc20
ISBN 3–8055–5555–5

Bibliographic Indices
This publication is listed in bibliographic services, including Current Contents® and Index Medicus.

Drug Dosage

The authors and the publisher have exerted every effort to ensure that drug selection and dosage set forth in this text are in accord with current recommendations and practice at the time of publication. However, in view of ongoing research, changes in government regulations, and the constant flow of information relating to drug therapy and drug reactions, the reader is urged to check the package insert for each drug for any change in indications and dosage and for added warnings and precautions. This is particularly important when the recommended agent is a new and/or infrequently employed drug.

Printed in Switzerland on acid-free paper by Thür AG Offsetdruck, Pratteln
ISBN 3–8055–5555–5

Contents

Preface

This is the 100th volume of the series *Contributions to Nephrology* which we founded 20 years ago as a vehicle of publishing meetings and symposia on all aspects of renal disease. Looking back over these volumes it is apparent that the majority of them recorded important milestones in the progress of understanding the kidney and kidney disease.

Since its inception the aim of the series has been to provide a convenient place for the rapid publication of state-of-the-art original work in the rapidly developing field of nephrology. To this end we have been aided by the wise counsel of an international editorial board whose advice has been crucial in making certain decisions, and as a result *Contributions to Nephrology* has fulfilled a role of presenting seminal data and ideas in the course of their being worked out, as presented at a symposium and before they have had the final polish put on them by the difficult process of multiple peer review resulting in their eventual appearance in the major nephrological journals. This can be verified by looking at the citation index in which *Contributions to Nephrology* has had remarkably high figures compared to similar but more review type publications in the same field. We have avoided the comprehensive and to our mind utterly boring reviews which are commonly published elsewhere, which are a year or more out of date by the time they are published, and are as intellectually stimulating as diazepam. Their main use consists of providing a list of references, and this is no longer necessary with the advent of Current Contents on floppy discs. We have also avoided purely local meetings as far as possible, so as to ensure a higher standard of original papers with the essentially international flavor of discussion so important to the stimulation of ideas, and the facilitation of excellence. Our choice of topics has been guided by two principles; first, no rehashing of old previously published well-known material disentombed for the conference by a celebrity who may have passed his best and most productive years, or who may talk at too many conferences, and second, that the subjects dealt with by the conference should be of either great clinical importance or be fundamental to our understanding of nephrology in its myriad aspects.

We conceived our purpose to be the presentation of clear advances and important new principles to the discerning nephrologist reading *Contributions to Nephrology*, and the fulfillment of an important continuing education role in this science. We have found that its articles are frequently chosen in Journal Clubs as the basis for not only widening the knowledge of the participants, but also for critical analysis of the facts presented. These relatively short articles are rapidly assimilated by the trainee nephrologist, whose fund of knowledge is improved pari passu with a sharpening of his critical faculties in reading scientific papers.

How far have we succeeded? The answer to this is purely subjective. The issues of *Contributions to Nephrology* are well received at major scientific meetings and are widely read and quoted. It is best left to the reader to decide which of the 100 volumes so far produced has been the most intellectually stimulating, or which article has been the best from the point of view of having succeeded in pushing the boundaries of knowledge further, or having influenced clinical nephrological practice.

For this 100th volume, we have invited authorities from around the world who have made serious and important contributions to our understanding of nephrology, to take part in this volume and write on a subject which they felt to be important. We have deliberately included controversial, refreshing and sometimes iconoclastic views in the articles which form this book. The chapters on acid base are particularly interesting in their diametrically opposed views. We hope you will enjoy the book and be stimulated by the wide variety of controversial topics covered.

We wish to thank Dr. T. Karger and the staff of S. Karger AG for much help in the production of this volume at all stages.

G.M. Berlyne
S. Giovannetti

Berlyne GM (ed): The Kidney Today. Selected Topics in Renal Science.
Contrib Nephrol. Basel, Karger, 1992, vol 100, pp 1–14

Renal Hypouricemia: Classification, Tubular Defect and Clinical Consequences

Oded Sperling

Department of Clinical Biochemistry, Beilinson Medical Center, Petah-Tikva, and Department of Chemical Pathology, Sackler Faculty of Medicine, Tel Aviv University, Israel

Hypouricemia in itself has no pathological consequences. Nevertheless, it may reflect pathological conditions, hereditary or acquired, some of which are severe or malignant. Etiologically, hypouricemia may represent two classes of underlying mechanisms: metabolic or renal. In metabolic hypouricemia, there is hereditary or drug-induced interference in uric acid production, more specifically in the degradative rather than the synthetic pathway of purine nucleotides. The two known hereditary conditions of metabolic hypouricemia are the deficiency of purine nucleoside phosphorylase, and the deficiency of xanthine oxidase (xanthinuria). Treatment by allopurinol (a potent inhibitor of xanthine oxidase) is the best example for drug-induced metabolic hypouricemia. In these conditions, both serum level of urate and urinary uric acid excretion are low (extremely low in the hereditary conditions), but renal urate clearance is relatively normal. In contrast, in renal hypouricemia, uric acid production is normal and urinary uric acid excretion is normal or elevated (see below), but serum urate is low, reflecting increased renal urate clearance, due to abnormal renal tubular handling of urate. Renal hypouricemia is the subject of this review. In the following, I shall elaborate on the various types of renal hypouricemia, on the nature of the renal tubular defect and on the clinical pathological manifestations.

The Various Types of Renal Hypouricemia

Hereditary Renal Hypouricemia

The term hereditary renal hypouricemia refers to the rare hereditary condition of increased renal urate clearance, caused by a specific (isolated) inborn error of membrane transport for urate in the renal proximal tubule, manifested in hypouricemia [1]. This defect is inherited in an autosomal-recessive mode. In homozygotes the defect is fully manifested in hypouricemia and increased renal urate clearance. Heterozygosity may be detected by moderately decreased serum urate levels and moderately, but significantly increased renal urate clearance. There are several other hereditary conditions of renal hypouricemia, which differ from the above in that the urate transport defect is only one component in a generalized disturbance of membrane transport. The most common of these is the Fanconi syndrome [2], including Wilson disease, cystinosis, galactosemia and hereditary fructose intolerance. Another such condition is the Hartnup syndrome [3]. In this group of diseases, uric acid reabsorption in the proximal tubule is decreased along with other crystalloid solutes.

Acquired Renal Hypouricemia

This type may accompany conditions of extracellular fluid volume expansion, such as in inappropriate antidiuretic hormone secretion [4], and in various malignancies, such as multiple myeloma [5], lymphomas [6] and pulmonary neoplasms [7], in heavy metal intoxication [8], following use of outdated tetracyclines [9], in liver diseases such as jaundice [10] and cirrhosis [11], and in patients with severe burns [12]. In some of the conditions of acquired renal hypouricemia, the renal abnormality exhibits a transientory nature. In Hodgkin's disease, the degree of renal urate clearance was found to correlate with the activity of the neoplastic process [7]. In jaundice and burns, the degree of renal urate clearance was shown to normalize with recovery [10, 12] and in cirrhosis, an inverse correlation was found between serum bilirubin and urate [11].

Renal Handling of Urate in Man

Experiments in man in order to clarify the renal handling of urate were limited to the study of the effects of urate loading [13, 14], and of various drugs [15, 16] on uric acid excretion. In addition, important information was furnished by the inborn defects in renal urate reabsorption in man. nevertheless, the exact nature of the renal handling of urate in man has not been clarified conclusively, mainly due to the lack of an animal model, in which the renal handling of urate is identical to that in man.

Glomerular Filtration of Uric Acid

Urate is probably bound to plasma proteins, the amount of binding being approximately 5–10% [17, 18]. Practically, urate is freely filtered at the glomerulus [19].

Reabsorption of Uric Acid

In man, the fractional clearance of urate, in relation to that of inulin or creatinine, is about 7–10%, indicating very efficient reabsorption. In all animals in which urate reabsorption has been localized, including the Cebus monkey [20], it was found to occur largely in the proximal tubules. Urate reabsorption is an active transport process. This is evident from the finding in free flow, micropuncture studies, that the ratio between urate concentration in the proximal tubule fluid to that in plasma is less than 1 [21]. There is additional experimental evidence for this [22]. Studies in the rat in vivo and in vesicles in vitro, demonstrated that urate reabsorption is inhibited by a number of substances, such as probenecid, furosemide, and other substances, indicating that urate reabsorption is mediated by an anion-exchange mechanism.

Secretion of Uric Acid

The first evidence for tubular urate secretion in man was found in 1950 [23], in the first subject with renal hypouricemia, in whom the fractional urate clearance (FC_{ur}) was 1.46. This early evidence for the bidirectional tubular transport of urate was subsequently augmented by studies in man, in which the FC_{ur} was increased artificially, by use of urate loading, mannitol diuresis and probenecid, to as high as 1.23 [14]. In man, the exact localization of urate secretion has not been conclusively established. In one study [24], the proximal nephron was found to transport urate to the tubular fluid, whereas the distal nephron was unable to do so. In vivo microperfusion and microinjection studies indicated that similar to the reabsorption of urate, the secretory process is a result of an anion-exchange mechanism [25]. An interesting finding in the rabbit is that urate secretion is modulated by a serum protein which affects the basolateral transporter by allosteric modification [26]. This finding may be taken to suggest that abnormalities in renal urate handling may also reflect primary defects in such modulators of secretion or reabsorption (see below).

Localization of the Various Transport Components in the Tubule

The development of the understanding of the renal handling of urate in man generated several models. The first model included two components, glomerular filtration followed by tubular reabsorption [13]. The second model included in addition the component of secretion [27]. More recently, a fourth component, that of postsecretory reabsorption, was added [28–31] in order to explain the effects of pyrazinamide in some patients or conditions [6, 27], which otherwise, according to the three-component model, could only be interpreted as enhanced tubular secretion of urate. The four-component model (fig. 1) includes glomerular filtration, early proximal reabsorption, later proximal secretion, and extensive post-secretory reabsorption, which could occur at the same location as the secretion, or separate and distal to it, or both. Recently, a similar model was proposed, including the same four components, but according to which, reabsorption and secretion of urate occur simultaneously along all the proximal tubule, each at different intensities at the different segments of the tubule [32].

Factors that Affect Renal Urate Clearance

Urate excretion becomes less efficient with age [33, 34], and the mean clearance of urate is 1.2–2.3 ml/min higher in females than males [35]. The extracellular fluid volume (or more precisely, the volume of the effective arterial circulation), is the most dominant factor in renal urate handling. Volume expansion increases urate clearance, whereas volume contraction decreases urate clearance. Presumably, these effects of changes in extracellular volume are mediated through alterations in urate reabsorption in the proximal tubule [36, 37].

Many drugs affect renal urate excretion [32, 38]. A biphasic effect was documented for some of the drugs and suggested for all others [38]. Pyrazinamide, probenecid, phenylbutazone and salicylate inhibit urate secretion at low doses, but inhibit urate reabsorption at high doses [27, 39]. Diuretic drugs are initially uricosuric by direct inhibition of urate reabsorption, but their chronic administration is associated with contraction of

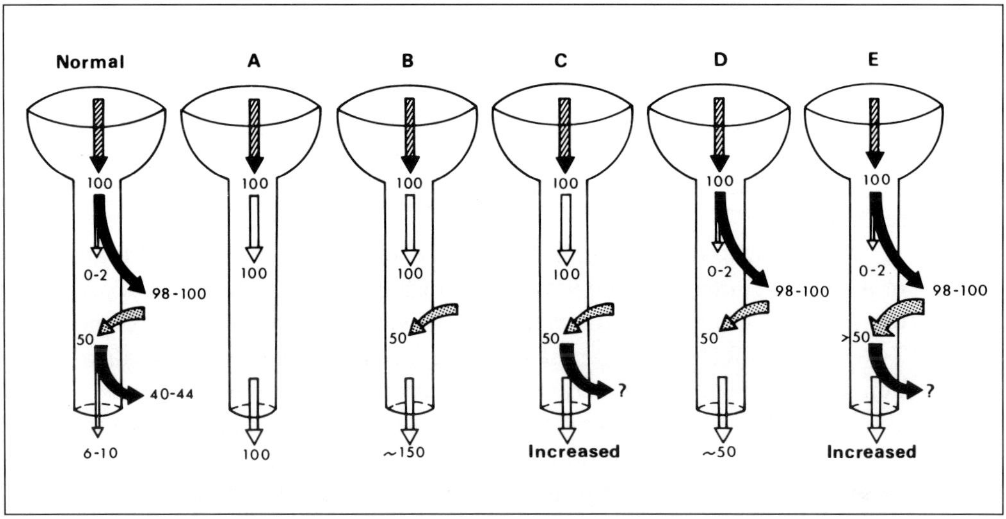

Fig. 1. The model for the renal handling of urate in normal man and the five possible defects (A–E) that may cause renal hypouricemia (see text for detailed explanation). The size and direction of arrows indicate intensity and direction of urate transport. The hatched arrows represent filtered urate; the solid arrows urate reabsorption; the dotted arrows urate secretion and the open arrows urate remaining in the tubular fluid after reabsorption. Numerical values indicate hypothetical order of magnitude of the transport process. The normal model is based on the model suggested by Rieselbach and Steele [31].

the extracellular fluid volume resulting in antiuricosuria [27]. An addition property of some of the diuretic drugs is their tendency to compete with tubular urate secretion, decreasing further urate excretion [40].

Nature of the Renal Tubular Defect in Renal Hypouricemia

The following discussion is based on the assumption that renal hypouricemia reflects a primary defect in one or more of the components of urate transport in the renal tubules. Nevertheless, the possibility that the renal tubular abnormality is secondary to a metabolite produced elsewhere in the body, should be kept in mind (see below). Increased renal urate clearance could be caused by defective reabsorption or by increased secretion. Based on the model of the renal handling of urate in man suggested by Rieselbach and Steele [31], five main possibilities of specific abnormalities in the active urate transport processes were considered as causing renal hypouricemia [1] (fig. 1), namely: (A) total transport defect (no reabsorption and no secretion); (B) total reabsorption defect; (C) presecretory reabsorption defect; (D) postsecretory reabsorption defect, and (E) increased secretion.

According to the four-component model of the renal handling of urate in man [28–31] (fig. 1), one may distinguish between the above types of transport defects by use of drugs inhibiting urate reabsorption or secretion. Two drugs, probenecid and pyrazinamide, are employed for the localization of the tubular defect in renal hypouricemia. Probenecid is employed as an inhibitor of the postsecretory reabsorption and pyrazinamide as an inhibitor or uric acid secretion. For detailed description of the effect of these drugs on uric acid transport in the renal tubule and of the reservations concerning the use of these drugs for the localization of the transport defect, the reader is referred to a previously published review [1]. The five types of renal hypouricemia would be expected to conform to the following responses of the drugs (fig. 1): In a total transport defect (type A), FC_{ur} should be about 1, and this value should not be altered by administration of pyrazinamide or probenecid. In a total reabsorption defect but presence of normal secretion (type B), FC_{ur} should be greater than 1. In such a defect, blocking secretion by pyrazinamide will result in an attenuated decrease in FC_{ur}, which should approach the value of 1. Administration of probenecid will not alter FC_{ur}. In a presecretory reabsorption defect (type C), administration of pyrazinamide will decrease FC_{ur} (which is probably greater than 1) and administration of probenecid will increase this parameter, but both effects will be attenuated. In defective reabsorption at the postsecretory site (type D), FC_{ur} will be smaller than 1, if the amount secreted is smaller than filtered. In presence of such a defect, administration of pyrazinamide will decrease FC_{ur} to normal level, but administration of probenecid will have no effect. In case of a defect manifested in increased secretion (type E), FC_{ur} will probably be smaller than 1 (unless the secretion is increased to such a level that its fraction escaping reabsorption will exceed the amounts of filtered urate). In such a defect, the administration of pyrazinamide and of probenecid will result in normal response.

Nineteen propositi with hereditary renal hypouricemia due to an isolated defect were studied for the effects of the drugs [1]. Twelve of these patients could definitely be classified with type C (presecretory reabsorption defect), 6 with type C or type A (total transport defect), and 1 with type B (total reabsorption defect). None of the propositi could be classified with type D (postsecretory reabsorption defect) or E (increased secretion).

The pyrazinamide and probenecid effects were studied also in patients with the other conditions of renal hypouricemia, such as the Fanconi's syndrome [37], Wilson's disease [42], Hodgkin's disease [6] and in a case of hyperparathyroidism [43]. Most if not all of these subjects could be classified as type D (postsecretory reabsorption defect) [1].

The above data may be taken to suggest that type C, the presecretory reabsorption defect, is the most common (if not the only type) among the subjects with hereditary (isolated) renal hypouricemia, whereas type D (postsecretory reabsorption defect) is the most common in patients with the acquired renal hypouricemia and in patients with hereditary renal hypouricemia, associated with a generalized renal tubular reabsorption defect.

The Possibility of a Nonrenal Primary Defect

The possibility that the renal tubular abnormalities in some or all types of renal hypouricemia may be secondary to an abnormal metabolite produced elsewhere in the body, or to qualitative or quantitative alterations in modulators affecting urate transport processes, should be taken into consideration. Indeed, the presence of a humoral uricosuric factor may be especially possible in some conditions with acquired renal hypouricemia, in which the transientory nature of the defect was demonstrated, such as severe burns [12] or Hodgkin's disease [6]. Nevertheless, until today, no experimental evidence could be obtained for the presence of a uricosuric agent in plasma of patients with acquired or hereditary renal hypouricemia [44].

Urate Transport in Nonrenal Tissue of Subjects with Hereditary Renal Hypouricemia

Uric acid transport into erythrocytes was studied in 5 hypouricemic subjects [45, 46]. In all these subjects, urate uptake by the erythrocytes, both total uptake as well as that inhibited by hypoxanthine (active transport), was normal. The intestinal absorption of uric acid was gauged in 1 hypouricemic subject and found similar to that obtained in 2 control subjects [45].

Clinical Consequences

The clinical consequences in renal hypouricemia are the manifestations of the hyperuricosuria, which appears to be a constant feature of this syndrome [1, 47]. The hyperuricosuria reflects diversion of intestinal urate elimination to urinary urate excretion, consequent to the hypouricemia. Uricolysis by intestinal bacteria is responsible for elimination of about 33% of the uric acid turnover each day [48]. The uric acid enters the alimentary tract form the blood in saliva, gastric juice, and bile. Accordingly, blood urate level influences the amount of uric acid secreted into the gut. Therefore, in hyperuricemia the fraction of uric acid produced, which is eliminated in the gut, is greater than in normal subjects, whereas in hypouricemia this fraction is markedly smaller. Accordingly a greater fraction of the uric acid produced leaves the body intact via the kidneys, manifested in hyperuricosuria. There is no evidence in the hypouricemic hyperuricosuric subjects for purine overproduction [49, 50]. The hyperuricosuria may be associated with uric acid lithiasis, or uric acid nephropathy.

Urolithiasis

Five out of the 21 propositi with inborn isolated renal hypouricemia [1] had urinary calculi: 3 uric acid stones, 1 calcium oxalate stone and 1 stone of unidentified composition. In 4 other propositi [51], urolithiasis was present in other family members. The high prevalence of urolithiasis among the subjects with hereditary renal hypouricemia may be explained by the high prevalence of hyperuricosuria. Hyperuricosuria is the most common cause for uric acid/urate supersaturation in urine, and thus it is considered as the underlying cause of the urolithiasis in renal hypouricemia [52].

Uric Acid Nephropathy

Uric acid nephropathy is distinguished from urate nephropathy in that in the former there is precipitation of uric acid in the collecting tubules, pelvis, or ureter, with subsequent impairment of urine flow, whereas in the latter, there is precipitation of monosodium urate in the renal interstitial tissue. Thus, uric acid nephropathy is associated with hyperuricosuria, whereas urate nephropathy is associated with hyperuricemia.

Acute uric acid nephropathy is relatively common in presence of massive uricosuria, such as in the tumor lysis syndrome, but it may occur also in some gouty subjects with excessive purine overproduction, especially in association with inborn errors of metabolism, such as the Lesch-Nyhan syndrome [53] and phosphoribosylpyrophosphate synthetase superactivity [54].

In view of the prevalence of hyperuricosuria in renal hypouricemia, the possibility of acute uric acid nephropathy in this syndrome should be taken into consideration. Indeed, one such case was reported recently [55], in a patient requiring hemodialysis due to oliguric renal failure. Renal biopsy showed amorphous uric acid crystals in some of the tubular lumina and mild to moderate interstitial inflammation. In this case the hyperuricosuria was attributed to the increased tubular uric acid secretion.

Very recently, Ishikawa et al. [56] reported mild acute renal failure induced by exercise, in 3 subjects with renal hypouricemia. The clinical manifestations in these patients resembled very much the acute renal failure syndrome described by this group in 1982 [57]. This syndrome, characterized by patchy renal vasoconstriction is usually mild, nonoliguric, appearing in young healthy subjects, and has a good prognosis. There is only mild to moderate rise in CPK, suggesting that massive rhabdomyolysis does not occur. In the 3 reported subjects, the renal failure was attributed to the precipitation of uric acid, which however could not be demonstrated in the affected patients.

Calcium Metabolism in Hereditary Renal Hypouricemia

Idiopathic hypercalciuria was found in 6 of the propositi with hereditary renal hypouricemia (isolated defect) [58–62], and in 3 of them [58, 60] it was proven to be of the hyperabsorptive type. The apparently frequent asssociation between the renal defect for urate handling and the intestinal calcium hyperabsorption is as yet unexplained.

Decreased bone density was found to be associated with the hypercalciuria in 2 of the renal hypouricemic sibs [61] and osteoporosis was found in a 4-year-old boy with hereditary renal hypouricemia [51]. Similar abnormalities have not been observed in the other reported hereditary hypouricemic families (isolated type). The association between renal hypouricemia and decreased bone density remains therefore unexplained.

Animal Model for Renal Hypouricemia

The Dalmatian coach hound is a breed of dogs which differs from other dogs in that it is relatively hyperuricemic and hyperuricosuric and exhibits increased renal urate clearance [63–68]. The relative hyperuricemia-hyperuricosuria in this breed of dogs was demonstrated to reflect a defective uric acid transport into hepatocytes, the uricase-containing tissue, but the reason for the increased renal urate clearance has not yet been clarified conclusively. It could reflect a urate transport defect, in the kidney, or in all tissues [63]. On the other hand, data are available (liver transplantation experiments performed by Kuster et al. [67, 68], suggesting that the increased renal urate clearance in this dog reflects the presence of a uricosuric substance produced in the Dalmatian liver.

Conclusion

The term renal hypouricemia refers to all conditions of low serum urate due to increased renal urate clearance. The increased clearance may reflect an hereditary or acquired renal tubular transport defect, isolated for uric acid or in combination with other transport defects. The present understanding of the renal handling of urate in man includes four components: glomerular filtration, early proximal reabsorption, later proximal secretion and postsecretory reabsorption. Studying the effect of drugs inhibiting urate reabsorption or secretion, revealed that in most patients with the hereditary isolated abnormality, the defect appears to be in the presecretory urate reabsorption site, whereas in the patients with hereditary, but generalized transport defect, as well as in the patients with acquired renal hypouricemia, the defect seems to be in the postsecretory urate reabsorption site. Renal hypouricemia is characterized by mild or massive uricosuria, reflecting diversion of intestinal urate elimination to urinary urate excretion, consequent to the hypouricemia. The hyperuricosuria may cause precipitation of uric acid in the kidneys and urinary tract, resulting in uric acid urolithiasis, or uric acid nephropathy. The latter may lead to acute renal failure.

References

1 Sperling O: Hereditary renal hypouricemia; in Scriver CR, Beaudet AK, Sly WS, Valle D (eds): The Metabolic Basis of Inherited Disease, ed 6. McGraw-Hill, 1989, p 2605.

2 Wallis IA, Eagle RI: The adult Fanconi syndrome. II. Review of eighteen cases. Am J Med 1957;22:13–23.

3 Baron DN, Dent CE, Harris H, Hard EW, Jebson JB: Hereditary Pellagra-like skin rash with temporary cerebellar ataxia, constant renal aminoaciduria and other bizarre biochemical features. Lancet 1956;271:421–428.
4 Beck IH: Hypouricemia in the syndrome of inappropriate secretion of antidiuretic hormone. N Engl J Med 1979;301:538.
5 Smithline N, Kassirer JP, Cohen JJ: Light-chain nephropathy. N Engl J Med 1976; 294:71–74.
6 Bennett JS, Bond J, Singer I, Gottlieb AJ: Hypouricemia in Hodgkin's disease. Ann Intern Med 1952;76:751–756.
7 Weinstein B, Irreverre F, Watkin DM: Lung carcinoma, hypouricemia and amino-aciduria. Am J Med 1965;39:520–526.
8 Chisholm JJ Jr, Harrison HAC, Everlein WR, Harrison HE: Amino aciduria, hypophosphatemia and rickets in lead poisoning; in Lawson RB, Weech AA, Smyth FS, Kennedy RIJ, Wilson JI, Zuelzer WW (eds): American Journal of Diseases of Children. Chicago, American Medical Association, 1955, p 159.
9 Gross JM: Fanconi syndrome (adult type) developing secondary to the ingestion of outdated tetracycline. Ann Intern Med 1963;48:523–528.
10 Schlosstein L, Kippen I, Bluestone R, Whitehouse MW, Klineberg JR: Association between hypouricemia and jaundice. Ann Rheum Dis 1974;33:308–312.
11 Michelis MF, Warms PC, Fusco RD, Davis BB: Hypouricemia and hyperuricosuria in Laënnec cirrhosis. Arch Intern Med 1974;134:681–683.
12 Weinberger A, Weinberger A, Sperling O, Ben-Bassat M, Kaplan I, Pinkhas J: Increased uric acid clearance in patients with burns. Biomed Express 1977;27:277–278.
13 Berliner RW, Hilton JG, Tu TF, Kennedy TJ Jr.: The renal mechanism for urate excretion in man. J Clin Invest 1950;29:396–401.
14 Gutman AB, Tu TF, Berger L: Tubular secretion of urate in man. J Clin Invest 1959; 38:1778–1781.
15 Yu TF, Berger L, Stone DJ, Wolf J, Gutman AB: Effect of pyrazinamide and pyrazinoic acid on urate clearance and other discrete renal functions. Proc Soc Exp Biol (NY) 1957;96:264–267.
16 Steele TH, Rieselbach RE: The renal mechanism for urate homeostasis in normal man. Am J Med 1967;43:868–875.
17 Abramson RG, Levitt MF: Micropuncture study of uric acid in rat kidney. Am J Physiol 1975;228:1597–1605.
18 Wyngaarden JB, Kelly WN: Gout; in Stanbury JB, Wyngaarden JB, Fredrickson DS, Goldstein JL, Brown MS (eds): The Metabolic Basis of Inherited Disease. New York, McGraw-Hill, 1983, p 1043.
19 Kahn AM, Weinman EJ: Urate transport in the proximal tubule: in-vivo and vesicle studies. Am J Physiol 1985;249:F789–F798.
20 Roch-Ramel F, Weiner IM: Excretion of urate by the kidney of Cebus monkeys: A micropuncture study. Am J Physiol 1973;224:1369–1374.
21 De Rougemont D, Henchoz M, Roch-Ramel F: Renal urate excretion at various plasma concentrations in the rat: a free-flow micropuncture study. Am J Physiol 1976;231:387–392.
22 Weiner IM, Fanelli GM Jr: Renal urate excretion in animal models. Nephron 1975; 14:33–47.

23 Praetorius E, Kirk JE: Hypouricemia with evidence for tubular elimination of uric acid. J Lab Clin Med 1950;35:865–868.
24 Podevin R, Ardaillou R, Paillard F, Fontannele J, Richet G: Etude chez l'homme de la cinetique d'apparition dans purine de l'acide urique 2-^{14}C. Nephron 1968;5:134–140.
25 Weinman EJ, Sansom SC, Steplock DA, Sheth AU, Knight TF, Senekjian HO: Secretion of urate in the proximal convoluted tubule of the rat. Am J Physiol 1980; 239:F393–F387.
26 Shimomura A, Chonko A, Tanner RM, Edwards R, Grantham JJ: Nature of urate transport in isolated rabbit proximal tubules. Am J Physiol 1981;241:F565–F578.
27 Gutman AB, Yu TF: A three-component system for regulation of renal excretion of uric acid in man. Trans Assoc Am Physicians 1961;74:353–364.
28 Steele TH, Boner G: Origins of the uricosuric response. J Clin Invest 1973;52:1368–1375.
29 Steele TH: Urate secretion in man: The pyrazinamide suppression test. Ann Intern Med 1973;79:734–737.
30 Diamond HS, Paolini JS: Evidence for a post-secretory reabsorptive site for uric acid in man. J Clin Invest 1973;52:1491–1499.
31 Rieselbach RE, Steele TH: Influence of the kidney upon urate homeostasis in health and disease. Am J Med 1974;56:665–675.
32 Grantham JJ, Chonko AM: Renal handling of organic anions and cations; metabolism and excretion of uric acid; in Brenner BM (ed): The Kidney, ed 3. Philadelphia, Saunders, 1986, pp 663–700.
33 Harkness RA, Nicol AD: Plasma uric acid levels in children. Arch Dis Child 1969; 44:773–778.
34 Stapelton FB: Renal uric acid clearance in human neonates. J Pediatr 1983;103: 290–294.
35 Wolfson WO, Hunt HJ, Levine E, Gutterman HS, Cohn C, Rosenberg EF, Huddlestun B, Kadota IC: The transport and excretion of uric acid in man. V. A sex differential in urate metabolism with a note on clinical and laboratory findings in gouty women. J Clin Endocrinol 1949;9:749–767.
36 Steele TH, Oppenheimer S: Factors affecting urate excretion following diuretic administration in man. Am J Med 1969;47:563–573.
37 Steele TH, Manuel MA, Boner G: Diuretics, urate excretion and sodium reabsorption: A test of acetazolamide and urinary alkalinization. Nephron 1975;14:48–61.
38 Emmerson BT: Abnormal urate excretion associated with renal and systemic disorders, drugs and toxins; in Kelley WN, Weiner IM (eds): Handbook of Experimental Pharmacology, Uric Acid. Berlin, Springer, 1978, vol 51, p 287.
39 Yu TF, Gutman AB: Study of the paradoxical effects of salicylate in low, intermediate and high dosage on the renal mechanisms for excretion of urate in man. J Clin Invest 1959;38:1298–1315.
40 Stewart RJ, Chonko AM: Pharmacologic inhibition of urate transport across perfused and non-perfused rabbit proximal straight tubules. Kidney Int 1981;19:258.
41 Meisel AD, Diamond HS: Hyperuricosuria in the Fanconi syndrome. Am J Med Sci 1977;273:109–115.
42 Wilson DB, Goldstein NP: Renal urate excretion in patients with Wilson's disease. Kidney Int 1973;4:331–336.

43 Gibson T, Sims HP, Jimenez SA: Hypouricemia and increased renal urate clearance associated with hyperparathyroidism. Ann Rheum Dis 1976;35:372–376.
44 Kay NE, Gotleib AJ: Hypouricemia in Hodgkin's disease: Report of an additional case. Cancer 1973;32:1508–1511.
45 Sperling O, Boer P, Weinberger A, De Vries A: Transport into erythrocytes and intestinal absorption of uric acid in hereditary renal hypouricemia. Biomedicine 1975;23:157–159.
46 Vinay P, Gatterean A, Moulin B, Gougoux A, Lemieux G: Normal urate transport into erythrocytes in familial renal hypouricemia and in Dalmatian dog. Can Med Assoc J 1983;128:545–549.
47 Sperling O: Urate deposition and stone formation in renal hypouricemia; in Zollner N, Gresser U (eds): Urate Deposition in Man and Its Clinical Consequences. Berlin, Springer, 1991.
48 Sorensen LB: Degradation of uric acid in man. Metabolism 1959;8:687–703.
49 Akaoka I, Nishizawa T, Yano E, Kamatani N, Nishida T, Sasaki S: Renal urate excretion in five cases of hypouricemia with an isolated renal defect of urate transport. J Rheumat 1977;4:86–94.
50 Kawabe K, Muryama T, Akaoka I: A case of uric acid renal stone with hypouricemia caused by tubular reabsorption defect of uric acid. J Urol 1976;116:690–692.
51 Takeda E, Kuroda T, Ito M, Toshima K, Watanabe T, Ito M, Naiko E, Yokota I, Huwang TJ, Miyao M: Hereditary renal hypouricemia in children. J Pediatr 1985; 107:71–74.
52 Sperling O: Uric acid nephrolithiasis; in Wickham JEA, Buck AC (eds): Renal Tract Stone: Metabolic Basis and Clinical Practice. London, Churchill Livingstone, 1990.
53 Stout JT, Caskey CT: Hypoxanthine-guanine phosphoribosyltransferase deficiency: The Lesch-Nyhan syndrome and gouty arthritis; In Scriver CR, Beandet AC, Sly WS, Valee D (eds): The Metabolic Basis of Inherited Disease, ed 6. New York, McGraw-Hill, 1989.
54 Sperling O, Boer P, Brosh S, Zoref E, De Vries A: Superactivity of phosphoribosylpyrophosphate synthetase due to feedback resistance, causing purine overproduction and gout; in Purine and Pyrimidine Metabolism. Ciba Found Symp 48. Amsterdam, Elsevier/Excerpta Medica/North-Holland, 1977.
55 Erley Ch MM, Hirschberg RR, Hoefer W, Schaefer K: Acute renal failure due to uric acid nephropathy in a patient with renal hypouricemia. Klin Wochenschr 1989;67: 308–312.
56 Ishikawa I, Sakurai Y, Masuzaki S, Sugishita N, Shinoda A, Shikura N: Exercise-induced acute renal failure in 3 patients with renal hypouricemia. Jap J Nephrol 1990;32:923–928.
57 Ishikawa I, Onoudi Z, Yuri T, et al: Acute renal failure with severe loin pain and patchy renal vasoconstrictions; in Eliahou HE (ed): Acute Renal Failure. London, Libbey, 1982, p 224.
58 Frank M, Many M, Sperling O: Familial renal hypouricemia: two additional cases with uric acid lithiasis. Br J Urol 1979;51:88–91.
59 Weitz R, Sperling O: Hereditary renal hypouricemia: Isolated tubular defect of urate reabsorption. J Pediatr 1980;96:850–853.

60 Greene ML, Marcus R, Aurbach GD, Kazam ES, Seegmiller JH: Hypouricemia due to isolated renal tubular defect. Am J Med 1972;53:361–367.
61 Sperling O, Weinberger A, Oliver I, Liberman UA, De Vries A: Hypouricemia, hypercalciuria and decreased bone density: a hereditary syndrome. Ann Intern Med 1974;80:482–487.
62 Garty BZ, Nitzan M, Sperling O: Inborn hypouricemia due to isolated defect in renal tubular uric acid transport. Israel J Med Sci 1981;17:295–297.
63 Briggs OM, Sperling O: Uric acid metabolism in the Dalmatian coach hound. J South Afr Vet Assoc 1982;53:201–204.
64 Friedman M, Byers SD: Observations concerning the causes of the excess excretion of uric acid in the Dalmatian dog. J Biol Chem 1948;175:727–735.
65 Kessler RH, Hierholzer K, Gurd RS: Localization of urate transport in the nephron of mongrel Dalmatian dog kidney. Am J Physiol Ther 1959;197:601–603.
66 Mudge GH, Gucchi J, Platts M, Brian O'Connell JM, Berndt WO: Renal excretion of uric acid in the dog. Am J Physiol 1968;215:404–410.
67 Kuster G, Shorter RG, Dawson B, Hallenbeck GA: Effect of allogenic hepatic transplantation between Dalmatian and mongrel dogs on urinary excretion of uric acid. Surg Forum 1967;18:360–362.
68 Kuster G, Shorter RG, Dawson B, Hallenbeck GA: Uric acid metabolism in Dalmatian and other dogs. Arch Intern 1972;129:492–496.

Oded Sperling, PhD, Department of Clinical Biochemistry,
Beilinson Medical Center, Petah-Tikva 49100 (Israel)

Berlyne GM (ed): The Kidney Today. Selected Topics in Renal Science.
Contrib Nephrol. Basel, Karger, 1992, vol 100, pp 15–24

Phagocytic Function in the Uremic Patient

S. Ringoir, R. Vanholder

Division of Nephrology, Department of Internal Medicine, University Hospital, Ghent, Belgium

Phagocytosis is the process whereby certain cells ingest foreign material, and enclose it within a cytoplasmatic vacuole named phagosome, which is subsequently destroyed. The role of phagocytosis in the immune response was underestimated for a long period of time. The importance of phagocytosis in the host defense against infection and cancer recently became more and more apparent, however. The uremic patient displays, as known, an enhanced susceptibility to infection as well as an increased incidence of cancer.

Phagocytic cells can be subdivided into polymorphonuclear granulocytes, monocytes, eosinophilic granulocytes, and macrophages. Whereas monocytes only circulate in the blood during a short period of time, they remain present for several consecutive months when they reside in the tissues: they then are called macrophages. Macrophages can be found in the liver (Kupffer cells), in the spleen, in the brain (glia cells), the lungs, the glomeruli (mesangial cells), the serosa, and the lymph nodes.

Phagocytes not only have the capacity to kill and destroy bacteria, but they also produce so-called cytokines that have the ability to activate other cells of the immune system such as T and B lymphocytes, or that may induce general reactions. Interleukin-1 (IL-1), produced by macrophages, provokes fever by the stimulation of the fever center. IL-6 is also produced. Tumor necrosis factor (TNF) provokes tumor necrosis, in addition to a whole series of other catabolic side effects. The macrophages also cause antigen processing in favor of the lymphocytic system.

During phagocytosis, glucose is metabolized via the hexose monophosphate pathway, which delivers the necessary energy for the production of free oxygen radicals, H_2O_2 and other compounds, that will provoke lysis of phagocytosed material. This process is called respiratory burst.

The modern methodology to investigate phagocytic function essentially consists of: (1) The follow-up of respiratory burst during phagocytosis by estimation of glucose metabolization after administration of labeled glucose and measurement of produced CO_2. Ribose-5-phosphate is indeed transformed to CO_2 in a series of consecutive steps. (2) During the chemical reaction of the respiratory burst, light is produced. This chemiluminescence can be measured in a Lumetron. In uremia one should take into account the possibility that the registration of chemiluminescence would be disturbed by the presence of uremic toxins, that act as scavengers of free radicals. (3) Follow-up of IL-1, IL-6 or TNF production.

For the investigations in our unit, we studied the glucose metabolization and production of CO_2 in the resting state and after stimulation with latex, zymosan or *Staphylococcus aureus*. These tests were performed on 50 μl of whole blood [2, 3]. There might be a slight bias in the results obtained with whole blood because red blood cells also have a glucose metabolism, but this bias exists only in the resting state and not after challenge, as erythrocytes do not display any phagocytosis.

Evaluation of Phagocytosis in Uremia

The following conditions of the uremic state were evaluated: (1) progressive chronic renal failure; (2) first weeks of hemodialysis treatment; (3) intradialytic evolution; (4) evolution during maintenance haemodialysis treatment; (5) continuous ambulatory peritoneal dialysis (CAPD) and renal transplantation, and (6) changes after stimulation of phagocytosis.

Normal Individuals

Normal individuals were first investigated as a control series. We consider as Δ latex, Δ zymosan or Δ *S. aureus* the increase after stimulation of radioactive disintegrations per minute (dpm), compared to the resting state.

Progressive Chronic Renal Failure

In a transversal study of an outpatient population of patients with increasing degree of renal failure, it appears that phagocytosis is only modified at a serum creatinine of more than 6 mg/dl, and even more with

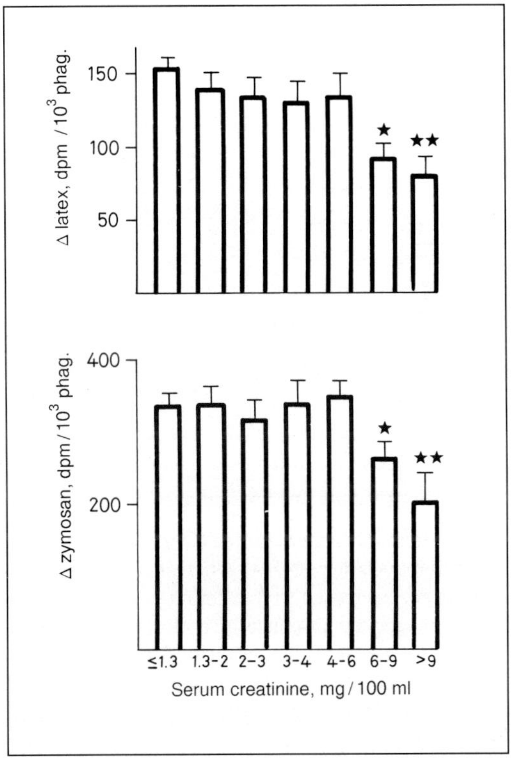

Fig. 1. Transversal evaluation of the glycolytic response towards latex and zymosan in an outpatient population (n = 119), with groups with different serum creatinine. The glycolytic response is only significantly modified from a serum creatinine about 6 mg/dl on. ★ $p < 0.05$; ★★ $p < 0.01$ vs. creatinine ≤ 1.3 mg/dl.

serum creatinine above 9 mg/dl (fig. 1). This is the period immediately before the start of renal replacement therapy. At a decrease of renal function as indicated by creatinine clearance, from 15 to 5 ml/min, in a prospective study, over a 5-month period, phagocytic function equally decreases significantly (fig. 2) [7].

Hemodialysis Patients

In a large group of hemodialysis patients, both Δ latex and Δ zymosan are considerably suppressed (fig. 3). Remarkably enough, phagocytic function decreases still further during the first weeks of hemodialysis treat-

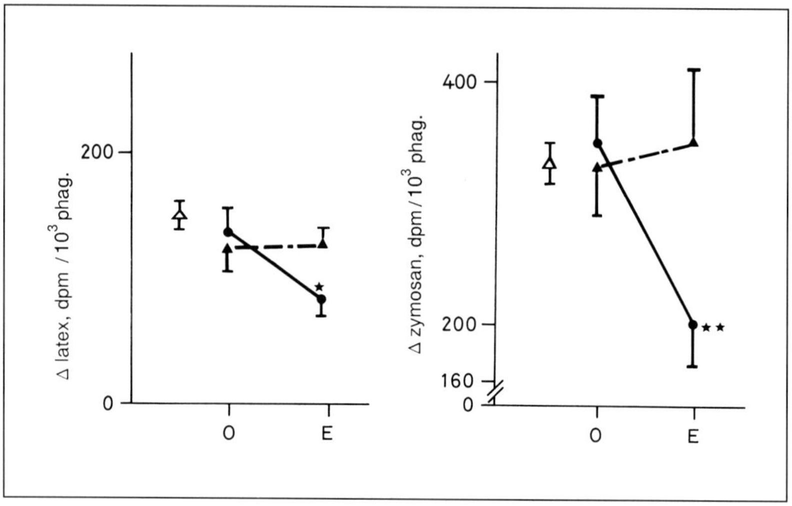

Fig. 2. Prospective study in 20 patients with a creatinine clearance of 15 ml/min. In 9 patients creatinine clearance remained unchanged (---), whereas in 11 patients over a mean period of 5.9 months, the clearance decreased to 5 ml/min (——). Normal control values as a reference (△). Whereas phagocytic response at a clearance of 15 ml/min remains unaltered versus the control, we see at an evolution to 5 ml/min a significant decrease. O = Start of follow-up; E = end of follow-up; ▲ = group with stable clearance; ● = group with decreasing clearance; ★ $p < 0.05$; ★★ $p < 0.01$ vs. normal control. [Figure from ref. 7, with permission.]

ment, especially when dialysis with a cuprophan membrane is performed (fig. 4). At this point we are probably not confronted with an inhibitory role played by uremic toxins, but with the influence of the extracorporeal circulation. The fact of bringing blood 3 times/week during 4 h at a blood flow of 250 ml/min or more into contact with dialysis machinery appears to be harmful, especially by the contact of blood with nonbiocompatible dialyzer surfaces or membranes. This bioincompatibility of membranes is also clearly illustrated by the well-known fact that at the 15th minute after submission to dialysis with cuprophan, the number of circulating polymorphonuclear cells has decreased drastically, together with complement activation [1, 5]. At contact with so-called biocompatible membranes (e.g. polysulfone, polyacrylonitrile, polymethylmethacrylate, hemophane) this phenomenon is much less pronounced or even absent [6, 8, 10].

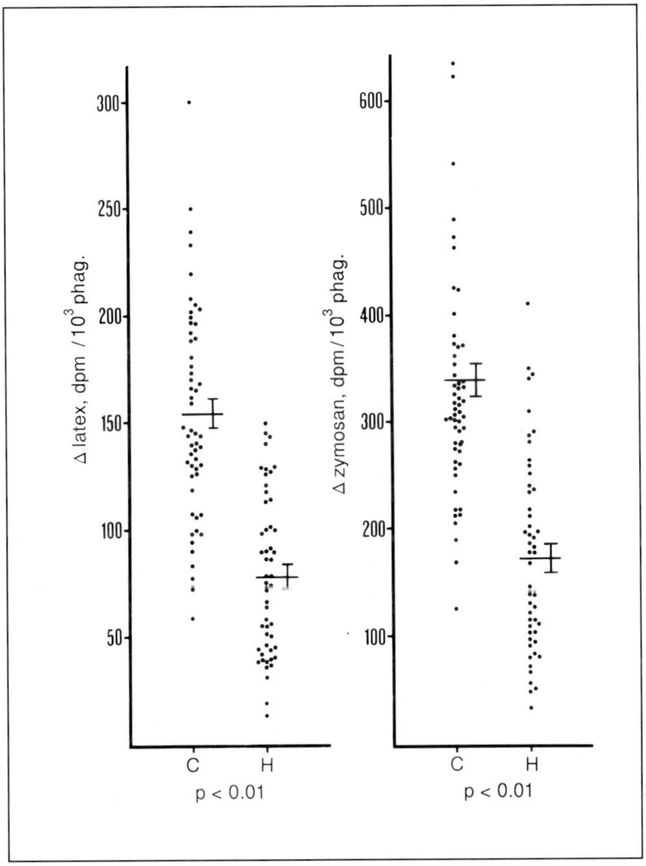

Fig. 3. Values obtained in hemodialysis patients (H) versus healthy controls (C). Studies in two groups of 50-year-old age- and sex-matched patients. The values in dialysis patients are significantly lower. [Figure from ref. 7, with permission.]

However, not only the number of polymorphonuclear cells is decreased at the 15th minute, but also the function of the remaining phagocytic cells is practically eliminated (fig. 5). Also after contact in vitro we observed a difference in reaction in the presence of biocompatible and nonbiocompatible membranes [8, 10]. It is quite conceivable that the risks for septicemia are considerably increased when in the period of this 15th minute of dialysis *S. aureus* or another bacteria is entering the circulation. In this context, we also studied the morbidity by infection or septicemia in

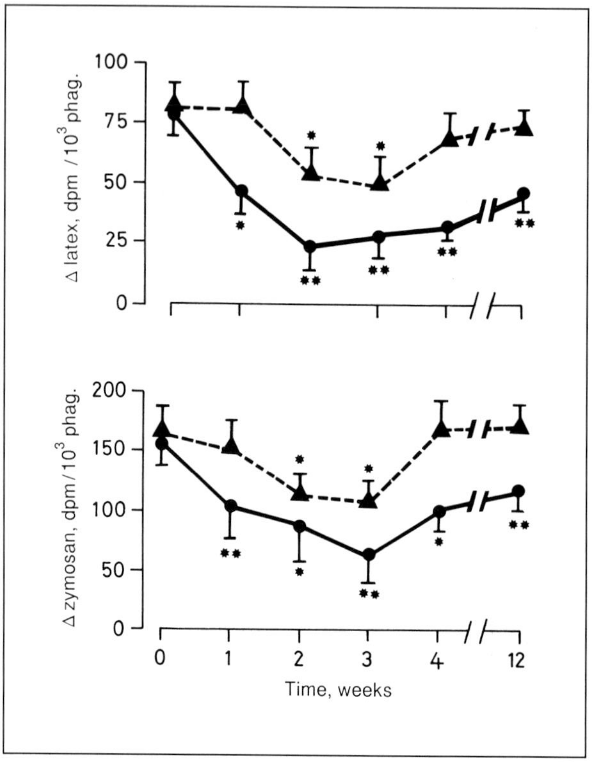
Δ latex, dpm / 10^3 phag.
100
75
50
25
0
Δ zymosan, dpm / 10^3 phag.
200
150
100
50
0
0
1
2
3
4
12
Time, weeks

4

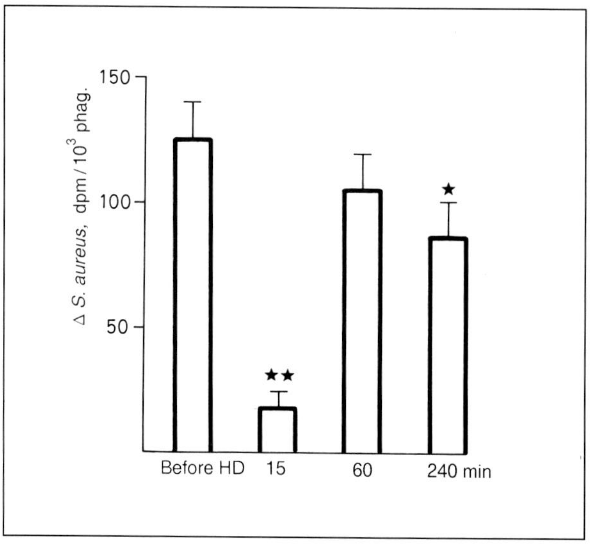
Δ *S. aureus*, dpm / 10^3 phag.
150
100
50
Before HD
15
60
240 min

5

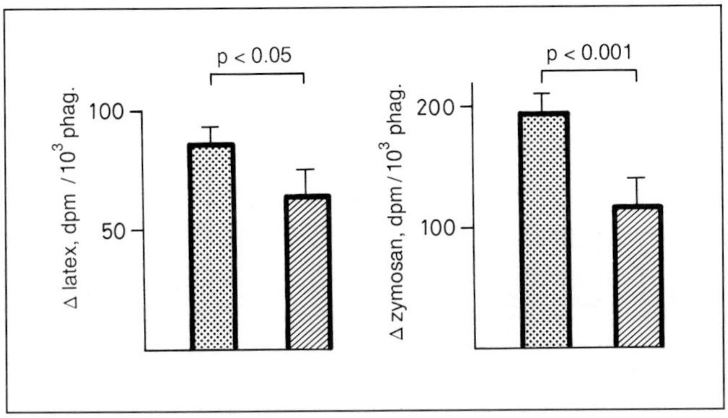

Fig. 6. Phagocytic response in a group (n = 35) with no major infections within 6 months after determination of phagocytic metabolic activity (▩) in comparison to a group that developed infections within the same time interval (n = 15) (▨). The results for the patients that will develop infection are lower.

hemodialysis patients, and observed that this morbidity was markedly higher when patients showed a decreased phagocytic glycolytic activity (fig. 6).

With the increase of the duration of the maintenance hemodialysis, phagocytic function of these patients appears to improve again. Our patients who have been on hemodialysis for more than 12 years show a virtually normal phagocytic function. Survival is probably better for those patients who show the most important resistance to aggression by the extracorporeal circulation [9].

Fig. 4. Evolution of phagocytic metabolic function during the first weeks after the start of hemodialysis with a cuprophan (——) or a polysulfone membrane (– – –) (n = 7 and 8 respectively). Study on blood samples taken predialysis. The values are markedly lower for cuprophan. * $p < 0.05$; ** $p < 0.01$ vs. 0 weeks. [Figure from ref. 7, with permission.]

Fig. 5. Phagocytic response during cuprophan dialysis, versus a standard quantity of *S. aureus*. Blood samples taken before dialysis, and after 15, 60 and 240 min. The response is clearly disturbed, especially after 15 min, which corresponds to the maximal activation of complement and leukocytes. * $p < 0.05$; ** $p < 0.01$. [Figure from ref. 7, with permission.]

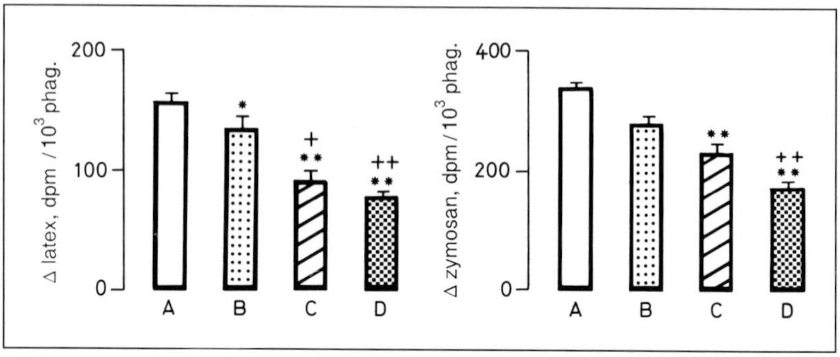

Fig. 7. Comparison between healthy controls (A), transplanted patients (B), CAPD patients (C), and hemodialysis patients (D). There is a slight decrease present for transplanted patients, but the decrease is most marked for dialysis (CAPD and hemodialysis). * $p < 0.05$; ** $p < 0.01$ vs. A; $^{+}p < 0.05$; $^{++}p < 0.05$ vs. B.

CAPD and Renal Transplantation

To make a comparison with other treatment modalities we have also evaluated phagocytic function in patients on CAPD and after renal transplantation (fig. 7). We demonstrated in 1980 the inhibitory role of a non-defined uremic serum retention fraction [3]. Whereas uremic serum suppresses phagocytosis, urea and creatinine as such are not inhibiting.

Stimulation of Phagocytosis

Stimulation of phagocytic function is possible. We have used for this purpose the new cephalosporin, Cefodizime®. Two grams of Cefodizime®, intravenously administered at the end of 5 consecutive dialysis sessions, caused in 5 of 6 patients a considerable increase of glycolytic response to latex and zymosan by the phagocytic system (fig. 8) [4]. This research is continuing with other antibiotics, which might also have a suppressing effect on phagocytic function.

Conclusion

We have reviewed in this contribution the ongoing investigations in our laboratory in the domain of phagocytosis. Some general conclusions may already be drawn: (1) uremia is a prototype of acquired immune de-

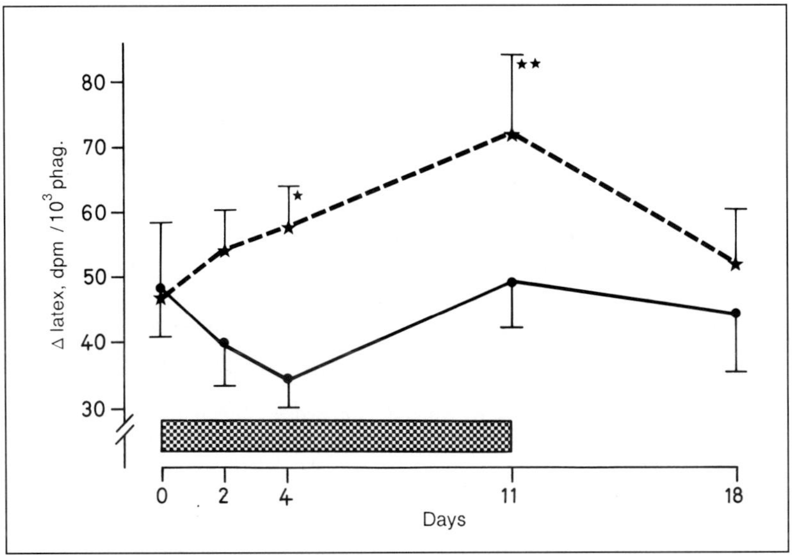

Fig. 8. Glycolytic response versus latex during treatment with cefodizime (---) versus placebo (——). The bar indicates the treatment period. There is a marked increase in activity during and after treatment with Cefodizime®. Similar results were obtained for the response towards zymosan. * $p < 0.05$; ** $p < 0.01$ vs. day 0.

ficiency; (2) contact with bioincompatible dialysis membranes is deleterious for phagocytosis; (3) other toxins than the classical urea and creatinine are apparently responsible for this disturbed phagocytosis; (4) stimulation of phagocytosis with medication is possible.

References

1 Kaplow LS, Goffinet JA: Profound neutropenia during the early phase of hemodialysis. JAMA 1968;203:1135–1137.
2 Ringoir S, Van Landschoot N, De Smet R: Influence of human urine on phagocytosis; in Dubach UC, Schmidt U (eds): Diagnostic Significance of Enzymes and Proteins in Urine. Bern, Huber, 1979, pp 358–367.
3 Ringoir S, Van Landschoot N, De Smet R: Inhibition of phagocytosis by a middle molecular fraction from ultrafiltrate. Clin Nephrol 1980;13:109–112.
4 Vanholder R, Van Landschoot N, Dagrosa E, Ringoir S: Cefodizime®: a new cephalosporin with apparent immune-stimulating properties in chronic renal failure. Nephrol Dial Transplant 1988;2:221–224.

5 Vanholder R, Ringoir S: Bioincompatibility: an overview. Int J Artif Organs 1989; 12:356–365.

6 Vanholder R, Dhondt A, Van Landschoot N, Waterloos MA, Ringoir S: Phagocytic activity of leukocytes induced by different dialyzer membranes. Contrib Nephrol. Basel, Karger, 1989, pp 86–90.

7 Vanholder R, Ringoir S, Dhondt A, Hakim R: Phagocytosis in uremic and hemodialysis patients: a prospective and cross-sectional study. Kidney Int 1991;39:320–327.

8 Vanholder R, Van Landschoot N, Waterloos MA, Delanghe J, Van Maele G, Ringoir S: Phagocyte metabolic activity during hemodialysis with different dialyzers not affecting the circulating number of phagocytes. Int J Artif Organs 1992;15:89–92.

9 Vanholder R, Ringoir S: Correction of phagocyte metabolic dysfunction after > 15 years of hemodialysis; in Giovanetti S (ed): Proc 16th Capri Conf on Uremia, in press.

10 Vanholder R, Dell'Aquila R, Ringoir S: Phagocyte function in uremia; in Hörl W, Schollmeyer E (eds): Recent Advances in Neutrophil Function. New York, Plenum Press, 1991, pp 193–205.

S. Ringoir, MD, Division of Nephrology, Department of Internal Medicine,
University Hospital, Ghent, De Pintelaan 185, B–9000 Ghent (Belgium)

Berlyne GM (ed): The Kidney Today. Selected Topics in Renal Science.
Contrib Nephrol. Basel, Karger, 1992, vol 100, pp 25–34

Lead and Hypertension

R. Nowack[a], *A. Wiecek*[b], *E. Ritz*[a]

[a] Department of Internal Medicine, University of Heidelberg, FRG, and
[b] Silesian School of Medicine, Katowice, Poland

Lead (Pb) poisoning was a serious problem in antiquity and in medieval times. In acute Pb poisoning, e.g. saturnine crisis, hypertension occurs, probably as a result of vasospasm. Acute Pb poisoning has been widely eliminated by way of strict control of Pb contamination of food and regulations for occupational Pb handling. Still, the persistingly high Pb burdens from nonoccupational sources continue to raise serious concern with respect to potential public health effects, particularly the effect on blood pressure. The Pb burden of the environment has continuously increased in the past centuries. As estimated from the Pb content of the Greenland shelf ice [1], today's Pb concentration of arctic snow has risen by a factor of 20 in comparison to ice deposited from snow in 800 B.C. The trend of the Pb burden to increase has even accelerated since 1940, when leaded gasoline was widely introduced.

The Pb-blood pressure hypothesis became popular when some authors proposed that lead smelter workers who were heavily Pb exposed had an excess mortality from cardiovascular and cerebrovascular disease [2]. Subsequently, epidemiologic evidence as well as experimental data have been provided to support the hypothesis of an increase of blood pressure by Pb exposure which does not cause acute Pb intoxication. At present, however, consensus has not been achieved. The controversy is well documented by the Proceedings of the Symposium on Lead-Blood Pressure Relationships in Chapel Hill, 1987 [3].

Epidemiologic Studies

The relationship of Pb exposure and blood pressure has been investigated in occupationally Pb-exposed populations as well as in the general population. Further studies examined the cardiovascular and cerebrovascular morbidity in occupationally Pb-exposed populations.

In surveys on the Pb-blood pressure relationships, two crucial problems have to be addressed:

(1) Confounding variables have to be carefully assessed, since smoking and alcohol consumption both increase body Pb burden and both influence blood pressure via alternative mechanisms.

(2) The question of what is the appropriate index for body Pb burden has to be resolved. Most of the investigators measured Pb concentration in whole blood with or without correction for hematocrit. Other authors [4] argued that blood Pb is a poor index of chronic Pb burden and rather reflects recent Pb exposure. Pb is stored in deeper compartments, especially in bone, and can be mobilized from these tissues by the chelating drug Na_2Ca-EDTA. Emmerson [4] suggested that the increment of Pb excretion over baseline after EDTA infusion reflects chronic Pb burden more accurately than does blood Pb.

When recently planning an epidemiologic study [5] on Pb and blood pressure by comparing hypertensive and normotensive subjects, we were careful to meet these specific problems by measuring both blood Pb and the increment of Pb excretion after EDTA infusion. Furthermore, confounding variables were carefully assessed. We studied 47 male patients with primary hypertension and 52 matched male normotensive controls from the same geographical area. Patients were approached by their physicians. In a structured interview, family history of hypertension, current smoking, alcohol consumption, history of gout, occupational or nonoccupational Pb exposure and residence in rural or inner-city areas was explored. Secondary hypertension was ruled out by urinalysis, renal sonography and measurement of plasma catecholamines. A blood sample for blood Pb was taken at the time of the examination before 1 g of Na_2Ca-EDTA was infused over 1 h. Basal urinary Pb excretion was measured in a 24-hour urine collection before the infusion and mobilizable Pb was measured in the 24-hour collection after the infusion.

Blood Pb levels in the hypertensive patients were higher than in the controls as in many previous studies. After correction for hematocrit, however, the difference disappeared since hypertensives had a higher hemato-

Table 1. Blood Pb, basal urinary Pb excretion and mobilizable Pb

	n	Blood Pb ng/ml	Basal Pb excretion nmol/24 h/1.73 m^2	Mobilizable Pb nmol/24 h/1.73 m^2
Controls	52	70 (51–94)	20 (12–38)	210 (125–396)*
Hypertension	47	78 (43–109)	47 (9–128)	255 (87–886)
Nonsmoking control	36	72 (52–93)	72 (52–93)	240 (125–304)
Nonsmoking hypertensives	26	68 (43–109)	68 (43–109)	244 (87–567)

* Difference between controls and hypertension $p < 0.05$.

crit. The increment of Pb excretion after EDTA infusion (table 1) was higher in hypertensive patients than in controls. Consideration of confounding variables accounted for this difference. Current smoking was more frequently reported by hypertensives (44 vs. 30%). There was no difference of Pb excretion between the nonsmoking hypertensives and the nonsmoking controls. We found a strong correlation between blood Pb levels and mobilizable Pb (fig. 1) which argues for blood Pb as a reliable index of Pb burden. This observation which differs from previous findings is most likely due to recent analytical progress and greater accuracy of blood Pb levels.

How must this negative finding be interpreted in view of previous studies reported in the literature? In the early observation of Beattie et al. [6] a high prevalence of hypertension and comparatively high blood Pb levels were found in an area of soft water supply. Soft water increases the solubility of heavy metals and may increase the Pb burden if water is supplied using old Pb water pipes. In the case-control study by Beevers et al. [7] a correlation was found between systolic blood pressure and blood Pb level. Both studies were conducted in a highly industrialized area and in a population presumably exposed to various risk factors for hypertension. However, important confounding variables had not been considered in the analysis. Pocock et al. [8] studied blood pressure-blood Pb relationships in 24 British cities taking into account confounding variables. In this study a correlation was found between blood Pb and systolic or diastolic blood pressure, respectively, but this correlation disappeared after adjusting for alcohol consumption. These studies exemplify that confounding factors

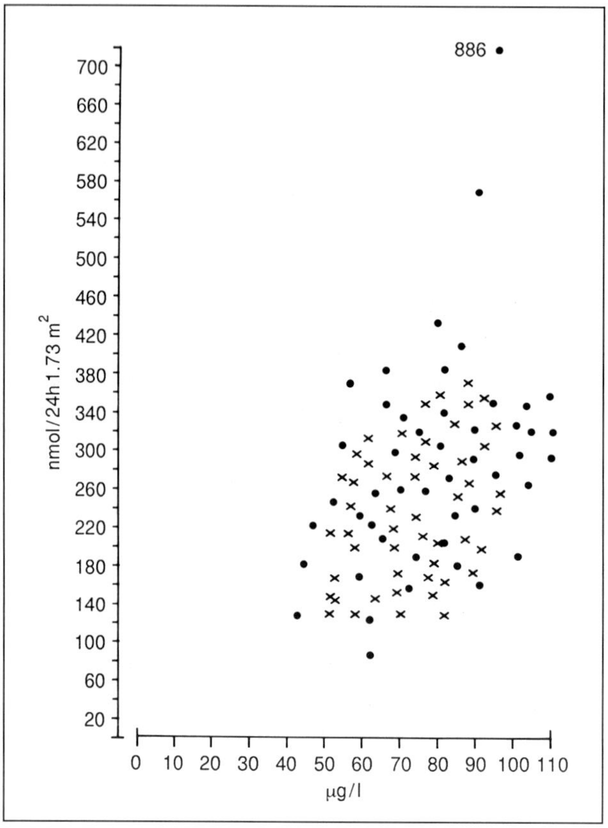

Fig. 1. Correlation between blood Pb and mobilizable Pb in hypertensive patients (47) and controls (52). ● = Hypertensives, $r = 0.77$, $p < 0.01$; × = normotensives, $r = 0.65$, $p < 0.01$.

may mimic a correlation between Pb exposure and blood pressure. The same was true in our own study.

The results of other studies suggest that the relationship between Pb and blood pressure is weak or may exist only in certain subpopulations. In the Zutphen Study [9] a correlation between blood Pb and systolic and diastolic blood pressure was found in 152 elderly men. After exclusion of 1 patient with excessive blood Pb levels, statistical significance of the difference was missed. Orssaud et al. [10] found a weak correlation between blood Pb and systolic blood pressure in young persons (24–34 years old);

in contrast no correlation exists in older persons from the same geographic area. Finally, no relationship of the Pb content in water and blood pressure of the general population in Boston was found by Sparrow et al. [11].

In contrast, however, impressive evidence in favor of the Pb-blood pressure hypothesis comes from the Second National Health and Nutrition Examination Survey (NHANES) [12, 13] in the United States from 1976 to 1980 conducted in 9,900 representative subjects. It showed a weak but statistically significant correlation between blood Pb and systolic and diastolic blood pressure using multiple regression analysis adjusting for age, hemoglobin, alcohol, nicotine consumption and numerous nutritional variables. The correlation existed in the range of comparatively low blood Pb levels (below 30 µg/dl) and was stronger for black subjects than for white subjects.

This conclusion has not been generally accepted, however. Gartside [14], analyzing the same data with a different statistical approach, failed to show a significant correlation. As of now, many still accept the interpretation that according to NHANES blood Pb is an independent predictor of blood pressure. In our own study, mentioned above, the hypertensive subjects do not differ from normotensive controls with respect to blood Pb or mobilizable Pb, when confounding factors like age, hemoconcentration, alcohol and nicotine consumption were considered.

In view of such uncertainty it is warranted to look at potential mechanisms by which Pb may raise blood pressure.

Experimental Studies on Blood Pressure Changes Induced by Pb

Reports on blood pressure changes induced by Pb in animals exposed to Pb, particularly in rats or dogs, are extremely variable. No change of blood pressure was found in some studies [15] while others noted a definite increase in systolic or mean blood pressure [16–18]. Cottier et al. [16] and Webb et al. [17] found a biphasic response to Pb exposure in rats, moderate Pb exposure increasing and more severe Pb exposure decreasing blood pressure. It may also be important whether adult animals or weanling animals are exposed to Pb. Aviv et al. [18] demonstrated that the blood pressure raising effect of Pb exposure is limited to a temporal window of a few weeks in weanling rats in whom hypertension develops only some time after discontinuation of Pb exposure. With respect to the mechanisms by which blood pressure is raised, several candidate systems must be considered: sympathetic nervous system, the renin-angiotensin-aldosterone system, sodium excretion via the kidney (or combinations of these).

Pb has important interactions with the sympathetic nerve system, particularly the central nervous system. Increased urinary catecholamine excretion has been found in Pb intoxicated mice [19] and this is reminiscent of the finding of increased urine catecholamine excretion in children with massive Pb exposure. It has been proposed that Pb acts by modulation of central sympathetic tone through modification of the calcium-sensitive neuron channels [20]. There may be also, however, a peripheral component of sympathetic dysfunction. Webb et al. [17] found an increased pressor response to methoxamine and noradrenaline in helical strips of tail arteries of Pb-exposed animals.

The effect of Pb on the renin system is also complex. Fleischer et al. [21] noted increased PRA in Pb-exposed rats whereas in Pb intoxicated men suppressed PRA was noted [22]. In vitro experiments showed increased renal renin concentrations in Pb-exposed rats and an exaggerated renin release after isoproterenol stimulation [23]. Finally, renal effects may play an important role. Aviv et al. [18] demonstrated a reduced single-nephron GFR in rats exposed to Pb in the weanling period. This should favor Na retention. In contrast, acute administration of Pb causes natriuresis and diuresis with unchanged renal hemodynamics in dogs [24]. Moreover, of interest in relation to recent findings of a circulating inhibitor of Na,K ATPase in sodium homeostasis, Tokashige et al. [25] found inhibition of Na,K ATPase of vascular smooth muscle cells in Pb-exposed animals.

Because of the contrasting reports on the effects of Pb in experimental studies, we decided to systematically address this issue [26]. We examined the following: dose-response relationship, time course, effects in different strains (WKY vs. SHR-sp) etc. Some measurements were also performed to look at volume control, concentrations of angiotensin II or catecholamines and pressor effects of these agonists. The fact that many authors found no blood pressure effects of Pb may be explained by the remarkable time course that we were able to delineate. In SD and in hypertensive prepubertal SHR-sp rats (fig. 2) Pb exposure over several weeks caused dose-dependent lowering of systolic pressure (by tail plethysmography) and mean arterial pressure (by intraarterial catheter in conscious animals). Blood pressure was lowered and in parallel animals gained less weight. Using a pair-feeding protocol, however, we were able to dissociate the effects on body weight from those on blood pressure, since lower blood

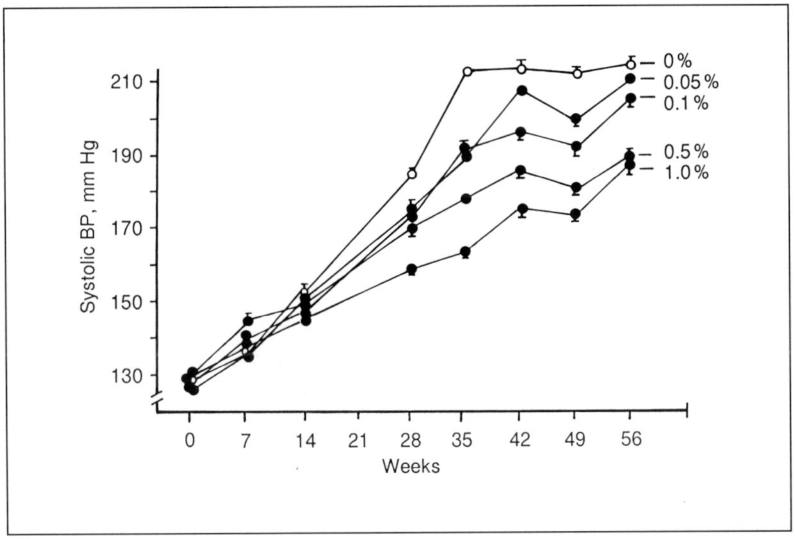

Fig. 2. Systolic blood pressure by tail plethysmography in male SHR-sp (150 g) exposed to different concentrations of Pb in drinking fluid. $\bar{x} \pm$ SD, n = 7 per group.

pressure was still found in Pb-exposed animals compared with control animals. Gender seems to be another confounding variable since the blood pressure lowering response to Pb is markedly attenuated in postpubertal male rats and was virtually absent in female animals. That androgens play a role was shown by an experiment in which surgical or pharmacological castration of prepubertal (but not of postpubertal) SHR-sp prevented the blood pressure changes. The Pb-exposed male SHR-sp had decreased plasma levels of testosterone, pointing to interference of Pb with synthesis of androgens ('pharmacological castration') and this could in part explain the mitigated development of hypertension. With respect to the acute blood pressure lowering, we wished to exclude a role of prostaglandins, the metabolism of which is affected by Pb. Cyclooxygenase inhibition did not attenuate the effect of Pb on blood pressure as shown in figure 3. While rats briefly exposed to Pb had similar plasma volume, total body sodium and plasma levels of angiotensin II or catecholamines, the pressor response to angiotensin II and noradrenaline were blunted. It therefore appears that reduced sensitivity to vasoconstrictors underlies the acute hypotensive effect of Pb.

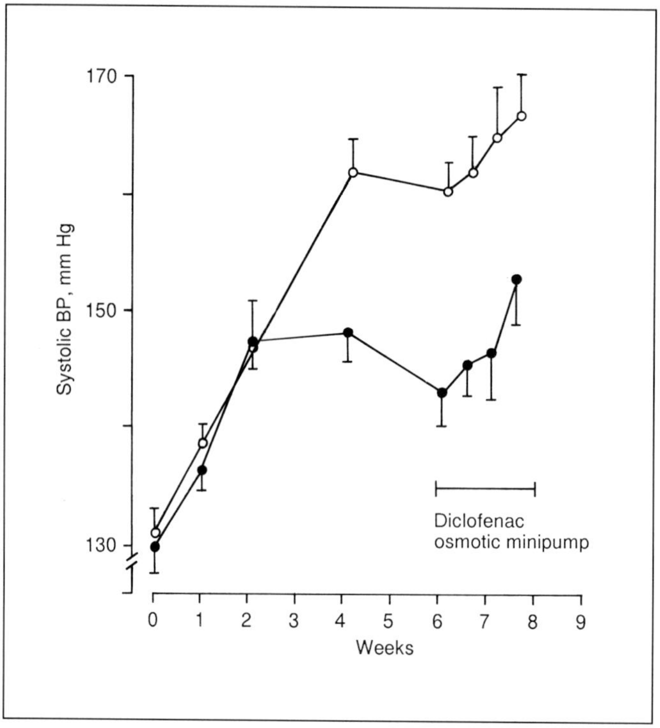

Fig. 3. Effect of diclofenac on systolic blood pressure (by tail plethysmography) of male SHR-sp rats exposed to 1% Pb acetate in drinking fluid. Controls received Na acetate. ● = Pb acetate; ○ = Na acetate.

Clear-cut hypertension developed, however, when normotensive rats were exposed to 1% of Pb acetate in drinking fluid over 18 months. Blood pressure elevation was not found before the 47th week of Pb exposure (tail plethysmography and intraarterial recording). After 18 months we found expanded plasma volume (by Evans blue dilution technique) and increased total body sodium. Interestingly, although renal ablation sensitizes for the development of volume-dependent hypertension, no effects of uninephrectomy on blood pressure development was seen in Pb-exposed animals. Volume retention did not result from renal failure, since GFR was found to be normal. Furthermore, plasma concentrations of vasoconstrictor hormones were unchanged as was the pressor response to noradrenaline or angiotensin II.

We conclude from these experiments that the effects of Pb on blood pressure in animals are complex and very much dependent on details of the experimental protocol; this may explain conflicting results in past literature. Nevertheless, very clear evidence for a chronic hypertensinogenic action of Pb by volume-dependent mechanisms was provided by the results of our study.

Conclusion

There can be no doubt that Pb may raise blood pressure either in patients with acute Pb intoxication (saturnine crisis) or patients with chronic Pb nephropathy. The evidence is much less clear whether a raised Pb burden which does not cause Pb intoxication is able to raise blood pressure. In this context it remains to be seen whether the recent increase in environmental Pb burden is relevant. While we provided clear experimental evidence for a hypertensinogenic action of Pb, our population study failed to confirm a major association of Pb with blood pressure (although for obvious reasons the sensitivity of our approach is limited). We conclude that if Pb has an effect on the blood pressure burden in the general population, it must be a minor one. The magnitude of the effect has probably been overestimated in some of the recent epidemiological studies.

References

1 Murozumi M, Chow TJ, Patternson C: Chemical concentrations of pollutant lead aerosols, terrestrial dusts and sea salts in Greenland and Antarctic snow strata. Geochim Cosmochim Acta 1969;38:1247–1294.
2 Selevan SG, Landrigan PJ, Stern FB, Jones JH: Mortality of lead smelter workers. Am J Epidemiol 1985;122:673–683.
3 Proceedings of the Symposium on Lead-Blood Pressure Relationships. Environ Health Perspect 1988;78:1–200.
4 Emmerson BT: Chronic lead nephropathy. Kidney Int 1973;4:1–5.
5 Nowack R, Schaumann HJ, Novak L, Stoeppler M, Ritz E: Is there a relation between body lead burden and blood pressure. Klin Wochenschr, in press.
6 Beattie AD, Moore MR, Devenay WT, Miller HR, Goldberg A: Environmental lead pollution in an urban soft water area. Br Med J 1972;ii:491–493.
7 Beevers DG, Erskine E, Robertson M, Beattie AD, Campbell BC, Goldberg A, Moore MR, Hawthorne VM: Blood lead and hypertension. Lancet 1976;ii:1–3.

8 Pocock SJ, Shaper AG, Ashby D, Delves T, Whitehead TP: Blood lead concentration. Blood pressure and renal function. Br Med J 1984;289:872–874.
9 Kromhout D, Wibowo AAE, Herber RFM, Dalderup M, Heerdink H, de Lezenne Coulander C, Zielhuis RL: Trace metals and coronary heart disease risk indicators in 152 elderly men (the Zutphen Study). Am J Epidemiol 1985;122:378–385.
10 Orssaud G, Claude JR, Moreau T, Lellouch J, Juguet B, Festy B: Blood lead concentration and blood pressure. Br Med J 1985;290:244.
11 Sparrow D, Sharrett AR, Garvey AJ: Trace metals in drinking water: Lack of influence on blood pressure. J Chronic Dis 1984;37:59–65.
12 Harlan WR, Landis R, Schmouder RL, Goldstein NG, Harlan LC: Blood lead and blood pressure. J Am Med Assoc 1985;253:530–534.
13 Pirkle JL, Swartz JR, Harlan L, Harlan WR: The relationship between blood lead levels and blood pressure and its cardiovascular risk implications. Am J Epidemiol 1985;121:246–285.
14 Gartside PS: The relationship of blood lead levels and blood pressure in NHANES II: Additional calculations. Environ Health Perspect 1988;78:31–34.
15 Skoczynska A, Szchinski J, Juzwa W, Smolik R, Behal FJ: Carotid sinus reflexes in rats given small doses of lead. Toxicology 1987;43:161–171.
16 Cottier P, Kunz HA, Zollinger HU: Experimenteller Beitrag zur Frage der Bleihypertonie. Helv Med Acta 1953;82:443–449.
17 Webb RC, Winquist RJ, Victery W, Vander A: In vivo and in vitro effects of lead on vascular reactivity in rats. Am J Physiol 1981;241:H211–H216.
18 Aviv A, John E, Berstein J, Goldsmith DI, Spitzer A: Lead intoxication during development: Its late effects on kidney function and blood pressure. Kidney Int 1980;17:430–437.
19 Silbergeld EK, Chisolm JJ: Lead poisoning: altered urinary catecholamine metabolites as indicators of intoxication in mice and children. Science 1976;192:153–155.
20 Rius J: Mechanisms of the effect of lead on brain neurotransmission, a calcium-mediated action. Sci Total Environ 1988;71:441–448.
21 Fleischer N, Mouw DR, Vander AJ: Chronic effects of lead on renin and renal sodium excretion. J Lab Clin Med 1980;95:759–770.
22 McAllister RG, Michelakis AM, Sandstead HH: Plasma renin activity in chronic plumbism. Arch Intern Med 1971;127:919–923.
23 Wilke L: Early and late effects of neonatal lead exposure on in vitro renin release (abstract). Thorofare (NJ), American Society of Nephrology, 1987.
24 Mouw D, Vander A, Cox J, Fleischer N: Acute effects of lead on renal electrolyte excretion and plasma renin activity. Toxicol Appl Pharmacol 1978;46:435–447.
25 Tokashige A, Higashino H, Bogden JD, Searle B, Bauman J, Aviv A: The effect of Pb, Mg, Cd and Ca on the vascular smooth muscle cell Na,K ATPase. Kidney Int 1983;23:177A.
26 Nowack R, Wiecek A, Exner B, Gretz N, Ritz E: Chronic lead exposure in rats: effects on blood pressure. Eur J Clin Invest, in press.

Prof. Dr. E. Ritz, Department of Internal Medicine,
Bergheimer Strasse 56a, D-W-6900 Heidelberg (FRG)

Berlyne GM (ed): The Kidney Today. Selected Topics in Renal Science.
Contrib Nephrol. Basel, Karger, 1992, vol 100, pp 35–47

Altered Cellular Calcium Metabolism in Hypertension: A Reassessment and a Hypothesis

Massimo Cirillo, Giovambattista Capasso, Natale Gaspare De Santo

Chair of Nephrology, Department of Pediatrics, University of Naples 'Federico II', Naples, Italy

The etiology of primary hypertension is still to be defined but it is generally recognized that this condition depends on the interaction of different factors [1]. High peripheral resistance to blood flow is the hemodynamic disorder underlying high blood pressure in the established phase of the disease [2]. This hemodynamic status could be caused by high vascular resistance and/or by high blood viscosity, which are the two major factors regulating total peripheral resistance [3]. Vascular resistance is elevated in human hypertension [4] and in experimental models of rat genetic hypertension [5] because of functional and structural changes of the resistance vessels. In fact, the vascular wall in hypertension is highly contractile and thickened [5, 6], suggesting that a vascular abnormality could be responsible at least for the maintenance of the disease.

Calcium (Ca^{2+}) metabolism plays a central role in many cellular functions including the contraction [7] and growth [8, 9] of vascular smooth muscle cells. The mechanisms responsible for the regulation of cellular Ca^{2+} metabolism (with particular attention to the vascular smooth muscle cell), the role of Ca^{2+} ions in the vascular smooth muscle cell, and the evidence of alterations of Ca^{2+} metabolism in hypertension will be discussed to support the hypothesis that different factors and mechanisms could interact at the cellular level inducing high blood pressure.

Regulation of Cellular Ca^{2+} Transport

The cytosolic free Ca^{2+} concentration is in the submicromolar range, while the extracellular ionized Ca^{2+} concentration is slightly higher than 1 mmol/l [10]. This gradient is maintained through a complex regulation of Ca^{2+} transport mainly based on the low rate of Ca^{2+} influx into the intracellular environment. This low influx of Ca^{2+} is maintained also in spite of the electrical gradient between the extracellular fluid and the intracellular electronegative environment [11]. In response to certain stimuli, the low permeability of the plasma membrane to Ca^{2+} is increased through specific high Ca^{2+} conductance membrane structures named Ca^{2+} channels. Three types of Ca^{2+} channels have been described in vascular smooth muscle cells: the receptor-operated Ca^{2+} channel; the T-type; and the L-type of voltage-dependent Ca^{2+} channel [12]. The first type responds to the stimulation with vasoconstrictor agonists while the voltage-dependent Ca^{2+} channels account for the high Ca^{2+} influx observed during electric depolarization. The T-type responds to minor depolarizations of the membrane potential and is quickly inactivated, while the L-type requires greater depolarizations and is inactivated slowly [13].

Active Ca^{2+} efflux from the cytosolic environment occurs thanks to the activity of the calcium- and magnesium-dependent adenosinetriphosphatase (Ca^{2+}-Mg^{2+} ATPase or Ca^{2+} pump) and of the sodium-calcium exchange (Na^+/Ca^{2+} exchange). The Ca^{2+} pump extrudes Ca^{2+} against the electrochemical gradient using the energy provided by ATP hydrolysis. It has been proposed that the Ca^{2+} efflux due to the Ca^{2+} pump activity is coupled with the simultaneous entry of two protons, raising the possibility that the Ca^{2+} pump is an electroneutral mechanism which contributes to the acidification of the intracellular fluid. The Ca^{2+} pump is a low capacity, high affinity transport system which accounts for most of the Ca^{2+} efflux at the normal low (100 nmol/l) cytoplasmic free Ca^{2+} concentration [14, 15]. The Na^+/Ca^{2+} exchange is an antiporter which induces the movement of Ca^{2+} and Na^+ in opposite directions. In presence of the normal negative membrane potential and of the normal low cytosolic concentration of Na^+ and Ca^{2+}, the Na^+/Ca^{2+} exchange is supposed to induce the efflux of one Ca^{2+} ion coupled with the influx of three Na^+ ions. This electrogenic exchanger has very high capacity but low affinity for cytosolic Ca^{2+}, accounting for only a negligible fraction (if any) of Ca^{2+} efflux in presence of the normal low cytoplasmic free Ca^{2+} concentration [14, 15].

Differently from other ions, the cytosolic concentration of free Ca^{2+} depends not only on the influx and efflux rates but also on the intracellular Ca^{2+} buffering capacity. Some of the cellular organelles – the sarcoplasmic reticulum, the nucleus, and the mitochondria – contain more Ca^{2+} than the cytosol [16]. The Ca^{2+} compartmentalization into these intracellular organelles occurs through active transport mechanisms located in the membrane of such organelles. The sarcoplasmic reticulum is the most important intracellular Ca^{2+} store in vascular smooth muscle [16]. Its membrane contains a Ca^{2+} pump which is similar to that of the plasma membrane and transports Ca^{2+} ions from the cytosol into the sarcoplasmic reticulum using energy provided by ATP hydrolysis [17]. The Ca^{2+} deposits stored in the reticulum are released in response to certain agonists which stimulate the inositol 1,4,5-triphosphate receptor located in the reticulum membrane leading to the increase of the Ca^{2+} conductance of the sarcoplasmic reticulum membrane and, therefore, to the release of the stored Ca^{2+} [18]. The mitochondria appear active only in the presence of an extremely high cytosolic Ca^{2+} concentration, playing only a minor role under normal conditions [19].

The cellular buffering capacity depends also on the capacity of certain proteins or other cellular molecules to bind Ca^{2+} ions. Proteins and phospholipids of the plasma membrane are some of the molecules capable of binding Ca^{2+} [20]. Even though the buffering capacity of these Ca^{2+} binding sites is small compared to that of the active transport systems, changes in Ca^{2+} binding can modify cytosolic free Ca^{2+} concentration [21]. Moreover, it has been proposed that membrane Ca^{2+} binding could be the signal through which high cytosolic Ca^{2+} reduces inward Ca^{2+} current [22].

In excitable cells, like vascular smooth muscle cells, the distribution of the systems regulating Ca^{2+} metabolism is almost homogeneous within the cell. In epithelial cells devoted to Ca^{2+} absorption these systems are differently distributed between the basolateral and the luminal surface according to the cell polarity. In fact, in intestinal and tubular renal cells the Ca^{2+} channels are mainly located in the luminal surface of the plasma membrane, while the active efflux systems are located on the basolateral surface. This particular arrangement is the basis for the unidirectional translocation of Ca^{2+} ions from the intestinal or tubular lumen to the basolateral environment which occurs in absorbing epithelia.

The complex and multiple regulation of Ca^{2+} transport in vascular smooth muscle cells makes it possible to modify the cytosolic Ca^{2+} concentration through a number of different pathways. Changes in the rate of

Ca^{2+} influx will induce parallel changes in the cytosolic Ca^{2+} concentration. Correspondingly, changes in the rate of Ca^{2+} efflux and in the membrane Ca^{2+} binding capacity will induce opposite changes in cytosolic Ca^{2+} concentration. Moreover, high levels of cytosolic Ca^{2+} can be transiently induced also by the release of Ca^{2+} ions from the sarcoplasmic reticulum. Therefore, the rate of transmembrane Ca^{2+} fluxes or of Ca^{2+} compartmentalization inside the cell allows for a continuous fine tuning of the cytosolic Ca^{2+} concentration.

Role of Ca^{2+} Concentrations in Vascular Smooth Muscle Cells

The contraction of vascular smooth muscle is regulated by the level of cytosolic Ca^{2+} [7] through an enzymatic cascade initiated by the binding of Ca^{2+} ions to calmodulin, a small peptide which mediates many Ca^{2+}-dependent functions. The Ca^{2+}-calmodulin complex activates the myosin light chain kinase stimulating the phosphorylation of myosin light chains. Only the phosphorylated myosin light chains can interact with actin to bring about contraction. When the cytosolic Ca^{2+} concentration returns to the normal low level, the reversible Ca^{2+} binding to calmodulin is reduced, and relaxation can occur due to rapid dephosphorylation of myosin light chain to its inactive form. A sustained contraction of vascular smooth muscle could be mediated by a different phosphorylation process depending on the activation of protein kinase C rather than that due to the Ca^{2+}-calmodulin complex [23]. However, the activation of protein kinase C is also induced and maintained by high Ca^{2+} concentration [23].

In addition to its role in the control and initiation of smooth muscle contraction, the cytosolic Ca^{2+} concentration appears to be a pivotal factor in the regulation of cell growth. An increase in the cytosolic Ca^{2+} concentration is induced by most of mitogens at the initiaion of the cell proliferative cycle [24] in different cell types, including vascular smooth muscle cells [25]. In these cells, when the cytosolic Ca^{2+} response to a growth factor is blunted by Ca^{2+}-channel blockers, the proliferative response is blunted too [9], suggesting a direct role for cytosolic Ca^{2+} level in this response. Finally, evidence suggests that in vascular smooth muscle cells Ca^{2+} and calmodulin are involved not only in the induction of the cell proliferative cycle [26] but also in the angiotensin II-induced protein synthesis [8].

Abnormalities of Ca^{2+} Metabolism in Hypertension

A number of studies have reported the existence of abnormalities in Ca^{2+} metabolism in human primary hypertension and in rat models of genetically determined hypertension. We will review: (1) the in vitro studies, investigating the cellular Ca^{2+} homeostasis in different cell types of hypertensive patients and animals; (2) the in vivo studies, providing evidence of altered cellular Ca^{2+} transport, and (3) the humoral and hormonal factors which modulate cellular Ca^{2+} transport and can be relevant to hypertensive disease.

In vitro Studies

In human and rat hypertension, increased cytosolic Ca^{2+} concentration has been reported in different types of cells, including platelets, leukocytes, and vascular smooth muscle cells [27–29]. Considering the role of cytosolic Ca^{2+} in smooth muscle contraction and growth, these data provide the evidence that an abnormality which is potentially responsible for increased vascular resistance is present at the cellular level in the hypertensive disease. Furthermore, it has been shown that in the so-called prehypertensive phase of rat genetic hypertension – i.e. in 3- to 4-week-old genetically hypertensive rats that have not yet developed hypertension – the cytosolic Ca^{2+} concentration is still normal [28], suggesting that the expression of this abnormality is affected to some degree by aging, like the appearance of hypertension in humans and in animal models.

Given the peculiar and complex pattern of cellular Ca^{2+} transport, the finding of high cytosolic Ca^{2+} in hypertension does not indicate the existence of a specific defect, since more than one mechanism could account for high cytosolic Ca^{2+}. In vascular smooth muscle cells of genetically hypertensive rats, high rate of Ca^{2+} influx has been reported through both the receptor-operated and the voltage-operated Ca^{2+} channels [13]. It is not yet established whether the high Ca^{2+} influx depends on a high number of channels or an altered modulation of their activity. The rate of active Ca^{2+} efflux in hypertension is altered too. In fact, the activity of Ca^{2+}-Mg^{2+}-dependent adenosine triphosphatase, and the ATP-dependent and calmodulin-stimulated Ca^{2+} efflux are reduced in hypertensive rats and in patients with primary hypertension compared to normotensive controls in erythrocytes [30, 31], smooth muscle cells [32], and in renal tubular cells [33]. A low maximal velocity of the Ca^{2+} pump appears to be the underlying mechanism, while its affinity for cytosolic Ca^{2+} is normal. The Ca^{2+} binding capacity of the cell membrane is reduced in experimental and human hypertension [30, 32, 34–36], because of a low number of membrane Ca^{2+} binding sites. This abnormality seems to be intrinsic to the cell membrane composition rather than explained by an abnormality of the membrane-attached cytoskeletal proteins [37]. Accordingly, the cell membrane of hypertensive rats contains a reduced amount of a vitamin D-dependent Ca^{2+} binding protein [38].

Overall, a number of studies suggest that the cytosolic Ca^{2+} concentration is slightly but significantly higher in human and rat hypertension likely because of several alterations of the mechanisms responsible for buffering the cytosolic free Ca^{2+}.

In vivo Studies

Hypertension is associated with an increased rate of urinary Ca^{2+} excretion. This finding has been consistently reported by clinical and epidemiological studies in humans [39, 40] and by experimental studies in genetically hypertensive rats [41]. In humans, the prevalence of true 'hypercalciuria' among hypertensives is rare despite the fact that the urinary Ca^{2+} excretion in hypertensives is on average greater than that of normotensives [39]. This evidence suggests that hypertension is associated to a moderate increase of urinary Ca^{2+} excretion which can be defined as 'relative hypercalciuria' to avoid confusion with clinically defined 'true hypercalciuria'. The high urinary Ca^{2+} excretion of hypertensives is not associated with a parallel increase in magnesium excretion, suggesting therefore that the risk of urinary stone formation in hypertension could be increased too [42]. Accordingly, preliminary epidemiologic data in humans [43] and previous observations in rat [44] indicate that hypertension is associated with high prevalence of urinary stone disease. In hypertensive rats, 'relative hypercalciuria' is present in an early phase of the disease [34] and is not corrected by the reduction of blood pressure [45], indicating that it is not secondary to high blood pressure.

The condition underlying this 'relative hypercalciuria' is still debated. Lau et al. [45] reported an increased intestinal Ca^{2+} absorption in genetically hypertensive rats of the Kyoto strain (SHR). That study is in sharp contrast with the data obtained by Schedl et al. [46] and Lucas et al. [47], who reported a reduced intestinal Ca^{2+} absorption in SHR. However, the evidence of an altered intestinal Ca^{2+} handling is not relevant in the assessment of the pathogenesis of hypercalciuria, given that a secondarily increased intestinal Ca^{2+} absorption can be observed also in renal hypercalciuria. In the study by Lau et al. [45], SHR had, under fasting conditions, an increased urinary Ca^{2+}/Na^{+} ratio despite low serum ultrafiltrable Ca^{2+}, a finding which suggests the existence of a reduced tubular Ca^{2+} reabsorption in SHR regardless of their intestinal Ca^{2+} handling. In the Milan model of genetically hypertensive rats the increased urinary Ca^{2+} excretion rate was observed both under fed and fasting conditions in the presence of a normal filtered load of Ca^{2+} at glomerular level [34]. Similar evidence was obtained in 20 hypertensive patients [39], suggesting that in that model of rat genetic hypertension and in human hypertension too a renal defect accounts for the high urinary Ca^{2+} excretion. Further evidence supporting the existence of renal hypercalciuria in hypertension comes from the observation of mild parathyroid hyperactivity [39] and slightly reduced serum ionized Ca^{2+} concentration described in rat and human hypertension [34, 48]. Moreover, we reported that genetically hypertensive rats not only have increased urinary Ca^{2+} excretion but also increased fecal Ca^{2+} excretion, leading to altered Ca^{2+} balance and reduced bone Ca^{2+} content [41]. This evidence increases the possibility that in hypertension the process of transepithelial Ca^{2+} absorption is abnormal at intestinal and renal levels.

We have also shown during experimentally induced hypercalcemia that the urinary Ca^{2+} excretion rate of 7 untreated hypertensives is higher than that of normotensives for any given level of glomerular filtered Ca^{2+} [39, 49], the difference between hypertensives and normotensives being greater with increasing levels of serum Ca^{2+}. According to a classic interpretation of studies based on experimentally-induced hypercalcemia [50], this finding should indicate an abnormality of tubular Ca^{2+} reabsorption located at a proximal level. The normal hypocalciuric response to thiazide treatment observed in hypertensives supports the possibility that the reduced tubular Ca^{2+} reabsorption is not dependent on an abnormality of the distal tubule [51]. Kinetic analysis of experimentally-induced hypercalcemia indicated that the rate of tissular Ca^{2+} uptake measured by this experimental approach was significantly higher in hypertensives [52, 53], suggesting the existence of an accelerated influx of Ca^{2+} ions into the cellular compartment of these patients. Accordingly, experiments based on the intra-arterial infusion of Ca^{2+}-channel blockers in hypertensives and healthy normotensives also suggest an enhanced function of Ca^{2+} channels in the hypertensive disease [54].

Humoral and Hormonal Modulation of Cellular Ca Metabolism

In vivo cellular Ca^{2+} transport is modulated by a number of factors and hormones which could be relevant to hypertensive disease. Insulin is able to inhibit the activity of the membrane Ca^{2+} pump in red blood cells [55]. If present at vascular level, such a mechanism could cause increased cytosolic Ca^{2+} concentration and account for the association between hyperinsulinemia and high blood pressure.

Parathyroid hormone increases cytosolic Ca^{2+} level through mobilization of intracellular Ca^{2+} stores and stimulation of Ca^{2+} influx [56]. Parathyroidectomy is known to blunt the development of rat hypertension [57, 58] and hyperparathyroidism is often associated with hypertension [59]. Moreover, a mild parathyroid hyperactivity has been observed in human and rat hypertension [39, 60].

Serum cholesterol level, which is a significant correlate of blood pressure and is often increased in hypertensives, appears to modulate the function of Ca^{2+} channels and the rate of Ca^{2+} influx probably through modification of the cholesterol content of the plasma membrane [61]. A similar effect could be produced by ethanol [62], a mechanism which could account for the known effect of alcohol intake on blood pressure level.

Finally, a previous hypothesis proposed that the Ca^{2+} efflux would be lowered in vivo by increased levels of a circulating natriuretic digitalis-like factor [63]. This factor, via the inhibition of the Na^+-K^+-adenosinetriphosphatase, would increase the cytosolic Na^+ concentration and, thus, it would reduce the activity of the Na^+-Ca^{2+} antiporter as a pathway for Ca^{2+} efflux. The existence of such a factor and its possible effects at vascular level are still to be demonstrated; however, this represents an intriguing bridge between the effect of high dietary salt intake on blood pressure and the cellular systems responsible for Ca^{2+} transport and vascular contractility.

Proposed Model

Primary hypertension in humans and genetic rat hypertension show similar abnormalities of Ca^{2+} metabolism. It is possible to propose a hypothetic model where hypertension is caused by an abnormality in cellular Ca^{2+} transport (fig. 1). According to such a model, the inherited and/or acquired alteration(s) able to increase the cytosolic Ca^{2+} concentration of vascular smooth muscle cells would also increase the vascular contractility. Given the role played by cytosolic Ca^{2+} also in the growth of smooth muscle, high

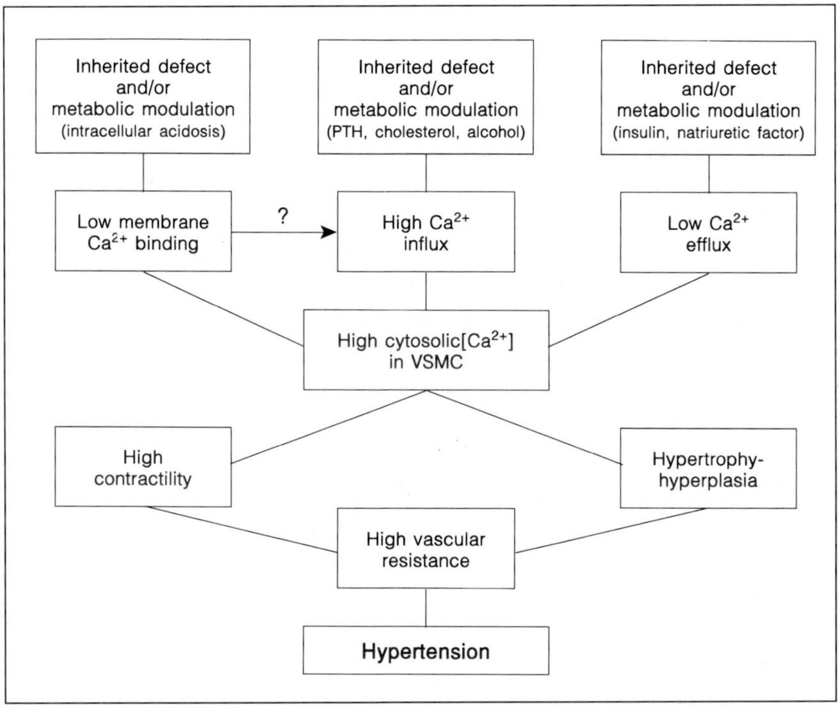

Fig. 1. Hypothetic model relating abnormalities of cellular Ca^{2+} metabolism to the etiopathogenesis of hypertension. Inherited abnormalities and/or metabolic modulation of the mechanisms involved in the control of cellular Ca^{2+} homeostasis could increase the cytosolic ionized Ca^{2+} concentration ($[Ca^{2+}]$) in vascular smooth muscle cells (VSMC). Such an increase in $[Ca^{2+}]$ could be responsible for high vascular resistance, and therefore hypertension, through its effects on contraction and growth of VSMC.

cytosolic Ca^{2+} could account at the vascular level not only for increased contractility but also for hyperplasia, both vascular changes which cause high vascular resistance and, therefore, hypertension. At different levels, such as the renal and the intestinal epithelia, the altered cellular Ca^{2+} transport could cause increased Ca^{2+} excretion and, secondarily, modifications in the Ca^{2+} balance, in the incidence of urinary stones, in the Ca^{2+} content of the bone, and in the parathyroid secretion.

Several points need to be further investigated to validate this working hypothesis. The existence and the independence of possible confounding factors for most of these abnormalities are still to be confirmed by epidemiologic studies in large series of individuals. Moreover, the predictive power of the presence of Ca^{2+} metabolism abnormalities for the development of hypertension is not yet known. Similarly, we do not know whether the correction of these abnormalities would lower blood pressure or prevent the development of hypertension and/or whether the reduction of high blood pressure would be followed by the correction of the abnormalities in Ca^{2+} metabolism.

Finally, the relative role of inherited and acquired factors in modulating cellular Ca^{2+} transport remains to be defined. Cellular Ca^{2+} metabolism is controlled by different systems: each one of them can theoretically be modulated by both genetic and environmental factors which could interact and cause an alteration in Ca^{2+} transport and cytosolic Ca^{2+} concentration. The clinical heterogeneity of primary hypertension and its multifactorial etiopathogenesis could reflect different cellular mechanisms, all resulting in an increased vascular resistance due to altered cellular Ca^{2+} transport at vascular level.

References

1 Page IH: The mosaic theory 32 years later. Hypertension 1983;4:177.
2 Pickering GW: The peripheral resistance in persistent arterial hypertension. Clin Sci 1936;2:209–235.
3 Chien S: Blood rheology in hypertension and cardiovascular disease. Cardiovasc Med 1977;2:356–360.
4 Dustan HP, Tarazi RC, Bravo EL: Physiologic characteristics of hypertension. Am J Med 1972;52:610–622.
5 Halloway ET, Bohr DF: Reactivity of vascular smooth muscle in hypertensive rats. Circ Res 1973;33:678–685.
6 Mulvany MJ, Hansen PK, Aalkjaer C: Direct evidence that the greater contractility of resistance vessels in spontaneously hypertensive rats is associated with a narrowed lumen, a thickened media, and an increased number of smooth muscle layers. Circ Res 1978;43:854–864.
7 Hai CM, Murphy RA: Ca^{2+}, crossbridge phosphorylation, and contraction. Annu Rev Physiol 1989;51:285–298.
8 Berk BC, Vekshtein V, Gordon HM, Tsuda T: Angiotensin II-stimulated synthesis in cultured vascular smooth muscle cells. Hypertension 1989;13:305–314.
9 Block LH, Emmons LR, Vogt E, Sachinidis A, Vetter W, Hoppe J: Ca^{2+}-channel blockers inhibit the action of recombinant platelet-derived growth factor in vascular smooth muscle cells. Proc Natl Acad Sci USA 1989;86:2388–2392.
10 Van Breemen C: Cellular mechanisms regulating $[Ca^{2+}]_i$ smooth muscle. Annu Rev Physiol 1989;51:315–329.
11 Hirst GDS, Edwards FR: Sympathetic neuroeffector transmission in arteries and arterioles. Physiol Rev 1989;69:546–604.

12 Van Breemen C, Cauvin C, Johns A, Leijten P, Yamamoto H: Ca^{2+} regulation of vascular smooth muscle. Fed Proc Fed Am Soc Exp Biol 1986;45:2746–2751.

13 Rusch N, Hemsmeyer K: Calcium currents are altered in the vascular muscle cell membrane of spontaneously hypertensive rats. Circ Res 1988;63:997–1002.

14 Furukawa KY, Tawada Y, Shigekawa M: Regulation of the plasma membrane Ca^{2+} pump by cyclic nucleotides in cultured vascular smooth muscle cells. J Biol Chem 1988;263:8058–8065.

15 Cirillo M, Canessa M: Activation of Ca^{2+}-Na^{+} exchange by platelet-derived growth factor in vascular smooth muscle cells. Ann NY Acad Sci, in press.

16 Somlyo AP, Himpens B: Cell calcium and its regulation in smooth muscle. FASEB J 1989;3:2266–2276.

17 Schatzman HJ: The calcium pump of the surface membrane and of the sarcoplasmic reticulum. Annu Rev Physiol 1989,51:473–485.

18 Ferris CD, Huganir RL, Supattapone S, Snyder SH: Purified inositol 1,4,5-triphosphate receptor mediates calcium flux in reconstituted lipid vesicles. Nature 1989; 342:87–89.

19 McCormack JG, Halestrap AP, Denton RM: Role of calcium ions in regulation of mammalian intramitochondrial metabolism. Physiol Rev 1990;70:391–425.

20 Grover AK, Kwan CY: High affinity Ca-binding to smooth muscle plasma membrane: inhibition by cations. Cell Calcium 1986;7:187–192.

21 Kohmoto O, Siptzer KW, Movsesian MA, Barry WH: Effects of intracellular acidosis on $[Ca^{2+}]_i$ transients, transsarcolemmal Ca^{2+} fluxes, and contraction in ventricular myocytes. Circ Res 1990;66:622–632.

22 Eckert R, Ewald D: Residual calcium ions depress activation of clacium-dependent current. Science 1982;216:730–733.

23 Rasmussen H, Takuwa Y, Park S: Protein kinase C in the regulation of smooth muscle contraction. FASEB J 1987;1:177–185.

24 Metcalfe JC, Moore JP, Smioth GA, Hesketh TR: Calcium and cell proliferation. Br Med Bull 1986;42:405–412.

25 Berk CB, Alexander RW, Brock TA, Gimbrone MA, Webb RC: Vasoconstriction: a new activity for platelet-derived growth factor. Science 1986;232:87–90.

26 Rasmussen CD, Means AR: Calmodulin, cell growth and gene expression. Trends Neurosci 1989;12:433–438.

27 Erne P, Bolli P, Burgisser E, Buhler FR: Correlation of platelet calcium with blood pressure. Effect of antihypertensive therapy. N Engl J Med 1984;310:1084–1088.

28 David-Dufilho M, Pernollet MG, Le Quang Sang H, Benlian P, De Mendonca M, Grichois ML, Cirillo M, Meyer P: Active Na^{+} and Ca^{2+} transport, Na^{+}-Ca^{2+} exchange and intracellular Na^{+} and Ca^{2+} content in young spontaneously hypertensive arts. J Cardiovasc Pharmacol 1986;8(suppl 8):S130–S135.

29 Socorro L, Vallega G, Nunn A, Moore TJ, Canessa M: Vascular smooth muscle cells from the Milan hypertensive rat exhibit decreased functional angiotensin II receptors. Hypertension 1990;15:591–599.

30 Cirillo M, Galletti F, Strazzullo P: Abnormalities of red cell membrane calcium handling in the Milan hypertensive strain of rat. Ann NY Acad Sci 1986;488:565–566.

31 Vincenzi FF, Morris CD, Kinsel LB, Kenny M, McCarron DA: Decreased calcium pump adenosine triphosphatase in red blood cells of hypertensive subjects. Hypertension 1986;8:1058–1066.

32 Orlov SN, Postnov YuV: Calcium accumulation and calcium binding by the cell membranes of cardiomyocytes and smooth muscle of aorta in spontaneously hypertensive rats. Clin Sci 1980;59:207S–209S.

33 Gmaj P, Bianchi G, Murer H: Calcium transport in basolateral plasma membranes from kidney cortex of Milan hypertensive rats. Biochim Biophys Acta 1988;941:187–197.

34 Cirillo M, Galletti F, Corrado MF, Strazzullo P: Disturbances of renal and erythrocyte calcium handling in rats of the Milan hypertensive strain. J Hypertens 1986;4:443–449.

35 Cirillo M, Trevisan M, Laurenzi M: Analysis of Ca binding to the inner side of the erythrocyte membrane in human hypertension. J Hypertens 1987;5(suppl 5):S291–S293.

36 Cirillo M, Trevisan M, Laurenzi M: Calcium binding capacity of erythrocyte membrane in human hypertension. Hypertension 1989;14:152–155.

37 Cirillo M: Reduced calcium binding capacity is an intrinsic abnormality of red blood cell membrane in spontaneously hypertensive rats. Am J Hypertens 1990;3:570–572.

38 Kowarski S, Cowen LA, Schachter D: Decreased content of integral membrane calcium-binding protein in tissues of the spontaneously hypertensive rat. Proc Natl Acad Sci USA 1986;83:1097–1100.

39 Strazzullo P, Nunziata V, Cirillo M, Giannattasio R, Ferrara LA, Mattioli PL, Mancini M: Abnormalities of calcium metabolism in essential hypertension. Clin Sci 1983;65:137–141.

40 Elliott P, Kesteloot H, Day S, Dyer A, for the INTERSALT Cooperative Research Group: Interrelationships of 24-hour urinary calcium and sodium excretion, and blood pressure: The INTERSALT Study (abstract). J Hypertens 1990;8(suppl 3):S128.

41 Cirillo M, Galletti F, Strazzullo P, Torielli L, Melloni MC: On the pathogenetic mechanism of hypercalciuria in genetically hypertensive rats of the Milan strain. Am J Hypertens 1989;2:741–746.

42 Cirillo M, Strazzullo P, Siani A, Mattioli PL, Giannattasio R, Nunziata V: Low urinary magnesium/calcium ratio due to increased urinary calcium excretion in essential hypertension; in Schwille PO, Smith LH, Robertson WG, Vahlensieck W (eds): Urolithiasis and Related Clinical Research. New York, Plenum Publishing, 1985, pp 245–247.

43 Cirillo M, Laurenzi M: Elevated blood pressure and positive history of kidney stones: results from a population based study. J Hypertens 1988;6(suppl 4):S485–S486.

44 Wexler BC, Murtry JP: Kidney and bladder calculi in spontaneously hypertensive rats. Br J Exp Pathol 1981;62:369–374.

45 Lau K, Zikos J, Spirnak J, Eby B: Evidence for an intestinal mechanism in hypercalciuria in spontaneously hypertensive rats. Am J Physiol 1984;247:E625–E633.

46 Schedl HP, Miller DL, Pape JM, Horst RL, Wilson JD: Calcium and sodium transport and vitamin D metabolism in the spontaneously hypertensive rat. J Clin Invest 1984;73:980–986.

47 Lucas PA, Brown RC, Drueke T, Lacour B, Metz JA, McCarron DA: Abnormal vitamin D metabolism, intestinal Ca absorption, and bone calcium status in the spontaneously hypertensive rat compared to its genetic control. J Clin Invest 1986;78:221–227.

48 McCarron DA: Low serum concentrations of ionized calcium in patients with hypertension. N Engl J Med 1982;307:226–228.
49 Cirillo M, Strazzullo P, Siani A, Galletti F, Nunziata V: Increased urinary calcium excretion of hypertensive patients is due to a reduced tubular calcium reabsorption. Kidney Int 1985;27:96–97.
50 Cirillo M, Strazzullo P Mattioli PL, Siani A, Cioffi G: Basal parathyroid activity and renal calcium handling during an intravenous calcium load. Nephron 1984;38:167–169.
51 Strazzullo P, Cirillo M, Turco S, Siani A, Mattioli PL, Ferrara LA, Scopacasa F, Mancini M: Increased urinary excretion of calcium and parathyroid hyperactivity in essential hypertension. Minerva Endocrinol 1983;8:151–156.
52 Cirillo M, Strazzullo P, Galletti F, Siani A, Nunziata V: The effect of an intravenous calcium load on serum total and ionized calcium in normotensive and hypertensive subjects. J Clin Hypertens 1985;1:30–34.
53 Strazzullo P, Galletti F, Cirillo M, Siani A, Nunziata V, Giannattasio R, Mancini M: Altered extracellular calcium homeostasis in essential hypertension: a consequence of abnormal cell calcium handling. Clin Sci 1986;71:239–244.
54 Robinson BF, Dobbs RJ, Bayley S: Response of forearm resistance vessels to verapamil and sodium nitroprusside in normotensive and hypertensive men: evidence for a functional abnormality of vascular smooth muscle in primary hypertension. Clin Sci 1982;63:33–42.
55 Pershadsingh HA, McDonald JM: Direct addition of insulin inhibits a high affinity Ca^{2+}-ATPase in isolated adipocyte plasma membrane. Nature 1979;281:495–497.
56 Kawashima H: Parathyroid hormone causes a transient rise in intracellular ionized calcium in vascular smooth muscle cells. Biochem Biophys Res Commun 1990;166:709–714.
57 Mann FE, Bommer J, Klooker B, Rambausek M, Ritx E: Permissive effect of parathormone on blood pressure development of spontaneously hypertensive rats (abstract). Eur J Clin Invest 1982;12:148.
58 Bethelot A, Gairard A: Parathyroid hormone and deoxycorticosterone acetate-induced hypertension in the rat. Clin Sci 1980;58:365–371.
59 Rosenthal FD, Roy S: Hypertension and hyperparathyroidism. Br Med J 1972;iv:396–397.
60 McCarron DA, Yung NN, Ugoretz BA, Kritzik S: Disturbances of calcium metabolism in spontaneously hypertensive rats. Hypertension 1981;3:I162–I167.
61 Bialecki RA, Tulenko TN: Excess membrane cholesterol alters calcium channels in arterial smooth muscle. Am J Physiol 1989;257:C306–C314.
62 Vasdev S, Sampson CA, Prabhakaran VM: Platelet-free calcium and vascular calcium uptake in ethanol-induced hypertensive rats. Hypertension 1991;18:116–122.
63 Blaustein MP: Sodium ions, calcium ions, blood pressure regulation and hypertension: a reassessment and a hypothesis. Am J Physiol 1977;232:C165–C173.

Prof. Natale G. De Santo, Nefrologia Pediatrica, Padiglione 17,
Nuovo Policlinico, Via Sergio Pansini 5, I–80131 Napoli (Italy)

Berlyne GM (ed): The Kidney Today. Selected Topics in Renal Science.
Contrib Nephrol. Basel, Karger, 1992, vol 100, pp 48–57

The Metabolic Acidosis of Chronic Renal Failure: Pathophysiology and Treatment

Sergio Giovannetti, Adamasco Cupisti, Giuliano Barsotti
Clinica Medica I, Università di Pisa, Italia

When the tubular capacity to excrete hydrogen ions becomes lower than their generation, some of them accumulate, and metabolic acidosis appears, which is indicated by low values of arterial blood pH and HCO_3^-. Several injurious effects are produced by uremic acidosis, especially if longlasting, and an adequate treatment should be instituted which, instead, is often omitted.

This paper is a short review of the pathophysiology of uremic acidosis and on its conservative therapy.

The Origin and the Destiny of Hydrogen Ions

The H^+ that enter the body fluids under normal conditions result mostly from the oxidation of sulfur-containing amino acids and, to a minor extent, from the nonmetabolizable organic acids (uric acid, oxalic acid, etc.). Hydrogen ions may be then regarded as an end-product of proteins, like urea; indeed, the urea and H^+ production are, in general, directly proportional to each other (fig. 1).

Both the exogenous and endogenous proteins are equally important; hence, a catabolic condition increases the H^+ production as well as a protein-rich diet. The ingested proteins are not, however, the only dietary factor modulating the H^+ generation; the amount of the ingested organic anions (tartrate, acetate, citrate, etc.) which are oxidized to HCO_3^-, is equally important. HCO_3^- plays, indeed, a key role in the physiological

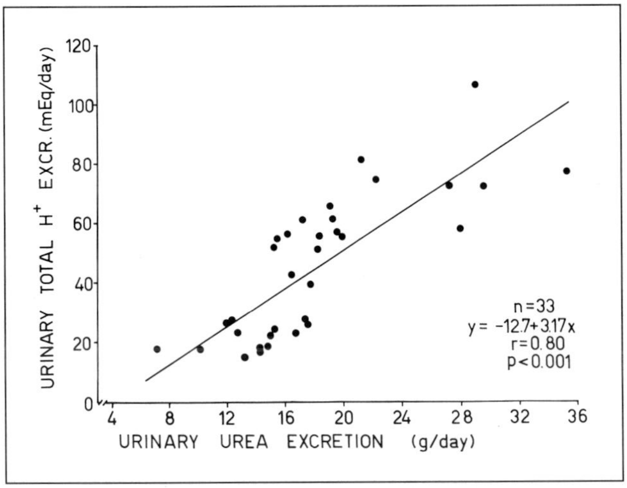

Fig. 1. The daily renal urea and the total H^+ secretion in normal volunteers at various protein intakes. The close direct relationship between the two parameters is well evident.

defence against acidosis, since it neutralizes equimolecular amounts of H^+. The biologically important amount of H^+ originating from foods is then the algebraic sum of H^+ and HCO_3^- precursors.

It is evident then that, when the ingestion of these organic anions exceeds that of H^+ precursors, as it occurs in herbivores, a tendency to alkalosis appears and alkaline urine may be excreted. This is not the case of omnivores, including humans, who always produce an excess of H^+ that have to be first buffered and then excreted.

The defence against acidosis has then two separate purposes, namely: (1) to maintain constant the pH of body fluids, and (2) to keep the H^+ balance in equilibrium.

(1) The constant value of the pH of body fluids is obtained through the buffering action of the buffer anions (proteins, hemoglobin, PO_4, creatinine and, particularly, HCO_3^-). For any acidifying process, HCO_3^- is converted into H_2CO_3, with a consequent increase of the blood CO_2, that stimulates the respiratory center. Hyperventilation takes place and the excess CO_2 is excreted, with the consequent fall of CO_2 and H_2CO_3 plasma concentrations. The pH of body fluids is thus maintained constant, in spite of a H^+ positive balance, at the price of a decreased buffering capacity of

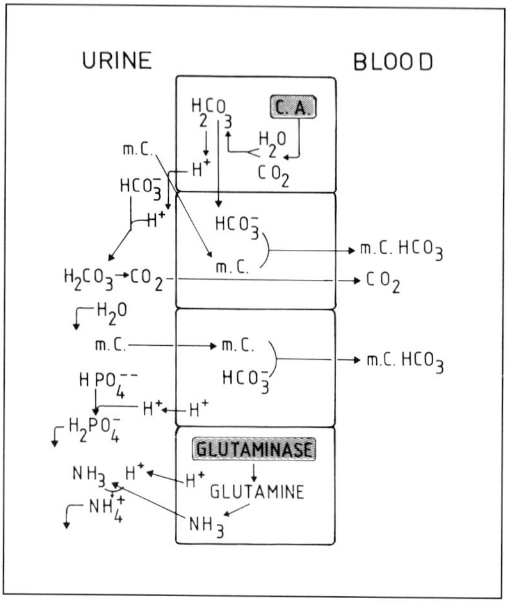

Fig. 2. A schematic representation of the tubular secretion of H^+. C.A. = Carbonic anhydrase; m.C. = metallic cation (mostly Na^+).

body fluids. This condition is metabolic acidosis, whose severity is indicated by the low values of plasma HCO_3^-.

(2) The return of H^+ balance to equilibrium, that is the physiological correction of metabolic acidosis, is exerted by the kidney. The tubular function operating such correction consists of the excretion of H^+ derived from the H_2CO_3 contained in the tubular cells. This is converted by carbonic anhydrase into H^+ and HCO_3^-; the former is excreted, and the latter is recovered to body fluids. The hydrogen ions thus excreted in the urine are captured by the buffer anions available there and by ammonia, which is converted into NH_4^+ (fig. 2).

In normal subjects, under normal conditions, the total capacity to excrete H^+ by renal tubular cells is only in part utilized. Indeed, less than 100 mEq of H^+ (as free acids plus ammonium) are excreted daily, but in the case of severe acidosis (e.g., diabetic), the total excretion of H^+ may reach values of several hundred milliequivalents per day, indicating the existence of a powerful secretory reserve.

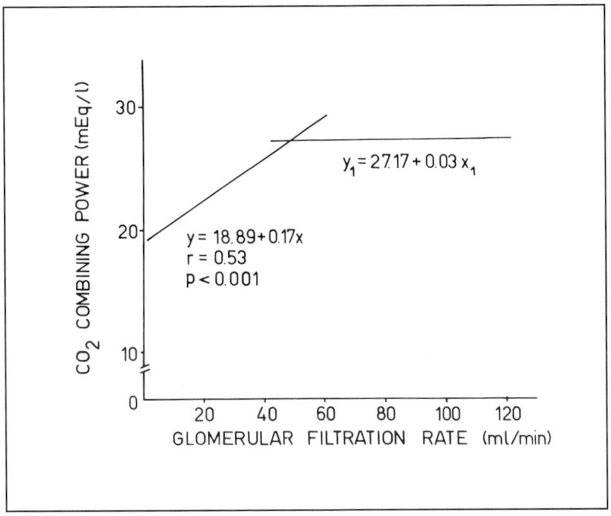

Fig. 3. The values of CO_2 combining power plotted against the rates of glomerular filtration (thiosulfate clearance). The curves have been calculated from the results of a study performed in 1956 [1], in this laboratory. Starting from a GFR of 40 ml/min and going to zero, the CO_2 combining power is directly proportional to the severity of renal failure.

In patients with GFR lower than 25 ml/min, this reserve is saturated and, in absence of an adequate dietary restriction, H^+ accumulate continuously. One would then expect the resulting acidosis to be progressive, and to be directly related to the duration of renal failure and not with its severity. This is not the case; severity of uremic acidosis is directly related to the severity of renal failure (fig. 3) [1], and it is constant in the time. This discrepancy has been generally attributed to the buffering action of bone tissue, and studies on its composition in acidotic uremics have shown that a decrease of $CaCO_3$ occurs, which is directly proportional to the duration of acidosis [2, 3]. This clearly demonstrates that a continuous demineralization occurs in bone tissue during uremic acidosis, but does not demonstrate that uremic acidosis remains constant in the time only for the buffering action of bone. On the contrary, the evaluation of its total buffering capacity has led to the conclusion that 'the contribution of the bone buffering in protracted metabolic acidosis seems negligible' [4].

Table 1. The known injurious effects of metabolic acidosis in untreated chronic renal failure

Injurious effect	Ref.
Intensification of the PTH action	7
Direct bone tissue demineralization	?
Inhibition of vitamin D_3 1α-hydroxylation	8
Hyperphosphatemia	9
Protein catabolism	10–12
Resistance to insulin	13
Hyperpotassemia	14
Hypercholesterolemia and hypertriglyceridemia	16
Decreased pulmonary vascular compliance	17
Depression of central nervous system	18

The hypothesis has been then put forward that gut may contribute in some way to the excretion of H^+ in presence of CRF acidosis [5], and some preliminary observations seem to support this view [6].

The Injurious Effects of Metabolic Acidosis

These deleterious effects are various and important, and some of them are difficult to reverse, particularly bone demineralization (table 1).

Acidosis increases the sensitivity of bone to PTH [7], that is probably one of the mechanisms through which it causes bone demineralization, while the inhibition of vitamin D_3 1α-hydroxylation may contribute to cause uremic osteodystrophy [8]. The hyperphosphatemia, also caused by acidosis [9], may be regarded as a consequence of bone demineralization.

It has been recently demonstrated [10, 11] that metabolic acidosis causes protein catabolism, and evidence suggesting this effect was also obtained in our laboratory several years ago [12].

Protein catabolism caused by uremic acidosis is extremely important, since it may be responsible for protein malnutrition in spite of an adequate dietary protein supply, and may trigger a vicious cycle, in which acidosis causes protein catabolism, and protein catabolism worsens acidosis. An increase of protein supply may, obviously, worsen this vicious cycle.

Table 2. Total urinary H^+ output (NH_4^+ plus titratable acids) and urea output in 3 normal volunteers who changed from a free diet to a low protein diet

	pH	NH_4^+ mEq/day	Titratable acids mEq/day	Total H^+ output mEq/day	Urinary urea g/day	Protein cat. rate g/day
Free mixed diet	5.7	38.2	35.4	73.6	22.9	87
Low protein diet (0.6 g/kg/day)	6.4	21.5	14.8	35.3	12.4	53

Resistance to insulin is another metabolic derangement caused by acidosis [13], and hyperkalemia, also caused by acidosis [14], is probably related to it. The improvement of carbohydrate tolerance occurring in diabetic patients with CRF following a protein-restricted diet [15] may be caused, in part at least, through the improvement of acidosis.

Elevated serum levels of cholesterol and triglycerides probably result from the abnormalities of carbohydrate metabolism [16] which may be corrected with a low protein diet that also corrects metabolic acidosis.

Other dangers related with uremic acidosis are a decreased compliance of pulmonary vessels with a high risk of pulmonary edema [17] and the depression of the central nervous system [18].

The Conservative Treatment of Chronic Uremic Acidosis

The conservative treatment of chronic uremic acidosis is based on the restriction of foods yielding H^+ and on the administration of the buffer anion HCO_3^- or precursors.

Since the urea and hydrogen ions production are directly correlated with each other, it is evident that a dietary treatment which reduces urea generation also reduces H^+ production. This is clearly shown in table 2, where the daily total urinary H^+ and urea output occurring on a normal free diet and on low protein diets, in normal subjects, are reported. It is then evident that in CRF patients the only restriction of protein intake which may reduce H^+ production below the capacity of kidney to excrete them, may prevent or correct metabolic acidosis.

Table 3. Urinary data of 4 normal volunteers who followed isonitrogen diets differing for the nature of foods: mixed and pure vegetarian

Origin of foods	Urine					
	protein cat. rate g/day	urinary urea g/day	pH	NH_4^+ mEq/day	TA mEq/day	total H^+ mEq/day
Mixed	70	17.8	5.8	35	33.1	68.1
Plant	67	16.8	6.8	26	7.0	39.4

TA = Titratable acidity.

Table 4. Mean values of urinary pH (pH), titratable acidity (TA), ammonium (NH_4^+) and total hydrogen ion output (total H^+) in 4 normal subjects during a steady-state period of low protein diet (0.6 g/kg/day), before and during the ingestion of calcium carbonate ($CaCO_3$ 4 g/day, for 4 days)

	pH	TA mEq/day	NH_4^+ mEq/day	Total H^+ mEq/day
Before $CaCO_3$	6.43	13.5	20.3	33.9
During $CaCO_3$	6.72	6.8	15.4	22.2

Another efficient dietary manipulation is the exclusive or largely prevalent use of foods of plant origin. All these foods cause a metabolic generation of H^+ lower than that caused by foods of animal origin per unit of weight and per unit of protein content [19]. This is accounted for by the precursors of HCO_3^- that are present in larger amounts in foods of plant origin than in those of animal origin. It follows that a pure vegetarian diet is much less acidifying than an isonitrogen diet supplying foods of animal origin (table 3) and here too acidosis may be corrected.

Both animal and plant origin foods have a protein content and an acidifying effect which do not directly correlate with each other. For instance, the protein content of beans is high though they exert an alkalinizing effect, whereas bread, having a lower protein content, exerts a mild acidifying action (fig. 4).

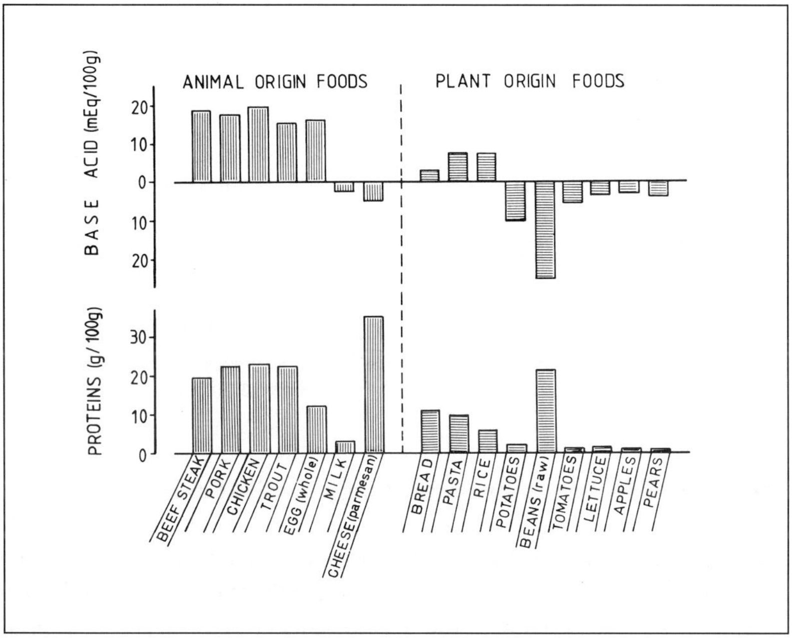

Fig. 4. The acid and the base resulting from the metabolization of 100 g of some foods of animal (left) and of vegetable origin, and their protein content.

It is well evident from what has been said that a low protein, pure vegetarian diet is efficient in the prevention and treatment of CRF acidosis, even if severe. However, the most largely used treatment of uremic acidosis is not the diet. The administration of alkalinizing agents is the most largely used treatment of uremic acidosis.

Sodium bicarbonate, given by mouth or intravenously, or its metabolic precursors (acetate, lactate, citrate) are usually employed to treat metabolic acidosis of CRF. Sodium bicarbonate is completely absorbed in the gastrointestinal tract, its efficiency is constant and the amount required to correct acidosis is easy to calculate [20]. The only, but very important, unwanted side effect of the treatments with $NaHCO_3$ is the risk of sodium and water retention, particularly in long-term treatments.

The calcium salts of CO_3^- and of its metabolic precursors (citrate, lactate, acetate) are also used as alkalinizing agents (table 4). The effect of calcium carbonate on acidosis was first documented in 1966 [21] and confirmed in 1969 [22]. It is due to the absorption of the buffer anion CO_3^{--}, which occurs when $CaCO_3$ is ionized. If this ionization is partially or totally prevented by physicochemical conditions of the gastrointestinal fluid, the alkalinizing action of $CaCO_3$ may be strongly reduced.

In spite of this drawback and of the frequent occurrence of stypsis or diarrhea, we think that the advantage to completely avoid the risk of sodium retention, and to obtain with the simultaneous correction of acidosis, hyperphosphatemia and of hypocalcemia, definitely prevails over the disadvantages.

The use of low protein vegan diets supplemented with $CaCO_3$ in adequate amounts, permits to correct or prevent acidosis even in the case of extremely severe CRF, when the capacity to excrete H^+ is reduced to a few milliequivalents per day.

While the correction of metabolic acidosis may be well evident in the severe cases for the disappearance of its well-known manifestations as hyperventilation, the beneficial effects of correcting the less severe cases may not be apparent. This does not mean, however, that no benefit is induced: protein hypercatabolism disappears [23, 24], and the negative calcium balance is changed into a zero balance [25].

References

1 Giovannetti S, Bigalli A: Sulla patogenesi dell'acidosi uremica. Minerva Nefrol 1956;3:1–16.
2 Pellegrino ED, Biltz RM: The composition of human bone in uremia. Medicine 1965;44:397–406.
3 Pellegrino ED, Biltz RM, Letteri JM: Interrelationship of carbonate, phosphate, monohydrogen phosphate, calcium, magnesium and sodium in uraemic bone: comparison of dialyzed and nondialyzed patients. Clin Sci Mol Med 1977;53:307–314.
4 Oh MS: Irrelevance of bone buffering to acid-base homeostasis in chronic metabolic acidosis. Nephron 1991;59:7–10.
5 Berlyne GM: Acid-base homeostasis in chronic renal failure. What happens to the H^+ produced in excess? In De Santo NG, Capasso GB (eds): Acid-Base Balance in Uremia. Cosenza, Editoriale Bios, 1991, pp 247–255.
6 Onori L, Giammaria U, De Meo F, et al: Is the large intestine an accessory pH-buffering organ in chronic uremic acidosis? In De Santo NG, Capasso GB (eds): Acid-Base Balance in Uremia. Cosenza, Editoriale Bios, 1991, pp 243–245.
7 Cohen JJ, Kassirer JP: Acid-Base. Boston, Little, Brown, 1982.
8 Reddy GS, Jones G, Koch SW, Fraser D: Inhibition of 25-hydroxyvitamin D_3 1α-hydroxylase in chronic metabolic acidosis. Am J Physiol 1982;243:E265–E271.

9 Barsotti G, Lazzeri M, Cristofano C, et al: The role of metabolic acidosis in causing hyperphosphatemia. Mineral Electrolyte Metab 1986;12:103–106.
10 May RC, Kelly RA, Mitch WE: Metabolic acidosis stimulates protein degradation in rat muscle by a glucocorticoid-dependent mechanism. J Clin Invest 1986;77:614–621.
11 May RC, Kelly RA, Mitch WE: Mechanisms for defects in muscle protein metabolism in rats with chronic uremia. Influence of metabolic acidosis. J Clin Invest 1987; 79:1099–1103.
12 Giovannetti S, Bigalli A, Della Santa M: Gli effetti dell'acidosi sperimentale sul metabolismo proteico. Minerva Nefrol 1957;48:1–5.
13 De Fronzo RA, Bekles AD: Glucose intolerance following chronic metabolic acidosis in man. Am J Physiol 1979;236:E328–E334.
14 Van Ypersele DE, Strinou C: Potassium homeostasis in renal failure. Kidney Int 1977;11:491–504.
15 Barsotti G, Ciardella F, Cupisti A, et al: Nutritional treatment of renal failure in type I diabetic nephropathy. Clin Nephrol 1988;29:281–287.
16 Chan MK, Varghese Z, Moorhead JF: Lipid abnormalities in uremia, dialysis and transplantation. Kidney Int 1981;19:625–632.
17 Kim SL, Shoemaker WC: Role of the acidosis in the development of increased pulmonary vascular resistance and shock lung in experimental hemorrhagic shock. Surgery 1973;73:723–730.
18 Winer MW, Epstein FH: Signs and symptoms of electrolyte disorders; in Maxwell MH, Kleeman CR (eds): Clinical Disorders of Fluid and Electrolyte Metabolism, ed 2. New York, McGraw-Hill, 1972.
19 McCance RA, Widdowson EM: The composition of foods. Medical Research Council Special Report Series, No 297. London, HMSO, 1960.
20 Klahr S, Weiner ID: Disorders of acid-base metabolism; in Chan JCM, Gill JR Jr (eds): Kidney-electrolyte disorders. New York, Churchill Livingstone, 1990, pp 46–47.
21 Clarkson EM, McDonald SJ, De Wordener HE: The effect of high intake of calcium carbonate in normal subjects and in patients with chronic renal failure. Clin Sci 1966;30:425–438.
22 Makoff DL, Franklin SS, Gordon A, et al: Chronic calcium carbonate therapy in uremia. Arch Intern Med 1969;123:15–21.
23 Lyon DH, Dunlop DM, Stewart CP: The alkaline treatment of chronic nephritis. Lancet 1931;ii:1009–1013.
24 Papadoyannakis NJ, Stefanides CJ, McGeown M: The effect of the correction of metabolic acidosis on nitrogen and potassium balance in patients with chronic renal failure. Am J Clin Nutr 1984;40:623–627.
25 Litzow JR, Lemann J, Lennon EJ: The effect of treatment of acidosis on calcium balance in patients with chronic azotemic renal failure. J Clin Invest 1967;46:280–286.

Prof. Sergio Giovannetti, MD, Clinica Medica I, University of Pisa,
Via Roma 67, I–56100 Pisa (Italy)

Berlyne GM (ed): The Kidney Today. Selected Topics in Renal Science.
Contrib Nephrol. Basel, Karger, 1992, vol 100, pp 58–88

pH Homeostasis: The Conceptual Change

Edmund Bourke[a], *Dieter Häussinger*[b]

[a] VA Medical Center and State University of New York Health Sciences Center, Brooklyn, N.Y., USA;
[b] Medical University Clinic, Department of Medicine, University of Freiburg, FRG

Since the 19th century, it has been known that in metabolic acidosis urinary urea decreases and urinary ammonium (NH_4^+) increases. (In the nomenclature used herein, NH_4^+ is referred to as ammonium and NH_3 is referred to as ammonia.) Until recently, studies on the causal relationships have focused on the latter observation [1]. Soon after the demonstration by Van Slyke et al. [2] that glutamine was the major source of urinary ammonium, Pitts [3] summarized his concept of the enhanced renal glutamine ammoniagenesis induced by metabolic acidosis (fig. 1), a concept that stimulated much work over subsequent decades. He concluded that NH_3 derived from glutamine in renal tubular cells combined with hydrogen ions in acidic urine as NH_4^+, thereby eliminating H^+ from the body. He envisioned the hydrogen ions as being derived from the hydration of CO_2 (H_2CO_3) and their removal as leaving behind a HCO_3^- which returned to the circulation to replete the depleted body stores, which was the cause of the adaptation in the first place. Many experimental observations are compatible with Pitts' hypothesis. Thus, in experimental metabolic acidosis in the dog, not only is there an increase in glutamine utilization but, at an approximately 2:1 ratio [4], an increase in total renal ammoniagenesis, resulting predominately from the sequential deamidation and deamination of glutamine.

When facts fit a theory, it does not mean that they prove the theory; much as it dominated thinking in the field for decades, the Pitts hypothesis was incompatible with physical chemistry.

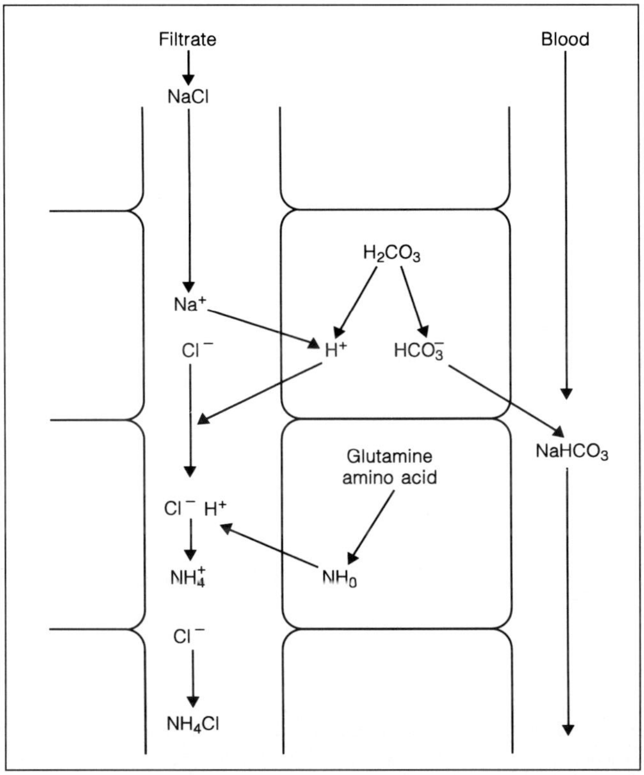

Fig. 1. Mechanisms of excretion of H^+ and NH_3 by the renal tubular cells as originally suggested by Pitts in 1948 [3].

Applications of Physical Chemistry

The misconcept of Pitts derived from writing the structural formula of glutamine in the unionized form as it had been frequently represented in biochemistry texts (fig. 2a). Were it to occur to any substantial extent in this form in physiological fluids, one could envisage that its deamidation to glutamate would yield NH_3 and its subsequent deamination to α-ketoglutarate would do likewise (fig. 2a). At physiological pH, however, the vast bulk of glutamine is zwitterionic [5] (fig. 2b), and its deamidation and deamination give rise to NH_4^+ (not NH_3) and α-ketoglutarate (not α-ketoglutaric acid) simply due to the pK values of these compounds. It is, thus,

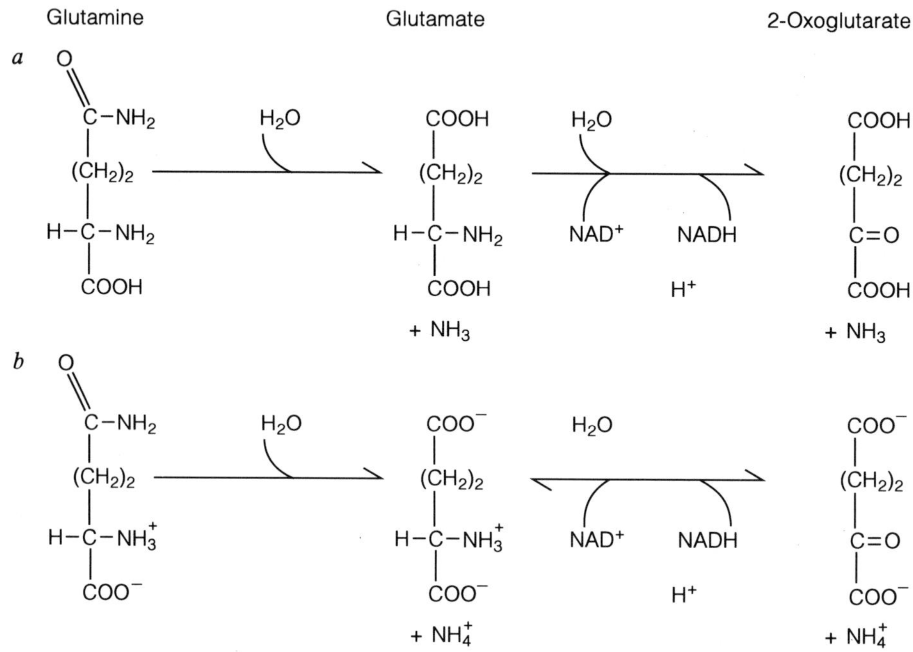

Fig. 2. Formation of 2-ketoglutarate (2-oxoglutarate) and ammonia (ammonium) from glutamine and glutamate in the unionized form (*a*) and in the zwitterionic form as occurs at physiological pH (*b*).

evident that glutamine does not give rise to NH_3 which can mop up hydrogen ions derived from the dissociation of carbonic acid, thereby leaving behind a newly generated bicarbonate to return to the body. When NH_4^+ is the product of glutamine degradation, its removal into the urine, where it also exists as NH_4^+, does not result in any uptake or release of hydrogen ions from the body's stores. The mechanisms whereby tubular NH_4^+ enters the urine are varied [6]. The extent to which there is direct transport of NH_4^+ across the luminal membrane – as distinct from its prior dissociation into H^+ and NH_3 which then recombine in acidic urine – will not, however, influence overall proton balance. The facts observed require an alternate explanation.

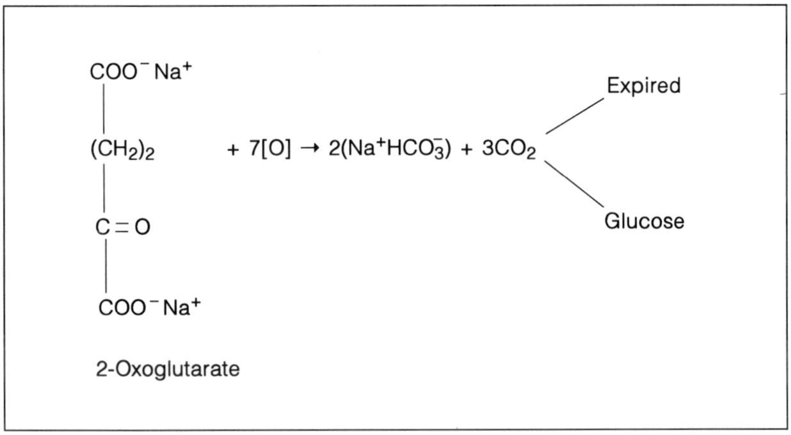

Fig. 3. Metabolism of 2-ketoglutarate (2-oxoglutarate).

Impact of Glutamine Carbon Oxidation on pH Homeostasis

Returning to the fate of the carbon skeleton of glutamine, α-ketoglutarate (fig. 2b), it carries two negative charges or carboxylate groups. Most of the increased production of α-ketoglutarate from glutamine by the kidney in metabolic acidosis is metabolized there, predominantly to glucose or CO_2. Whatever its fate, however, balancing the equation with respect to charge requires that it generates two bicarbonates [5] from each of the two carboxylate groups (fig. 3).

We will subsequently expand on this conclusion to a more general one, that the carboxylate groups of all amino acids must, on metabolism, give rise to bicarbonate [7]. But focusing at this stage on the bicarbonate generated in apparently increased quantities from the carbon skeleton of glutamine, α-ketoglutarate, in metabolic acidosis: could this represent an alternative explanation of the observed renal adaptation? Moreover, although the renal elimination of NH_4^+ may not mean proton elimination, are not two new bicarbonates generated from the resultant carbon skeleton of glutamine? Although the foregoing proposal was initially forwarded as an alternate interpretation of Pitts' hypothesis [5, 8–10], the shortcomings of such an interpretation were already implicit in Pitts' own experimental data in the dog kidney in vivo [11] on the renal utilization of lactate in metabolic acidosis and alkalosis. It is, as it were a mirror image of gluta-

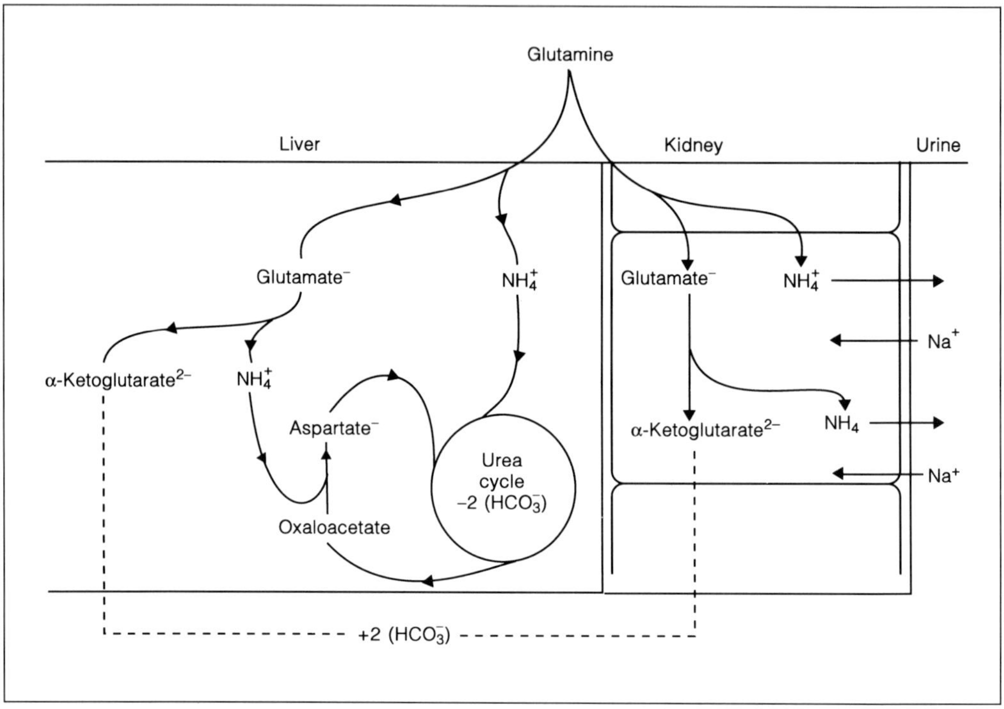

Fig. 4. Schematic representation of glutamine catabolism in the liver and kidney. The subsequent metabolism of the carbon skeleton gives rise to equal amounts of bicarbonate irrespective of where it is catabolized. For bicarbonate utilization in the urea cycle, see text. [From ref. 5.]

mine, decreased lactate utilization in acidosis. The interrelationships between these metabolic substrates are complex and have been elaborated on in rat kidney in vitro [12]. For acid-base purposes, however, by changing from one metabolic fuel to another, the kidney is changing from one bicarbonate precursor to another because the carboxylate group of lactate will also yield bicarbonate. Total renal CO_2 production is the same in metabolic acidosis and alkalosis [11]. Thus, the shift in renal substrate metabolism in metabolic acidosis could be seen as a shift in the utilization of bicarbonate precursors. The conclusion that the segregation into the urine of ammonium derived from glutamine, in metabolic acidosis, leads to a stoichiometric addition of new bicarbonate to the body is simply not substantiated by experimental data nor is it to be expected from physical

$$CH_3-\underset{\displaystyle NH_3^+}{\underset{|}{CH}}-COO^- + 3O_2 \rightarrow 2CO_2 + HCO_3^- + NH_4^+ + H_2O$$

Fig. 5. Oxidation of alanine.

chemistry. More cogently, when viewed in the context of total body metabolism, the renal generation of bicarbonate from the metabolism of carboxylate-containing substrates is not significant in the context of overall acid-base homeostasis. As already demonstrated in the case of glutamine and α-ketoglutarate, and alluded to in more general terms, the carboxylate groups of all amino acids and all other carboxylate-containing substrates including lactate will give rise to bicarbonate [7] irrespective of the organ system where they are metabolized. In so far as it is true that HCO_3^- generated in the kidney from the oxidation of the carbon skeleton of glutamine and entering the blood is alkalinizing, it is so only to the same extent as the same oxidation occurring elsewhere. And, if that HCO_3^- had not been generated in the kidney, the parent compound would have been oxidized elsewhere and exactly the same amount of HCO_3^- would have been supplied to the blood. Specifically in the case of glutamine, metabolic acidosis, in addition to increasing its renal utilization, decreases its hepatic utilization and increases its hepatic synthesis [13]. But whether it is utilized in the liver or transported to the kidney and utilized there, the same two bicarbonates are generated in both instances (fig. 4). Moreover, in parallel with the shift in α-ketoglutarate oxidation from the liver to the kidney in acidosis, the contribution of the kidney to whole-body gluconeogenesis increases whereas that of the liver decreases [14].

Protein Breakdown Is an Alkalinizing Process

Extrapolating from the metabolism of glutamine to amino acid metabolism in general, it can be seen that, as in the case of alanine (fig. 5), in aqueous solutions at physiological pH a simple amino acid contains one negatively charged carboxylate group and one positively charged substituted ammonium group. The oxidative metabolism of an amino acid necessarily leads to as many bicarbonates as the number of carboxylate groups on the amino acid and an approximately equal number of ammonium ions. It is self-evident and readily demonstrable that ammonium bicarbonate is an alkalinizing salt. This insight led Atkinson and Camien [7] to the inescapable conclusion that, in stark contrast to traditional thinking, protein catabolism is an alkalinizing process. When the amide (peptide) bonds of a protein are hydrolyzed there is no net liberation or consumption of protons but the resulting amino acids are dipolar ions at physiological pH. The subsequent metabolism of the amino acids of a typical human daily protein intake of 100 g yields about 1,000 mmol of bicarbonate [15, 16]. This is not to deny the liberation of protons that accompanies the metabolism of sulfur-containing amino acids such as cysteine and methionine. In quantitative terms, however, this would titrate only about 5% of the bicarbonate produced in overall protein catabolism [16]. As predicted from physical chemistry, the large alkalinizing load of ammonium bicarbonate that derives from the catabolism of protein differs negligibly in terms of its direct effects on pH from an equimolar load of sodium bicarbonate. Accordingly, NH_4^+ and Na^+ are interchangeable in aqueous solutions with respect to pH constancy in that solution. By analogy with this, excretion of NH_4^+ into urine as for instance by tubular sodium/ammonium exchange cannot per se be expected to affect the pH inside the body. In our marine ancestors, disposal of the metabolic load of bicarbonate (and ammonium) occurs through the gills. The alligator, during the periods when it is primarily aquatic, excretes the end products of protein catabolism, ammonium and bicarbonate, through the kidneys via a prodigious flow of urine [17]. For air-breathing land animals to emerge, an alternate evolutionary development was required for bicarbonate disposal. The disposal mechanism in mammals is ureagenesis [15, 16, 18].

Relations between Protein Oxidation, Ureagenesis and Renal Ammoniagenesis

Although decreased urea excretion in metabolic acidosis is as old an observation as the reciprocal rise in urine ammonium [1], it was the subject of less attention until recently. When experiments were subsequently repeated under more controlled conditions, the finding was essentially confirmed [5]. For instance, during chronic hydrochloric acidosis in the rat, the predicted rise in urinary ammonium was accompanied by an equimolar decrease in urinary urea [5]. Similar findings have been reported in man [19] where a fall in blood urea nitrogen was also observed. Furthermore, in the isolated perfused rat liver, lowering of medium pH is accompanied by a significant decrease in the rate of ureagenesis [20, 21], accounting for the decreased excretion observed in vivo. Studies on glutamine metabolism in the perfused liver tend to mirror those of urea, with an increase in net glutamine release at lower medium pH [18, 21–23]. It is of further interest that, during its passage through the isolated rat liver, the perfusing medium containing ammonium chloride becomes more acidic by about 0.2 pH units. When the effluent is exposed to the action of urease, however, its pH returns to that of perfusion input, demonstrating that the change results from urea synthesis rather than from other metabolic activities of the liver [13]. The relevance of these observations to acid-base homeostasis is apparent from review of the metabolic cycle of hepatic ureagenesis [5, 7, 13, 16, 22] (fig. 6). Ammonium and intramitochondrial bicarbonate are joined to form carbamylphosphate, with the self-evident consumption of a bicarbonate. In the subsequent formation of citrulline, a proton is released which in turn titrates a second bicarbonate. The second nitrogen is incorporated into the cycle from aspartate and the cycle ends in the production of urea. Thus, two bicarbonates are consumed with each revolution of the urea cycle. The balanced equation for ureagenesis is summarized in figure 7. Accordingly, urea synthesis is nothing more than an irreversible and energy-driven neutralization reaction of the strong base HCO_3^- (pK = 6.1) by the weak acid NH_4^+ (pK = 9.3). This important aspect of urea synthesis was overlooked in the past because attention focused mainly on ammonium detoxication. Hydrochloric acidosis, by decreasing ureagenesis, slows down that process relative to the rate of protein breakdown with resultant conservation of bicarbonate. Expressed in another way, as we shall further exemplify subsequently, in metabolic acidosis a considerable amount of the substituted ammonium of amino acids, ordinarily incorporated into urea with protonation of bicarbonate, is diverted via glutamine to be extracted by the kidney into the urine as NH_4^+ by a mechanism that does not utilize bicarbonate while simultaneously maintaining ammonium homeostasis. The overall effect is to mitigate the acidotic challenge. So we are no longer talking about a single organ, the kidney, as the metabolic regulator of systemic pH but of the coordinated action of at least two organs (fig. 8) in the overall response to metabolic acid-base disturbances. Adaptations in other organs such as muscle, bone and the gastrointestinal tract have also been referred to in the literature [7, 16, 21, 24–28], but are outside the scope of this review. This does not carry

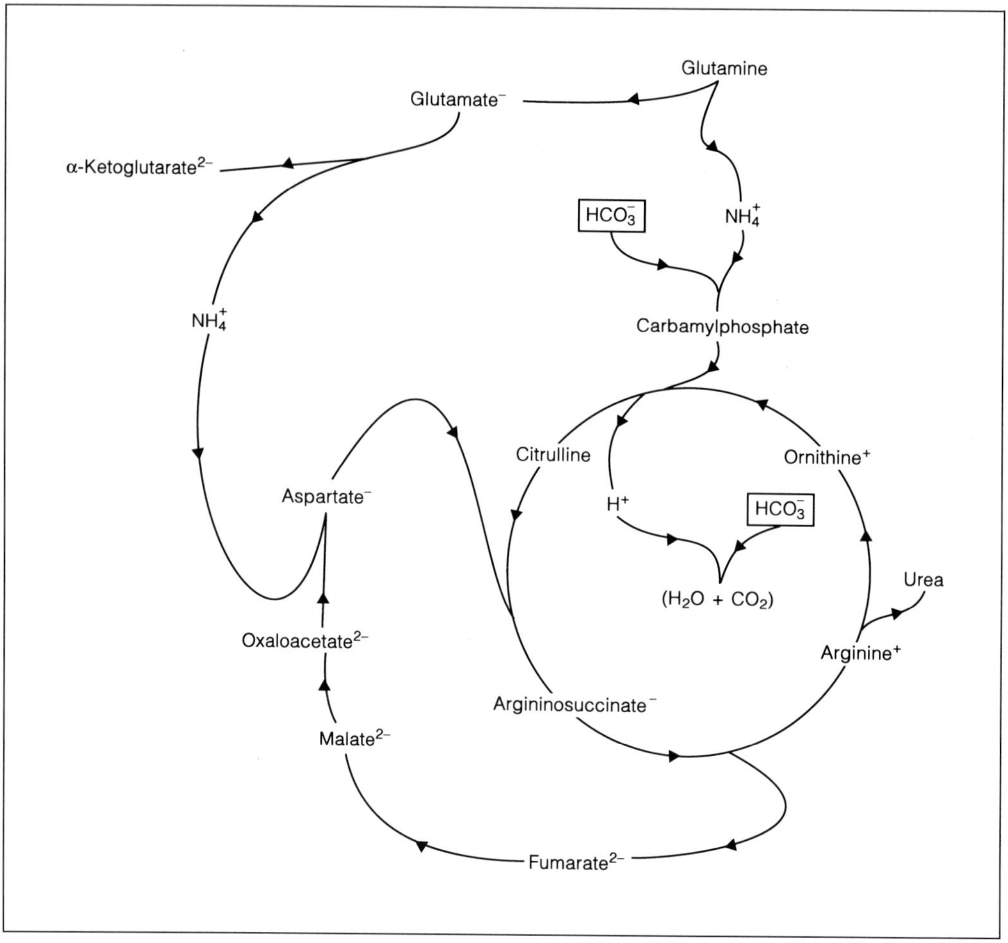

Fig. 6. Schematic representation of the urea cycle showing that two bicarbonates are effectively consumed with each turn of the cycle. [From ref. 5.]

$$2NH_4^+ + 2HCO_3^- \rightarrow NH_2\text{—}CO\text{—}NH_2 + CO_2 + 3H_2O$$

Fig. 7. Balanced equation for the synthesis of urea.

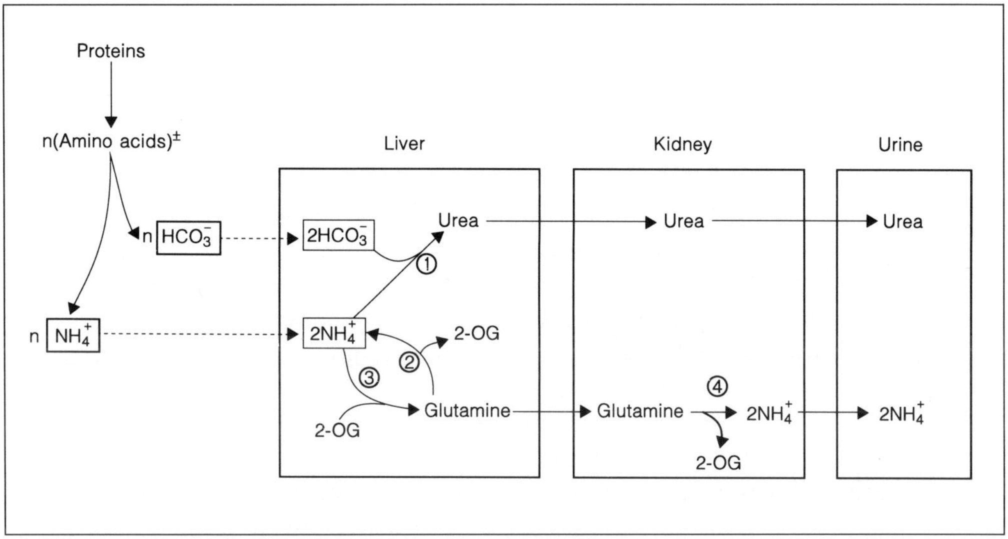

Fig. 8. Ammonium metabolism and bicarbonate homeostasis. NH_4^+ and HCO_3^- generation are ultimately linked in a 1:1 stoichiometry during protein catabolism as is the irreversible elimination of both compounds via hepatic urea synthesis. Flux through the urea cycle is sensitively controlled by the extracellular acid-base status. The mechanisms involved adjust bicarbonate-consuming urea synthesis to the requirements of acid-base homeostasis. When urea synthesis decreases relative to the rate of protein catabolism in acidosis, bicarbonate is spared and NH_4^+ is excreted as such in the urine (renal ammoniagenesis), with glutamine serving as a nontoxic transport form of ammonium from liver to kidney. When NH_4^+ is excreted as such in the urine there is no net production or consumption of α-ketoglutarate (2-oxoglutarate; 2-OG) in the organism. Numbers in circles refer to major points of flux control by the acid-base status. In metabolic acidosis, flux through the urea cycle (reaction 1) and hepatic glutaminase (reaction 2) is decreased, whereas flux through hepatic glutamine synthetase (reaction 3) and renal glutaminase (reaction 4) is increased. This interorgan team effort between the liver and the kidney results in NH_4^+ disposal without concomitant HCO_3^- removal from the organism. [From ref. 13.]

any implication for downplaying the importance of the kidney in this homeostatic response. The kidney plays a prerequisite role through the elimination of ammonium. The renal metabolic pathways of glutamine deamidation and deamination to ammonium have been extensively studied [for reviews, see ref. 29, 30 and 30a] and remain a field for productive future investigation since the enhanced glutamine ammoniagenesis induced by metabolic acidosis is still incompletely understood. Failure of the kidney to eliminate ammonium could have undesired side effects including overriding, by ammonium, of the pH-sensitive regulation of ureagenesis with resultant bicarbonate consumption and metabolic acidosis [13, 16]. The experimental acidosis induced by oral ammonium chloride administration is a classic example of such an override. This is readily illustrated in the rat in vivo by comparing dietary supplementation with ammonium chloride or ammonium bicarbonate [5]. Ammonium bicarbonate administration is accompanied by a marked increase in urinary urea with only a negligible change in urinary ammonium. In this instance, the bicarbonate utilization induced by ureagenesis is balanced by that administered in the salt with no effect on systemic pH. Conversely, following ammonium chloride, where the incorporation of ammonium into urea causes net bicarbonate utilization with metabolic acidosis, most of the additional dietary nitrogen intake is accounted for as increased urinary ammonium. Urinary urea increases significantly but to a much lesser extent than following ammonium bicarbonate. The increase, nonetheless, indicates override of the pH-sensitive regulation of ureagenesis by the ammonium load. It appears that, when ammonium chloride is converted to urea with a consequent fall in systemic pH, the incorporation of the substituted ammonium of amino acids into urea is decreased and diverted via glutamine synthesis to be subsequently released and excreted in the urine as NH_4^+. It is important in this regard to point out that the decrease in urea synthesis does not have to be absolute in order for hepatic bicarbonate conservation to occur [22]. Thus, in the case of ammonium chloride administration, there is an increase in urea excretion, but in the context of the total dietary nitrogen load there is an actual decrease [5]. The fall in systemic pH or some function thereof has, in relative terms, decreased the rate of protonation of bicarbonate which, in due course, would restore acid-base homeostasis.

Renal Ammonium Handling

The complexity of ammonium excretion [for review, see ref. 6] attests to the importance of this renal function even though it does not represent excretion of protons. In the proximal convoluted tubule where the bulk of the ammonium is generated, its transport across the luminal membrane is mainly by secondary active countertransport of NH_4^+ which substitutes for

H^+ on the luminal amiloride-sensitive Na^+/H^+ exchanger [31]. The ammonium is then delivered to the renal medulla via the loop of Henle in whose thick ascending limb it is substantially reabsorbed into the interstitium. This again occurs predominantly by secondary active transport, this time substituting for potassium on the luminal membrane Na^+-K^+-$2Cl^-$ cotransporter [32]. The associated medullary countercurrent multiplication favors secretion into the collecting duct. This 'shunt' results in substantial bypassing of the distal convoluted tubule [6]. In the collecting duct, NH_3 permeability is increased and it preferentially diffuses into the lumen in parallel with active H^+ secretion via a luminal proton translocating ATPase [33, 34]. The result is formation of NH_4^+ in the acidic milieu, facilitating enhanced outward diffusion of NH_3 down a concentration gradient. This 'diffusion trapping' is further augmented in parts of the collecting duct by the absence of luminal carbonic anhydrase with a resultant luminal acidic disequilibrium pH [35].

In summary, renal ammonium generation [29, 30, 30a] and elimination [6] are complex renal functions of substantial importance to the overall regulation of the homeostatic response to metabolic acidosis without invoking the need for the kidney to generate new bicarbonate. As already pointed out, failure of ammonium excretion may lead to bicarbonate consumption in ureagenesis [13, 16]. Its enhanced excretion, on the other hand, may be a reflection of bicarbonate conservation, a conservation that has already occurred by decreased ureagenesis in the liver.

Hepatic Ammonium Handling

Viewed from the perspective of total body metabolism, it is not so difficult to eschew the traditional concepts and realize that there is really no need for the kidney to generate new bicarbonate: since the metabolism of protein produces 1,000 mmol of bicarbonate daily [7], all that is necessary is to decrease its consumption. This is what the liver does in metabolic acidosis. In relative or in absolute terms, a decrease in hepatic ureagenesis slows the rate of bicarbonate consumption. From a teleological perspective it would make little sense to continue to consume bicarbonate unabated and at the same time to generate new bicarbonate. What happens instead, in the coordinated hepatorenal adaptation to an acid challenge, is that the liver consumes less bicarbonate leaving more behind to titrate the additional protons. In this context, ammonium is used for ureagenesis to the extent needed for disposal of bicarbonate with any excess being incorporated into glutamine for transport to and release in the kidney (fig. 8).

Arguments have been forwarded in the literature regarding what might be a primacy of organs in the regulated adaptations of these two processes, hepatic ureagenesis and renal ammoniagenesis, in metabolic acidosis [10, 24, 36, 37], proposing a primary adjustment in one organ resulting in secondary changes in the other. This is not a core issue since the coordination of both processes, hepatic conservation of bicarbonate and urinary excretion of ammonium, is essential to the overall effective regulation of bicarbonate and ammonium homeostasis. Moreover, available evidence indicates that both organs can adjust their function independently. The changes in the isolated perfused liver have already been alluded to [18, 21–23], and it has also been shown that acidosis-induced enhancement of glutamine ammoniagenesis is seen in the perfused kidney, independently of any possible hepatic influence [29]. What the evidence indicates is that hepatic and renal nitrogen metabolism are linked by an interorgan glutamine flux, coupling both renal ammoniagenesis and hepatic ureagenesis to systemic acid-base regulation.

A hepatic role in this interorgan 'team effort' is based upon three features. (1) The presence of a quantitatively important and liver-specific pathway for irreversible removal of metabolically generated bicarbonate, i.e. urea synthesis. (2) A structural-functional organization which uncouples urea cycle flux from the need to maintain ammonium homeostasis. (3) A sensitive and complex control of ureagenesis by the extracellular acid-base status, suggestive of a feedback control loop between the acid-base status and the rate of bicarbonate consumption, i.e. a bicarbonate homeostatic response in the liver.

The structural and functional organization of ammonium-metabolizing pathways in the liver is crucial for the effectiveness of the hepatic role in acid-base regulation. The functional unit of the liver is the acinus (fig. 9).

Each acinus extends from a terminal portal venule along a sinusoid to a terminal hepatic venule [38]. The hepatocytes near the sinusoidal inflow are termed periportal and those near the sinusoidal outflow are termed perivenous hepatocytes. A remarkable functional hepatocyte heterogeneity with respect to nitrogen metabolism (fig. 9) has been shown in comparative studies of antegrade and retrograde perfusion of the intact liver [39, 40]. These findings have now been amply confirmed in experiments on zonal liver damage, by immunohistochemistry and by mRNA in situ hybridiza-

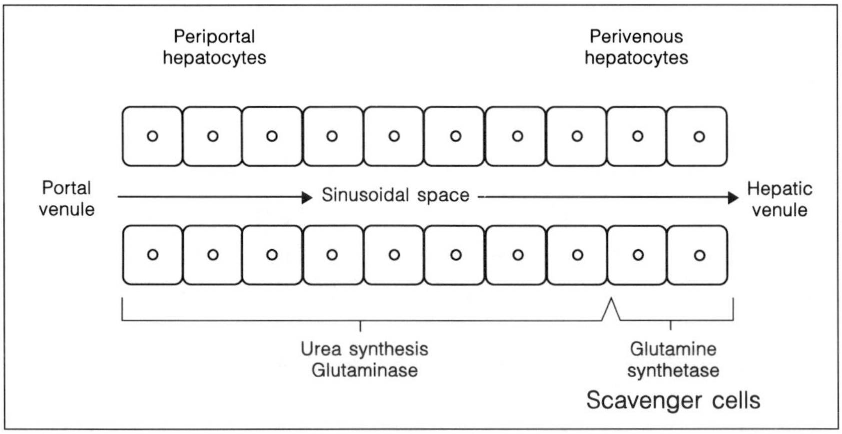

Fig. 9. The liver acinus. Periportal hepatocytes contain urea cycle enzymes and glutaminase but no glutamine synthetase. The latter enzyme is found exclusively in a small perivenous hepatocyte population near the outflow of the sinusoidal bed which is virtually free of urea cycle enzymes. Urea and glutamine synthesis, respectively, are thus structurally and functionally arranged in sequence, the periportal high-capacity but low-affinity system being followed by the perivenous high-affinity system for ammonium scavenging or detoxication. [From ref. 39.]

tion [for review, see ref. 22]. What the studies demonstrate is metabolic zonation of ureagenesis and glutamine synthesis, respectively attributable to a different distribution of key enzymes between the periportal (urea cycle enzymes, glutaminase) and perivenous (glutamine synthetase) hepatocytes of the hepatic acinus [for reviews, see ref. 22 and 23]. Accordingly, along the sinusoid the pathways of urea and glutamine synthesis are arranged in sequence. This organization prevents competition for ammonium between ureagenesis and glutamine synthesis. Rather there exists an established priority for ureagenesis (fig. 10). Thus, from the portal venule, throughout a substantial length of the hepatic acinus are found the enzyme glutaminase which contributes to the supply of ammonium for ureagenesis and also the enzymes of the urea cycle itself, sharing as it were a common hepatic compartment and giving prevalence to ureagenesis. Downstream, separately compartmentalized in the last rung of cells of the acinus surrounding the hepatic venule, is the enzyme glutamine synthetase, acting as a high-affinity scavenger for the ammonium which was not used in the periportal synthesis of urea [22]. This subacinar reciprocal distribution is

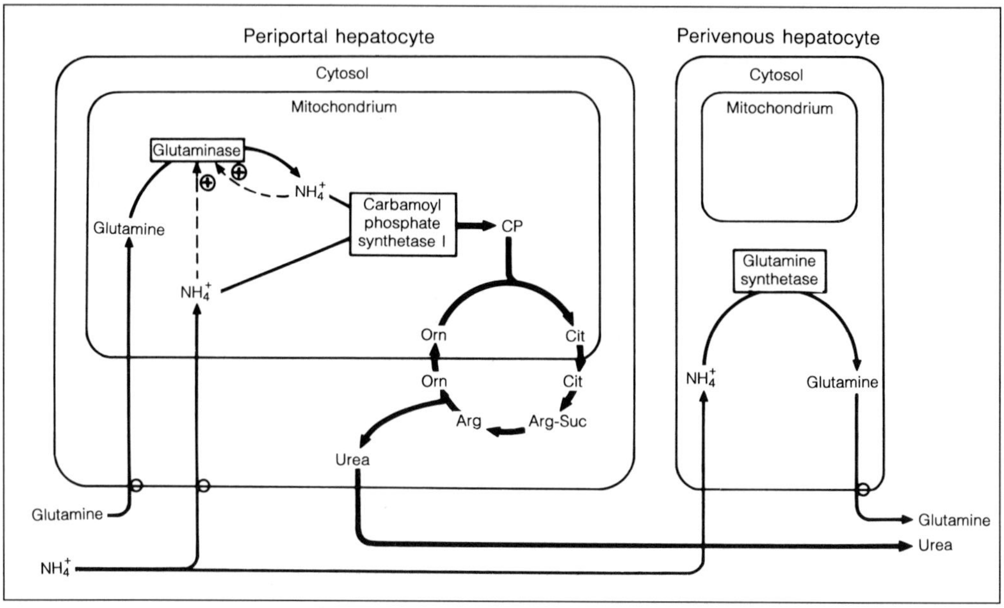

Fig. 10. Intercellular glutamine cycling and ureagenesis. Periportal glutaminase is activated by ammonium and acts as a pH- and hormone-modulated ammonium amplifier inside the mitochondria. The activity of this amplifier determines flux through the urea cycle. Glutamine synthetase in perivenous cells acts as a scavenger for the ammonium escaping periportal urea synthesis. This anatomic sequence uncouples urea synthesis from the vital need to ensure nontoxic ammonium levels and provides the basis for acid-base control of urea synthesis without the threat of hyperammonemia. [From ref. 23.]

regulated at a pretranslational level, has recently been demonstrated in man and seems to be unique to mammalian liver [19, 20, 41]. Furthermore, plasma membrane transport systems have a zonal distribution that appears complementary to the enzymes. Thus, plasma glutamate and α-ketoglutarate are transported into the perivenous hepatocytes whereas glutamine is transported predominantly into the periportal hepatocytes [22, 42]. Consequently, as the blood moves downstream through the sinusoid from the portal towards the hepatic venule (fig. 10), there is, initially, the formation of urea from the consumption both of bicarbonate derived from the metabolism of carboxylate groups of ketoacids and of ammonium derived in part from the portal blood but kept in adequate supply by the action of glutaminase on glutamine. Acidosis reduces this bicarbonate-utilizing process. As a consequence, there is ammonium left over which is not required for ureagenesis. This leftover ammonium is taken up by the last rung of cells of the acinus where glutamine synthetase incorporates it into glutamine, thereby controlling the blood ammonium concentration, preventing hyperammonemia and serving as a transport mechanism for ammonium to, for instance, the kidney.

Thus, urea synthesis is uncoupled from the vital need to dispose of potentially toxic ammonium because perivenous scavenger cells act as a 'backup system' for ammonium detoxication, even when urea synthesis is inhibited as for example in acidosis. This uncoupling without the threat of hyperammonemia is essential for urea cycle flux control by the acid-base status [for reviews, see ref. 22 and 23]. In a balanced acid-base situation, the rate of bicarbonate removal from the organism (urea synthesis) must match the rate of bicarbonate production (protein catabolism); in line with this, urea cycle flux is normally adjusted to the rate of protein breakdown. In acidosis, urea synthesis decreases relative to the rate of protein catabolism, resulting in retention of HCO_3^- as a pH homeostatic response by the liver. Several mechanisms have been identified which facilitate this sensitive adaptation of ureagenesis to the acid-base status [for review, see ref. 22 and 23]. These include liver glutaminase and carbonic anhydrase v, which play major regulatory roles: flux through both enzymes is inhibited during acidosis thereby regulating the extremely pH-sensitive input of ammonium and bicarbonate to carbamylphosphate synthetase, the rate-controlling enzyme of the urea cycle. Other factors contributing to this regulation include pH-dependent changes in the NH_3/NH_4^+ ratio and possibly also the formation of N-acetylglutamate, an activator of carbamylphosphate synthetase [22, 23]. Thus, the periportal acinar compartment seems to be primarily engaged in maintaining bicarbonate homeostasis by consuming as much ammonium as is required for bicarbonate elimination via urea synthesis, whereas the perivenous compartment serves the function of ammonium homeostasis, controlling blood ammonium levels by incorporating into glutamine the ammonium left over after urea synthesis. Final elimination of the ammonium, however, is achieved by renal ammoniagenesis, thereby reflecting the coordinated interorgan team effort (fig. 8).

In summary, the complex and sensitive control of ureagenesis by pH, HCO_3^- and CO_2 in experimental systems such as the perfused rat liver [for review, see ref. 22] seems well designed for an important contribution by the liver to the maintenance of acid-base homeostasis in vivo: whenever the pH and/or HCO_3^- and CO_2 concentrations in the extracellular space fall, the liver responds with an inhibition of bicarbonate-consuming ureagenesis.

Consequences of the Foregoing Considerations

A number of issues which have been raised in the recent literature will be addressed in light of the consequences of the foregoing considerations.

(1) Arguing that toxic levels of ammonium would have otherwise fatal consequences before life-threatening alkalosis would have time to develop [10, 43], it has been proposed that a role for ureagenesis in the regulation of systemic pH is, therefore, outside the bounds of normal physiology. Such an argument, however, does not necessarily hold validity. The acuteness of a potential consequence has no logical relation to the mode of regulation of a physiological process. In that ureagenesis clearly has a dual role, disposal of ammonium and of bicarbonate, there is some parallel here with the regulation of respiration. As exemplified above, an ammonium load (such as the nonphysiological administration of NH_4Cl) can stimulate ureagenesis, even overriding its regulation by pH. Similarly, in certain pathophysiological states, hypoxia may constitute the respiratory drive [44], and indeed the hypoxic consequences of strangulation are obvious even before severe respiratory acidosis would have time to develop. Yet this by no means negates the major role of pH and/or carbon dioxide concentration in the regulation of respiration and the role of the lungs in the maintenance of pH homeostasis [44].

Homeostatic regulatory mechanisms have evolved to deal with ongoing challenges to stability, not with impending doom. So the consequences of terminal liver failure and strangulation do not necessarily have relevance to the analysis of pH homeostatic mechanisms. More importantly, however, this viewpoint overlooks the fact that there are two major ammonium 'detoxicating' systems in the liver acinus which operate under physiological conditions and which, as emphasized, are represented in sequence (fig. 10) [16, 22, 39, 45, 46]. The $[K_{0.5}(NH_4^+)]$ for the more proximally located synthesis of urea is 3.6 m*M* compared to 0.11 m*M* for the distally located synthesis of glutamine, indicating a dramatically higher affinity of ammonium for glutamine synthesis [47]. In the perfused liver, roughly one third of the normal portal ammonium gets beyond the low-affinity high-capacity urea-synthesizing segment of the hepatic acinus, to be subsequently scavenged by the perivenous hepatocytes for high-affinity glutamine synthesis and ammonium and/or glutamate 'detoxication'. The importance of this 'perivenous scavenger cell hypothesis' [22], namely the distal acinar ammonium utilization process for the regulation of blood ammonium, is apparent from the hyperammonemia that results from inhibition of glutamine synthetase with methionine sulfoximine [39, 40, 48] or selective destruction of the glutamine-synthetase-containing perivenous hepatocytes by carbon tetrachloride administration, despite the fact that ureagenesis is unimpaired [46]. The regulatory advantage of this structural and functional backup system is that periportal ureagenesis can respond to acid-base challenge without the threat of hyperammonemia. This uncoupling [22] of ureagenesis from the prevention of hyperammonemia is an important prerequisite for efficient and sensitive feedback control of ureagenesis by systemic pH. Particularly cogent in this regard are studies in liver slices from cirrhotic patients [49]. The hyperammonemia of the hepatocellular decompensation of liver cirrhosis has classically been attributed to a decreased capacity to convert ammonium into urea. These studies point to a failure of the perivenous high-affinity system for ammonium detoxication, namely glutamine synthesis as a likely major factor in the hyperammonemia of cirrhotic liver failure, in addition to portal-systemic shunting.

(2) The acidosis that frequently accompanies renal disease has been interpreted to indicate a primacy of the kidney in the regulation of metabolic acid-base homeostasis. The traditional view that protein catabolism yields an excess of protons, particularly from sulfur-containing amino acids and phosphoproteins [37, 43, 50], and that homeostasis is maintained by their stoichiometric elimination in the form of urinary ammonium plus titratable acid [43, 51], might seem to find support in the metabolic acidosis of both generalized renal failure and specific renal tubular defects associated with decreased urinary ammonium and titratable acid (renal tubular acidosis) [52]. That excretion of ammonium does not represent excretion of acid has already been addressed and is discussed further in section 5 below. In the formation of urinary titratable acid, however, protons are unequivocally eliminated from the body [53, 54], accompanied by a net renal generation of bicarbonate [55]. The titratable acidity of the urine appears to be reflected in the amount of sulfate and phosphate that would arise from catabolism. It is conventionally viewed as primarily a means of preservation of acid-base balance by eliminating the excess protons generated with the release of these anions [56, 57]. From the perspec-

tive of total bicarbonate production from overall protein catabolism, however, this interpretation becomes highly questionable. As previously noted, when a sulfate ion and two protons are generated in the oxidation of cysteine and methionine, the protons will titrate two bicarbonate ions; in effect, two $NaHCO_3$ moles will be replaced by one Na_2SO_4. The roughly 70 milliequivalents of dietary-derived protons released daily would be readily titrated by the 1,000 mmol of bicarbonate also released by dietary catabolism. It does not require the elaboration of a urine of low pH, containing titratable acid, to achieve these ends. The process does, however, serve other crucial functions. It is an effective mechanism of excreting poorly reabsorbable anions while permitting reabsorption of what would otherwise be an obligatory loss of sodium and/or potassium [58]. It also represents an effective mechanism of trapping ammonium in the urine [6], facilitating its removal from the body under conditions of enhanced renal production, e.g. metabolic acidosis. The greater relevance of urinary acidification to ammonium excretion than to proton excretion is apparent from the physicochemical consideration of the consequences of lowering urine pH, a predominantly collecting-duct function. Some 75% of urinary titratable acid is attributable to $H_2PO_4^-$, which is already mostly formed by the end of the proximal tubule where luminal pH approximates 6.7 [59]. The elaboration of a urine pH which is roughly two orders of magnitude lower than that in the collecting duct results in only a modest further increase in titratable acid. Secretion of ammonium into the collecting duct (in contrast to the proximal tubule) is predominantly as NH_3; H^+, derived from peritubular NH_4^+, is separately pumped into the urine, to recombine as NH_4^+ at the lower luminal pH [6]. Highly acidic urine would thereby greatly accelerate NH_3 movement down a concentration gradient, facilitating net removal of NH_4^+.

Although a detailed treatment of the acidosis of renal disease is beyond the scope of this review, a few points are in order. Proximal (type 2) renal tubular acidosis is predominantly a bicarbonate reclamation defect [60] of diverse etiologies. Its consequence, a lowered plasma bicarbonate, is as predictable as the bicarbonate wastage from the gastrointestinal tract in certain diarrheal states. Classic distal (type 1) renal tubular acidosis is characterized by impaired acidification of the collecting-duct lumen and hence the final urine. Since urine pH in this condition generally reaches below the pK of the phosphate buffer system,

$$\frac{HPO_4^{2-}}{H_2PO_4^-}$$

(pK = 6.8) [59], the reduced excretion of titratable acid is not substantial and in light of the concepts presented here would not, per se, affect systemic pH. The restricted capacity to lower urine pH, one the other hand, impairs the renal elimination of ammonium [52, 60]. When distal renal tubular acidosis is induced acutely in the rat by amiloride administration, there is no change in total renal ammonium production, but its decreased urinary excretion is accompanied by a reciprocal increase in its renal venous return [61]. Decreased excretion of ammonium would be predicted to have a similar effect to its excessive ingestion, override of the pH-sensitive regulation of ureagenesis with increased hepatic consumption of bicarbonate and metabolic acidosis. Another form of renal tubular acidosis (type 4) is accompanied by impaired potassium excretion and hyperkalemia [60]. Hyperkalemia reduces renal ammoniagenesis and renal tubular ammonium transport [6], but further studies are required to delineate the pathogenesis of the acidosis. A striking acidosis may complicate diversion of the ureters into the colon in certain states of severe bladder dysfunction. There is an increased colonic load of ammonium derived from both its urinary excretion into the colon and the formation of ammonium bicarbonate from urinary urea by colonic bacterial urease. This is absorbed into the portal circulation [62]. There is also enhanced mucosal Cl^-/HCO_3^- exchange [63]. The net result is increased entry of NH_4Cl into the portal vein with the same consequences as an oral load of NH_4Cl, namely metabolic acidosis. Finally, although the metabolic acidosis of generalized renal failure is not yet entirely understood and is probably multifactorial in etiology [64–68], it is again characterized by a significant reduction in ammonium excretion with the same sequela as outlined above.

(3) Although a role for hepatic ureagenesis in acid-base homeostasis is gaining increasing acceptance [68–70], certain clinical and experimental observations have been interpreted as not favoring any such role. A sodium bicarbonate load was found to have no effect on urea synthesis in the rat [5] or man [71, 72]. Such would not have been anticipated, however, from our analysis. The large release of alkalinizing ammonium bicarbonate from protein catabolism is due to the release of approximately equal amounts of both ions, and ureagenesis is the process by which protons are irreversibly unlocked from ammonium to neutralize bicarbonate. It cannot occur if an additional excess of bicarbonate is not accompanied by an additional ammonium source such as, for instance, the simultaneous administration of arginine or lysine with bicarbonate. More relevant to our thesis is the potential role of metabolic alkalosis as an amplifier for hepatic

glutamine deamidation and consequently enhanced urea cycle flux under conditions of decreased ureagenic capacity [22, 49] (see below). Another study [37] concluded that urea formation was not regulated primarily by acid-base balance in vivo, based on the authors' interpretation of the consequences of a large ammonium chloride load: following its sustained administration, a constant increase in ureagenesis was observed over the time interval studied and systemic pH was noted to fall. This, from our analysis, is scarcely surprising. The authors' interpretation of their data, however, could be misleading [16, 25]: plotting systemic pH against urea synthesis, it appeared that pH was progressively falling while ureagenesis was remaining constant. Displaying the functional relationships in this way results in a misrepresentation. What was happening in reality was that systemic pH was being driven down by enhanced ureagenesis due to a massive NH_4Cl load and not a failure of a decreased pH to reduce ureagenesis. Another argument in this area has been related to the effects of hepatectomy. Because urea synthesis is a major pathway for irreversible disposal of bicarbonate, selective impairment of urea synthesis in liver disease might predictably diminish hepatic bicarbonate disposal with resultant metabolic alkalosis. Experimental hepatectomy did not appear to support this predication [73]. In acute studies in the rat, the failure of partial hepatectomy (85%) to influence acid-base homeostasis, despite a significant decrease in ureagenesis, led the authors to the conclusion that a major role of the liver in the regulation of acid-base homeostasis was unlikely. Importantly, however, bicarbonate conservation occurs only when urea synthesis decreases relative to protein breakdown. This apparently did not occur in this experimental design because neither bicarbonate nor ammonium accumulated to a significant extent; it seems that partial hepatectomy in these studies diminished both ureagenesis and protein breakdown to the same extent as might be predicted from the known major role of the liver in amino acid breakdown. In chronic liver disease in man, however, the picture becomes fully evident [49]: in a series of patients with histologically stratified liver disease, the in vivo plasma bicarbonate increased with progressive loss of urea cycle capacity. In these patients, other causes of metabolic alkalosis such as diuretics or antacids or vomiting, hyperaldosteronism or renal disease were rigorously excluded. Because alkalosis in turn is a potent stimulus for the ammonia amplifier, glutaminase [49], a feedback circuit between urea synthesis, bicarbonate accumulation and amplification of hepatic ammoniagenesis via glutaminase can be hypothesized as follows [49, 74, 75]: a decrease of urea cycle capacity leads to hyperbicarbonatemia and alkalosis which, in turn, activates glutaminase. Glutaminase activation augments urea synthesis and restores a normal urea flux despite diminished capacity to synthesize urea [22]. Since the net consequence of this metabolic shift is stimulation of hepatic bicarbonate consumption, metabolic alkalosis may have a self-limiting effect in the stable cirrhotic patient which would predictably increase to a new steady state with progressive loss of hepatic functional reserve. Interestingly, the progressive loss of urea cycle capacity in cirrhosis is paralleled by an increase in renal ammonium excretion, despite coexisting metabolic alkalosis, indicating that the kidney undertakes the task of eliminating ammonium when urea synthesis fails [49, 75]. This increase is not attributable to the small rise in plasma NH_4^+ concentration (which remains within the normal range), but requires de novo renal generation. This opens up a new facet on the regulation of renal ammoniagenesis. It appears that renal ammonium production is activated when ureagenesis decreases; this occurs not only in acidosis when urea synthesis is inhibited due to homeostatic regulation, but also during alkalosis when urea synthesis fails due to liver disease.

(4) Some investigators have seen the metabolism of protein to urea simply as one metabolic sequence resulting in no net bicarbonate turnover and without any impact on bicarbonate concentration. Thus, protein catabolism is not seen as a source of alkali. As a consequence of this formulation, the traditional view is upheld that protein catabolism predominately due to sulfur-containing amino acids leads to a net excess production of protons which requires the action of the kidney alone to prevent metabolic acidosis. In such a view, bicarbonate is seen to behave like pyruvate as an intermediate in the conversion of glucose to CO_2 [57, 76]. This view, however, is incorrect because urea is not necessarily the only end product of protein catabolism. For example, the process may stop at the level of ammonium and bicarbonate which, as mentioned above, are excreted by the gills of fish and the kidneys of the aquatic alligator. Nor can protein degradation be seen as a metabolic sequence [77, 78], but rather as proteolysis followed by the degradation of 20 amino acids along individual pathways, each with its own regulatory interactions. On the other hand, ureagenesis can be more accurately described as a typical biosynthetic pathway, converting simple starting materials into a more complex product at the expense of 4 ATP equivalents for every revolution of the cycle and a substantial utilization of hepatic energy metabolism [25], and responding to regulatory signals quite different from those of amino acid catabolic pathways. Bicarbonate is, therefore, an end product of metabo-

lism, eliminated as such, for instance through the gills of fish, into the surrounding oceans throughout countless millions of years of evolution. In this regard, the adaptation to urea excretion of a teleost fish, *Oreochromis alcalicus grahami* [79, 80], is noteworthy as a strategy for survival in the extremely alkaline environment of Lake Magadi in the Great Rift Valley, Kenya (pH = 10, total CO_2 = 180 mmol/l). This is the only known instance of complete ureotelism in an entirely aquatic teleost fish. Urea output was inhibited completely at pH 7. A closely related species of fish without this capacity for hepatic ureagenesis dies within an hour in such an alkaline milieu. Ureagenesis in this instance is an adaptation to the inability to adequately excrete bicarbonate and consequently the need to remove it by titration. Likewise, the emergence of air-breathing land animals is similarly faced with the need to dispose of bicarbonate, and in mammals, as outlined above, the mechanism evolved to do this is ureagenesis.

(5) Some investigators persist in the misunderstanding that NH_4^+ excretion is equivalent to proton excretion. There are two separate defenses for this viewpoint.

(a) The first relates to the assumption that NH_3 is produced in metabolism and that any NH_4^+ that is excreted in the urine may be considered to be an NH_3 molecule carrying a proton that it has picked up along the way [51, 56, 57, 81]. That would be a valid conclusion if NH_3 were provided to the kidney (or anywhere else in the body for that matter), for example by infusing NH_3 gas. But nothing of this kind happens in the body. Amino acids are dipolar ions, each bearing a carboxylate group, $-COO^-$, and a substituted ammonium group, $-NH_3^+$. Thus, the N is already protonated in the amino acid. When NH_4^+ is liberated in the course of catabolism of the amino acid, it merely retains the proton that it already had in the amino acid. If the NH_4^+ turns up in the urine, it cannot have picked up a proton from the body (including the kidney). Movement of NH_4^+ from its site of generation in the body to the urine obviously cannot involve any net transfer of proton. The actual pathway taken makes no difference. The overall process is merely $R\text{-}NH_3^+$ (intracellular) $\rightarrow$ NH_4^+ (urine) and there is no change in proton balance. The same applies to the amide nitrogen of glutamine. Apparently, it is the positive charge on the NH_4^+ that has frequently led to the conventional view in renal physiology that it should be considered to be an NH_3 molecule carrying a proton piggyback and that this proton is somehow different from protons in other molecules or ions. As has recently been emphasized [82], charge is irrelevant in such contexts. Thus, despite its negative charge, $H_2PO_4^-$ is about 100 times as strong an acid as NH_4^+; that is, it donates a proton about 100 times as readily as NH_4^+. Yet, no one says that the proton available from $H_2PO_4^-$ should be added to the titratable acidity in estimating acid excretion. For precisely the same reason that such an addition would be chemically absurd, it is not valid to consider the proton in the much weaker conjugate acid NH_4^+ to represent strong acid nor to add it to the titratable acidity in determining net renal acid excretion [82]. The point at issue can be further illustrated with the example of the normal urinary constituents, methylammonium, $CH_3NH_3^+$, derived from the endogenous metabolism of sarcosine [83] and ethylammonium, $C_2H_5NH_3^+$, derived from theanine, a constituent of tea leaves [84]. Unlike ammonium these substituted ammoniums are not significantly utilized in the body. Their oral administration does not affect systemic pH [E. Bourke, unpubl. observ.] and their urinary excretion could hardly be conceived of as representing significant proton excretion. Finally, if for arguments' sake NH_3 were somehow produced from R-NH_3^+, where would the proton go? This conventional view not only abandons the Bronsted [85] concept of acidity but also violates simple stoichiometry or conservation of mass and charge!

(b) A second group of investigators who hold that NH_4^+ excretion is equivalent to proton excretion have a somewhat different perspective. Arguing that, whereas NH_4^+ may not be an acid in a beaker, the definition of an acid in a living organism is more complex and NH_4^+ is an acid in a physiological context [86]. The proponents of this view conclude that because NH_4^+ can donate a proton in ureagenesis, albeit metabolically, it is an effective and important acid in the body. Consequently, its renal excretion is equivalent to the loss of acid from the body since it allows equal amounts of bicarbonate to persist. Such a view, on the one hand, acknowledges the important role of hepatic ureagenesis in controlling bicarbonate homeostasis, i.e. one hallmark of our considerations, but, on the other hand, compounds are seen as physiologically relevant acids when they possess the inherent potential to generate a proton during their metabolism. In such a case, also methanol (see below), ethylene glycol and even glucose (which can yield pyruvic and lactic acid) would fulfill the criteria of an acid, and the discrepancy with the acid-base definition given by Bronsted [85] and Lowry [87] becomes obvious. However, there is no physicochemical nor physiological justification for this departure from the Bronsted-Lowry definition. The fact that a compound can be used in a metabolic sequence in which protons are generated does not justify it being labeled an acid. This applies as much to ammonium chloride administration as to, for instance, methanol administration [88]. Neither can donate protons directly in significant amounts under physiological conditions. It is not the presence of either compound but specifically their conversion to urea and formate, respectively, that liberates protons and consumes bicarbonate. Should either compound not be metabolized at all or be metabolized by potential alternate routes such as to methyltetrahydrofolate in the case of methanol or to amino acids in the case of ammonium, bicarbonate consumption would not result. If one were to accept the capacity to donate a proton through metabolism as an appropriate definition of an acid, then methanol would be included and the elimination of methanol from the body could be perceived as allowing bicarbonate to persist as has been alluded to above in the case of ammonium. Since the metabolic steps liberating the proton from NH_4^+ or methanol are subject to metabolic regulation, this nonconventional acid-base terminology implies that the strength of an acid can also be regulated and thereby varied. Such thinking is neither meaningful nor helpful. In aqueous solutions, acids and bases react with each other spontaneously and instantaneously to reach equilibria which are determined by their respective pK values and the ambient pH. In the physiological context the only special situation is an ambient pH that fluctuates narrowly around 7.4 rather than between 1 and 14.

(6) Sometimes coupled with the proposal that NH_4^+ excretion represents acid excretion [51] and sometimes as a separate proposal [6, 10, 89, 90], the argument persists that renal metabolism of α-ketoglutarate derived from glutamine breakdown in acidosis will, via the oxidative or the gluconeogenic pathway, result in HCO_3^- formation or proton consumption by the kidney. As previously alluded to, however, this cannot affect HCO_3^- homeostasis in the whole organism, because α-ketoglutarate formation in the kidney is in effect accompanied by α-ketoglutarate consumption in the glutamine-producing organs such as the liver (fig. 8) and muscle. In metabolic acidosis, flux through urea synthesis and hepatic glutaminase is decreased, whereas flux through glutamine synthesis and renal glutaminase is increased, resulting in NH_4^+ disposal without concomitant HCO_3^- removal (fig. 8). When NH_4^+ ions are excreted as such in the urine, there is no net production or consumption of α-ketoglutarate in the whole organism. Expressed in another way, the NH_4^+ that the liver processes is partitioned between urea and glutamine. In the synthesis of urea a proton is released from each NH_4^+ and is available to titrate HCO_3^-. In the synthesis of glutamine, NH_4^+ is consumed, balancing the negative charge of the γ-carboxylate group of glutamate. The glutamine is not excreted, and in the first step of

its metabolism, whether in the liver, the kidney or elsewhere, NH_4^+ is regenerated along with the carboxylate group. Thus, the partitioning of NH_4^+ between urea and glutamine is of critical importance to pH homeostasis. Each ammonium that is used in ureagenesis liberates a proton permanently and thus removes a bicarbonate by titration. Incorporation of NH_4^+ into glutamine involves no donation of protons to the surroundings and so has no net effect on systemic acid-base balance. The cycling between glutamate and glutamine, and by the same token between α-ketoglutarate and glutamate, can neither yield nor consume a proton and is without effect on the bicarbonate buffer system or on blood pH. The real homeostatic response to acidosis has actually taken place when the bicarbonate, already formed from protein catabolism, is conserved through inhibition of ureagenesis. The variety of ways in which a 5-carbon compound may be cycled through different organs need not complicate this simple fact.

Conclusion

The metabolic component of acid-base homeostasis has traditionally been seen to involve only one organ, the kidney. Recent conceptual developments, however, point to an involvement of the liver as a major pH homeostatic organ. The catabolism of proteins generates large amounts of bicarbonate and thus a potential threat of alkalosis. Nature has developed different strategies for its disposal in different organisms. In mammals, there is a liver-specific pathway for irreversible removal of metabolically generated bicarbonate, namely urea synthesis. There is a sensitive and complex control of bicarbonate disposal via this pathway by the extracellular acid-base status. A structural-functional sequence for nitrogen disposal in the hepatic acinus which uncouples urea synthesis from the need to prevent hyperammonemia permits the effective regulation of systemic pH via ureagenesis. Thus, in acidosis, urea synthesis decreases relative to the rate of protein catabolism, resulting in a retention of bicarbonate as a pH homeostatic response by the liver. This sparing of bicarbonate in acidosis occurs without hyperammonemia because perivenous scavenger cells downstream from the acinar site of ureagenesis act as an efficient backup system for ammonium removal via an alternate route, namely its incorporation into glutamine. Renal glutaminolysis and subsequently urinary ammonium excretion provide a final sink for this leftover ammonium. Both hepatic ureagenesis and renal ammoniagenesis are subject to complex and sensitive short- and long-term regulation under such conditions as metabolic acidosis. Renal ammoniagenesis can be seen as a finely tuned spillover for the ammonium requiring elimination without concomitant bicarbonate consumption, and renal ammoniagenesis may therefore represent a primary ammonium homeostatic response rather than a bicarbonate homeostatic one. In fact, the bicarbonate homeostatic response antedates and in a sense leads to the ammoniagenic response by virtue of the structural-functional arrangement in the hepatic acinus. In this regard, ureagenesis is seen as an ATP-driven proton pump in which protons are forcibly transferred from NH_4^+ to HCO_3^- at the expense of 2 ATP equivalents each, a process that is responsive to the needs of systemic pH. In this shift of emphasis from a single organ regulator to a two or 'two plus' organ overview of the metabolic component of acid-base homeostasis, an important role is seen for hepatic ureagenesis. Clinical disorders of kidney or liver function are associated with acid-base disturbances which can be readily understood in terms of the conceptual changes in pH homeostasis outlined in this review.

References

1 Bourke E, Atkinson DE: The development of acid-base concepts and their application to mammalian pH homeostasis; in Häussinger D (ed): pH Homeostasis: Mechanisms and Control. New York, Academic Press, 1988, pp 163–179.

2 Van Slyke DD, Phillips RA, Hamilton PB, Archiballd RM, Fructor PH, Hiller A: Glutamine as source material of urinary ammonia. J Biol Chem 1943;150:481–483.

3 Pitts RF: The renal excretion of acid. Fed Proc 1948;7:418–426.

4 Bourke E, Frindt G, Paez-Rubio D, Schreiner GE: Effects of chronic alkalosis and acidosis on the glutaminase II pathway in the dog kidney in vivo. Am J Physiol 1971;220:1033–1036.

5 Oliver J, Bourke E: Adaptations in urea and ammonium excretion in metabolic acidosis in the rat: A reinterpretation. Clin Sci Mol Med 1975;48:515–520.

6 DuBose TD Jr, Good DW, Hamm LL, Wall SM: Ammonium transport in the kidney: New physiological concepts and their clinical implications. J Am Soc Nephrol 1991;1:1193–1203.

7 Atkinson DE, Camien MN: The role of urea synthesis in the removal of metabolic bicarbonate and the regulation of blood pH. Cur Top Cell Regul 1982;21:261–302.

8 Bourke E: Studies on metabolic aspects of acid-base regulation. Ir J Med Sci 1977; 146:119–129.

9 Halperin ML, Jungus RL: Metabolic production and renal disposal of hydrogen ions. Kidney Int 1983;24:709–713.

10 Walser M: Roles of urea production, ammonium excretion and amino acid oxidation in acid-base balance. Am J Physiol 1986;250:F181–F188.
11 Pitts RF: Production of CO_2 by the intact functioning kidney of the dog. Med Clin North Am 1975;59:507–518.
12 Preuss HG, Roxe DM, Bourke E: Acidotic alterations in oxidative metabolism influences rat renal slice ammoniagenesis. Life Sci 1987;41:1695–1702.
13 Häussinger D, Gerok W, Sies H: The effect of urea synthesis on extracellular pH in isolated perfused rat liver. Biochem J 1986;236:261–265.
14 Yudkin J, Cohen RD: The contribution of the kidney to the removal of a lactic acid load under normal and acidotic conditions in the conscious rat. Clin Sci Mol Med 1975;48:121–131.
15 Atkinson DE, Bourke E: The role of ureagenesis in pH homeostasis. Trends Biochem Sci 1984;9:297–300.
16 Atkinson DE, Bourke E: Metabolic aspects of the regulation of systemic pH. Am J Physiol 1987;252:F947–F956.
17 Coulson RA, Hernandez T: Alligator metabolism: Studies on chemical reactions in vivo. Comp Biochem Biophys [B] 1983;74:1–182.
18 Häussinger D, Gerok W, Sies H: Hepatic role in pH regulation: Role of the intercellular glutamine cycle. Trends Biochem Sci 1984;9:300–302.
19 Fine A, Carlyle JE, Bourke E: The effects of administrations of HCl, NH_4Cl and NH_4HCO_3 on the excretion of urea and ammonium in man. Eur J Clin Invest 1977; 7:587–589.
20 Lueck JD, Miller LL: The effects of perfusate pH on glutamine metabolism in the isolated perfused rat liver. J Biol Chem 1970;245:5491–5497.
21 Oliver J, Koelz AM, Costello J, Bourke E: Acid-base alterations in glutamine metabolism and ureagenesis in perfused muscle and liver of the rat. Eur J Clin Invest 1977; 7:445–449.
22 Häussinger D: Nitrogen metabolism in liver: Structural and functional organization and physiological relevance. Biochem J 1990;267:281–290.
23 Häussinger D, Meijer AJ, Gerok W, Sies H: Hepatic nitrogen metabolism and acid-base homeostasis; in Häussinger D (ed): pH Homeostasis: Mechanisms and Control. New York, Academic Press, 1988, pp 337–377.
24 Welbourne TC: Interorgan glutamine fluxes in acid-base disorders; in Häussinger D (ed): pH Homeostasis: Mechanisms and Control. New York, Academic Press, 1988, pp 379–401.
25 Atkinson DE: Functional roles of urea synthesis in vertebrates. Physiol Zool, in press.
26 Foley WJ, Hume ID: Nitrogen requirements and urea metabolism in two arboreal marsupials, the greater glider *(Petauroides volans)* and the brushtail possum *(Trichosurus volpecula)* fed *Eucalyptus* foliage. Physiol Zool 1987;60:241–250.
27 Bushinsky DA, Lechleider RJ: Mechanism of proton-induced bone calcium release: Calcium carbonate dissolution. Am J Physiol 1987;253:F998–F1005.
28 Guppy M: The hybernating bear: Why is it so hot and why does it cycle urea through its gut? Trends Biochem Sci 1986;11:274–276.
29 Tannen RL: Ammonia and acid base homeostasis. Med Clin North Am 1983;67: 781–798.
30 Tannen RL, Sastrasinh S: Response of ammonia metabolism to acute acidosis. Kidney Int 1984;25:1–10.

30a Schoolwerth AC: Regulation of renal ammoniagenesis in metabolic acidosis. Kidney Int 1991;40:961–973.
31 Nagami GT: Luminal secretion of ammonia in the mouse proximal tubule perfused in vitro. J Clin Invest 1988;81:159–164.
32 Watts BA III, Good DW: Effects of NH_4^+ transport on intracellular pH (pH_i) in rat medullary thick ascending limb (MTAL). J Am Soc Nephrol 1990;1:660A.
33 Good DW, Caflisch CR, DuBose TD Jr: Transepithelial gradients in inner medulla of the rat. Am J Physiol 1987;252:F491–F500.
34 Gluck S, Kelly S, Al-Awqati Q: The proton translocating ATPase responsible for urinary acidification. J Biol Chem 1982;257:9230–9233.
35 Starr RA, Burg MB, Knepper MA: Luminal disequilibrium pH and ammonia transport in outer medullary collecting duct. Am J Physiol 1987;252:F1148–F1157.
36 Phromphetcharat V, Jackson A, Dass PD, Welbourne TC: Ammonia partitioning between glutamine and urea: Interorgan participation in metabolic acidosis. Kidney Int 1981;20:598–605.
37 Halperin ML, Chen CB, Cheema-Dhadli S, West ML, Jungas RL: Is urea formation regulated primarily by acid-base balance in vivo? Am J Physiol 1986;250:F605–F612.
38 Lamers WH, Hilberts A, Furt E, Smith J, Jonges GN, Van Noorden CJF, Gaasbeek Janzen JW, Charles R, Moorman AFM: Hepatic enzyme zonation: A reevaluation of the concept of the liver acinus. Hepatology 1989;10:72–76.
39 Häussinger D: Hepatocyte heterogeneity in glutamine and ammonia metabolism and the role of an intercellular glutamine cycle during ureagenesis in perfused rat liver. Eur J Biochem 1983;133:269–275.
40 Häussinger D, Gerok W: Hepatocyte heterogeneity in glutamate uptake in isolated perfused rat liver. Eur J Biochem 1983;136:421–425.
41 Campbell SW: Excretory nitrogen metabolism; in Prosser CL (ed): Comparative Animal Physiology. New York, Wiley, 1991, vol 1, pp 277–324.
42 Stoll B, NcNelly S, Buscher HP, Häussinger D: Functional hepatocyte heterogeneity in glutamate, aspartate and alpha-ketoglutarate uptake: A histoautoradiographical study. Hepatology 1991;13:247–253.
43 Knepper MA, Burg MB, Orloff J, Berliner RW, Rector FC: Ammonium, urea and systemic pH regulation. Am J Physiol 1987;253:F199–F202.
44 Piiper J: Pulmonary and circulatory carbon dioxide transport and acid-base homeostasis; in Häussinger D (ed): pH Homeostasis: Mechanisms and Control. New York, Academic Press, 1988, pp 181–202.
45 Häussinger D, Sies H, Gerok W: Functional hepatocyte heterogeneity in ammonia metabolism: The intercellular glutamine cycle. J Hepatol 1984;1:3–14.
46 Häussinger D, Gerok W: Hepatocyte heterogeneity in ammonia metabolism: Impairment of glutamine synthesis in CCl_4 induced liver cell necrosis with no effect on urea synthesis. Chem Biol Interact 1984;48:191–194.
47 Kaiser S, Gerok W, Häussinger D: Ammonia and glutamine metabolism in human liver slices: New aspects on the pathogenesis of hyperammonemia in chronic liver disease. Eur J Clin Invest 1988;18:535–542.
48 Oliver J, Bourke E: Adaptations in metabolic acidosis: A reinterpretation. Proc Eur Dial Transplant Assoc 1975;11:411–416.
49 Häussinger D, Steeb R, Gerok W: Ammonium and bicarbonate homeostasis in chronic liver disease. Klin Wochenschr 1990;68:175–182.

50 Schwartz WB, Relman AS: Renal acidosis; in Richet G (ed): Proc 1st Int Congr Nephrol. Basel, Karger, 1961, pp 478–483.
51 Knepper MA, Packer R, Good DW: Ammonium transportation in the kidney. Physiol Rev 1989;69:179–249.
52 Wrong OM, Davies HEF: The excretion of acid in renal disease. Q J Med 1959;28: 259–313.
53 Henderson LJ: A critical study of the process of acid excretion. J Biol Chem 1911;9: 403–424.
54 Henderson LJ, Palmer WW: On the several factors of acid excretion. J Biol Chem 1914;17:305–315.
55 Pitts RF, Alexander RS: The nature of the renal tubular mechanism for acidifying the urine. Am J Physiol 1945;144:239–254.
56 Sullivan LP, Grantham JJ: The role of the kidney in acid-base regulation; in Sullivan LP, Grantham JJ (eds): Physiology of the Kidney. Philadelphia, Lea & Febiger, 1982, pp 196–200.
57 Carroll HJ, Oh MS: Role of urea synthesis in acid base balance; in Carrol HJ, Oh MS (eds): Water, Electrolyte and Acid-Base Metabolism: Diagnosis and Management, ed 2. Philadelphia, Lippincott, 1989, pp 64–65.
58 Schwartz WB, Cohen JJ: The nature of the renal response to chronic disorders of acid base equilibrium. Am J Med 1978;64:417–428.
59 Halperin ML: How much 'new' bicarbonate is formed in the distal nephron in the process of net acid excretion? Kidney Int 1989;35:1277–1281.
60 Sebastian A, Morris RC Jr: Renal tubular acidosis. Clin Nephrol 1977;7:216–230.
61 Kurtz I, Dass PD, Cramer S: The importance of renal ammonium metabolism to whole body acid-base balance: A reanalysis of the pathophysiology of renal tubular acidosis. Miner Electrolyte Metab 1990;16:331–340.
62 McConnell JB, Murison J, Stewart WK: The role of the colon in the pathogenesis of hyperchloremic acidosis in ureterosigmoid anastomosis. Clin Sci 1979;57:305–312.
63 Davis GR, Morawski SG, Santa Ana CA, Fordtran JS: Evaluation of chloride/bicarbonate exchange in the human colon in vivo. J Clin Invest 1983;71:201–207.
64 Schwartz WB, Hall PW, Hays RM, Relman AS: On the mechanism of acidosis in chronic renal disease. J Clin Invest 1959;38:39–52.
65 Muldowney FP, Donahoe JF, Carroll DV, Powell D, Freeney R: Parathyroid acidosis and uremia. Q J Med 1972;163:321–342.
66 Warnock DG: Uremic acidosis. Kidney Int 1988;34:278–287.
67 Buerkert J, Martin D, Trigg D, Simon E: Effect of reduced renal mass on ammonia handling and net acid formation by the superficial and juxtamedullary nephron of the rat: Evidence of impaired re-entrapment rather than decreased production of ammonia in the acidosis of uremia. J Clin Invest 1983;71:1661–1675.
68 Cohen RD: The renal tubular acidoses; in Weatherall DJ, Ledingham JGG, Warrell DA (eds): The Oxford Textbook of Medicine, ed 2. New York, Oxford University Press, 1987, vol 18, pp 171–174.
69 Meijer AJ, Lamars WH, Chamuleau RAF M: Nitrogen metabolism and ornithine cycle function. Physiol Rev 1990;70:701–748.
70 Cohen RD, Woods HF: Disturbances of acid-base homeostasis; in Weatherall DJ, Ledingham JGG, Warrell DA (eds): The Oxford Textbook of Medicine, ed 2. New York, Oxford University Press, 1987, vol 9, pp 164–175.

71 Gougeon-Reyburn R, Marliss EB: Effects of sodium bicarbonate on nitrogen metabolism and ketone bodies during very low energy protein diets in obese subjects. Metabolism 1989;12:1222–1230.
72 Bjerrum K, Vilstrup H, Almdal TP, Kristensen LO: No effect of bicarbonate induced alkalosis on urea synthesis in normal man. Scand J Clin Lab Invest 1990;50: 137–141.
73 Almdal T, Vilstrup H, Bjerrum K, Kristensen LO: Decrease in ureagenesis by partial hepatectomy does not influence acid-base balance. Am J Physiol 1989;257:F696–F699.
74 Häussinger D: Organization of hepatic nitrogen metabolism and its relation to acid-base homeostasis. Klin Wochenschr 1990;68:1096–1101.
75 Häussinger D: Hepatic nitrogen metabolism and systemic pH control: New concepts and pathophysiological aspects. Clinica Laboratorio 1991;15:77–86.
76 Halperin ML, Jungas RL, Cheema-Dhadli J, Brosnan JT: Disposal of the daily acid load: An integrated function of the liver, lungs and kidneys. Trends Biochem Sci 1987;12:197–199.
77 Atkinson DE, Bourke E: pH balance. Trends Biochem Sci 1987;12:179–180.
78 Häussinger D: pH balance. Trends Biochem Sci 1987;12:179–180.
79 Randall DJ, Wood CM, Perry SF, Bergman H, Maloiy GMO, Mommsen TP, Wright PA: Urea excretion as a strategy for survival in a very alkaline environment. Nature 1989;337:165–166.
80 Wood CM, Perry SF, Wright PA, Bergman HL, Randall DJ: Ammonia and urea dynamics in the Lake Magadi tilapia, a ureotelic teleost fish adapted to an extremely alkaline environment. Respir Physiol 1989;77:1–20.
81 Toto RD: Metabolic acid-base disorders; in Kokko JP, Tannen RL (eds): Fluids and Electrolytes, ed 2. Philadelphia, Saunders, 1990, pp 301–390.
82 Atkinson DE: Urinary NH_4^+ and pH homeostasis. Respir Physiol, in press.
83 Davis EJ, DeRopp RS: Metabolic origin of urinary methylamine in the rat. Nature 1961;283:162–167.
84 Asatoor AM: Tea as a source of urinary ethylamine. Nature 1966;210:1358–1360.
85 Bronsted JN: Einige Bemerkungen über den Begriff der Säuren und Basen. Recueil Trav Chim 1923;42:718–728.
86 Alpern RJ, Starr R, Seldin DW: Hepatic renal interrelations in acid-base regulation. Am J Physiol 1988;255:F807–F808.
87 Lowry TM: The uniqueness of hydrogen. Chem Ind 1923;43:43–47.
88 Atkinson DE, Bourke E: Hepatic renal interrelations in acid-base regulation. Am J Physiol 1988;255:F808–F809.
89 Good DW: New concepts in renal ammonium excretion; in Seldin DW, Giebisch G (eds): The Regulation of Acid-Base Balance. New York, Raven Press, 1989, pp 170–171.
90 Brosnan JT, Vinay P, Gougoux A, Halperin ML: Renal ammonium production and its implications for acid-base balance; in Häussinger D (ed): pH Homeostasis: Mechanisms and Control. New York, Academic Press, 1988, pp 281–304.

Edmund Bourke, MD, VA Medical Center, 800 Poly Place,
Brooklyn, NY 11209 (USA)

Berlyne GM (ed): The Kidney Today. Selected Topics in Renal Science.
Contrib Nephrol. Basel, Karger, 1992, vol 100, pp 89–104

Whole Body Acid-Base Balance

Man S. Oh, Hugh J. Carroll

Department of Medicine, SUNY, Health Science Center at Brooklyn, N.Y., USA

Under steady-state condition, net acid production must equal net acid excretion. The concept currently accepted for the balance between acid production and acid excretion is as follows. There are three major metabolic sources of acid: sulfuric acid, incompletely metabolized organic acid, and acid (or alkali) absorbed from the gastrointestinal (GI) tract [1–7]. Urinary excretion of sulfate and organic anions measures the first two components of net acid production, and the analysis of food and feces has been used to estimate the third [1]. Acid is excreted in urine either as ammonium or titratable acid. The sum of these two minus urinary bicarbonate is net acid excretion, which represents net renal generation of bicarbonate. The evidence most widely cited in support of the aforementioned techniques of measuring net acid production and net acid excretion is the observation that acid production measured from urinary sulfate and urinary organic acid was exactly equal to net acid excretion when the influence of the GI absorption of alkali was eliminated by placing the experimental subjects on a liquid formula diet containing a net alkali content of zero [2]. However, these techniques have not been rigorously examined, and further careful analysis of them and of data from other sources casts considerable doubt on the validity of such conclusions. Furthermore, the demonstration of an apparent parity between net acid production and net acid excretion as defined above does not prove that no other sources of acid production and acid excretion exist that might have escaped detection by these techniques. The purpose of this presentation is to evaluate the techniques that are presently used to measure net acid production and net acid excretion. Particular emphasis will be placed on various methods of measuring net alkali absorption from the GI tract, net acid excretion, and urinary organic anions.

Uncertainty of Net Acid Excretion as the Measure of Net Renal Production of Alkali

At the present time, net acid excretion is measured as urinary ammonia + titratable acid – bicarbonate. The sum of urinary ammonium and titratable acid is acid excretion and represents production of alkali by the kidney. When urine bicarbonate, which represents alkali loss, is subtracted from acid excretion, the result is termed net acid excretion, which represents net renal generation of bicarbonate. The concept of titratable acid is reasonably straightforward. The pH of urine at the glomerulus (the ultrafiltrate) is the same as that of plasma. Urine pH falls as H^+ is added by its tubular secretion. Some H^+ is buffered by HCO_3^- with formation of H_2CO_3, and consequent reabsorption of bicarbonate. H^+ accepted by other urinary buffers is excreted in the urine. Back titration of urine to the level of the plasma pH releases an amount of H^+ equal to that titrated by urinary buffers other than NH_3, and the quantity of strong alkali used to titrate urine to the plasma pH is called titratable acid. Titration of urine to the pH of the plasma does not release H^+ that is associated with NH_4+, and urinary NH_4+ is separately measured.

The main question to be addressed here is, does net acid excretion as currently measured accurately quantify net acid excretion by the kidney? First, there is a minor technical problem with the measurement of titratable acid, which has been addressed by Lemann et al. [8]. When urine is titrated to the plasma pH, urinary calcium tends to be precipitated with phosphate with liberation of additional H^+, which leads to overestimation of titratable acid. In addition, titration of urine to the plasma pH results in titration of about 1/40 of urinary ammonium, again leading to overestimation of titratable acid. The latter error is usually minor, but could be substantial if urine contained a large amount of ammonium.

A potentially more serious source of error is the possible presence of some other buffered H^+ in the urine that is not represented by ammonium, and that escapes titration of urine to pH 7.4. The method for the measurement of net acid excretion assumes that the only buffered H^+ that escapes detection in the measurement of titratable acid is that associated with ammonium. However, this assumption has not been the result of a negative finding after a careful search. In fact, there is some evidence that urine, in normal as well as in abnormal conditions, contains buffers that titrate H^+ but escape detection when urine is titrated to pH 7.4. For example, normal urine contains various amines such as ethanolamine, phosphoethanolamine, methylamine, and dimethylamine [9]. The normal concentration in urine of each of these substances is less than 1 mmol [9], but it is possible that many other amines besides those mentioned above could be excreted in small amounts individually, but in substantial amounts collectively.

The excretion of amino acid in urine is considered to be neutral with respect to acid-base balance, because most amino acids contain both $-NH_3^+$ and $-COO^-$ groups. Metabolism of amino group releases a proton, whereas metabolism of carboxyl group produces alkali. However, excretion of basic amino acids (which contain more amino group than carboxyl group) in urine represents the excretion of acid, whereas excretion of acidic amino acids (which contain more carboxyl group than amino group) represents the loss of alkali. Normal urine contains more basic amino acids such as histidine and arginine than acidic amino acid [10]. The total quantity of amino acid excreted in urine is normally insignificant, but this may change in certain disease states. The preliminary results of an investigation performed in our laboratory indicates that normal urine contains substantial amounts of buffer which can be identified in the pH range between 7.4 and 10.5 after ammonium, phosphate, and bicarbonate have been removed. The amount ranged from 13.3 to 40.0 mEq/g of creatinine [Oh and Carroll, unpubl. data]. At pH 10.5, the amounts of calcium and magnesium that can be titrated are insignificant. Since the amino groups of amino acids are mostly titratable in the pH range of 7.4 and 10.5, and since 8–10 mmol of amino acids are excreted in urine daily, part of this buffer undoubtedly represents titration of amino acids. However, the quantity of these unidentified buffers far exceeds the quantity of amino acid usually found in urine.

Measurement of Acid Production

Among the various mechanisms of acid production, that of sulfuric acid is best understood. It results from metabolism of sulfur-containing amino acids: $R\text{-}S \rightarrow CO_2 + \text{urea} + H_2SO_4$. The resulting H_2SO_4 is promptly buffered by a base such as $NaHCO_3$: $H_2SO_4 + 2\ NaHCO_3 \rightarrow Na_2SO_4 +$

H_2CO_3. Since sulfate does not accumulate in the body but is excreted promptly in urine, and since sulfate usually is not ingested as a salt, urinary sulfate fairly accurately represents sulfuric acid production [1]. Empirically, administration of sulfur-containing amino acids also has been shown to produce predictable reduction in serum bicarbonate [1].

Organic acid consumes bicarbonate in the reaction: organic acid + $HCO_3^- \rightarrow$ organic anion + CO_2. When the retained organic anion is metabolized, bicarbonate is regenerated as shown in the reaction: organic anion + $H_2CO_3 \rightarrow HCO_3^-$ + organic acid $\rightarrow HCO_3^- + CO_2 + H_2O$.

When organic anion is not metabolized, either because it cannot be metabolized, e.g., urate, or escapes metabolism, e.g. citrate, it is excreted in the urine, and it represents loss of alkali. Urinary organic anion is measured collectively by the Van Slyke-Palmer method [11], which requires titration of urine between pH 2.7 and 7.4 after removal of phosphate.[1] The result is then corrected for the presence of creatinine, which is titrated in the same pH range but is not an organic anion.

The measurement of organic acid by the Van Slyke and Palmer method has several pitfalls. First, urine may contain neutral substances besides creatinine that are titrated in the pH range of 2.7 and 7.4. An example is creatine. Normally the amount of creatine excreted daily in the urine is less than 2 mmol, but that amount may be increased in certain abnormal conditions. Other such substances may also be excreted in small quantities individually but in substantial amounts collectively. Leaf and co-workers [12] demonstrated that certain patients with renal tubular acidosis excreted a large quantity of substances that were titrated by the Van Slyke-Palmer method and removed by Lloyd's reagent, but were not creatinine. Second, the removal of phosphate with calcium hydroxide is known to remove some of the organic anions. The well-known examples are citrate and oxalate. The latter is insignificant quantitatively, but the

[1] Only the loss of organic anions in urine represents loss of alkali, but titration of urine between pH 2.7 and 7.4 by the Van Slyke-Palmer method measures total organic acid in urine (acids as well as anions), and hence slightly overestimates the actual alkali loss. However, the overestimation of alkali loss by this method is exactly cancelled out by the overestimation of titratable acid when urine is titrated from the pH of the voided urine to 7.4. Titration of urine to pH 7.4 titrates, in addition to phosphate and creatinine, any organic acid in the undissociated form. Excretion of an organic acid in urine does not represent gain of alkali, since organic acids originate from neutral substances such as fat or carbohydrate, and thus excretion of organic acids is no different from excretion, for example, of glucose.

former is excreted in substantial amounts. Furthermore, since urinary citrate excretion increases with alkalosis and decreases with acidosis [13–15], the amount of citrate removed by addition of calcium hydroxide will not remain constant but vary with systemic pH. Some other organic anions that behave like citrate and oxalate might also be removed by calcium hydroxide. Third, not all organic anions are titrated when urine is titrated to pH 2.7. For example, about 20% of an acid with pK of 3.3 would still remain untitrated at pH 2.7. It is not known what percentage of the urinary organic acids have a pK sufficiently low to permit them to remain substantially untitrated at pH 2.7. Although a large number of organic anions have been identified in urine, still the sum of these known organic anions is far less than the amounts that are presumed to be measured by the van Slyke and Palmer technique [16]. Finally, some neutral amino acids, whose excretion should not be included as loss of alkali, would also be partially titrated by the method. Although the carboxyl groups of amino acids usually have pKs in the range of 2.7 and 1.5, still about 20% of an amino acid with pK of 2.1 would be titrated at pH 2.7 [17]. With normal amounts of amino acid excretion in urine, the amount of amino acid titrated by the Van Slyke-Palmer method would be insignificant, but it could be substantial in subjects excreting large amounts of amino acids, such as those with aminoaciduria.

An additional source of acid is phosphoric acid produced from the metabolism of phosphoesters and phosphoproteins [2, 6]: $R\text{-}PO_4 \rightarrow CO_2 + H_3PO_4$; $5\ H_3PO_4 + 9\ NaHCO_3 \rightarrow 4\ Na_2HPO_4 + 1\ NaH_2PO_4 + 9\ H_2CO_3$. Phosphates produced in these reactions would ultimately be excreted in the urine. However, unlike sulfate, urinary phosphate cannot be equated with phosphoric acid production, since most of the phosphate excreted in urine was ingested as neutral phosphate. Another source of acid is metabolism of cationic amino acids. Acid is generated only when such cationic amino acid is balanced by noncombustible anion such as chloride: $R\text{-}NH_3{+}\ Cl^- \rightarrow$ urea $+ CO_2 + HCl$. The resulting HCl will be quickly buffered, and chloride ultimately excreted in the urine. Again, since much of the chloride appearing in urine is chloride ingested as dietary salts, NaCl and KCl, measurement of urinary chloride cannot be used to quantify the acid produced from metabolism of cationic amino acids balanced by chloride. For these reasons, the contribution of any mineral acids other than sulfuric acid cannot be assessed through excretion of individual anions, but can be measured collectively as part of net GI absorption of alkali, which is discussed in detail in the next section.

GI Absorption of Alkali and Acids

The normal diet contains large quantities of alkali, acids, potential alkali, and potential acids, but their ultimate impact on the body's acid-base balance depends primarily on their absorption. The pattern of absorption depends not only on the nature of the substance but also on the interaction with other chemicals, both endogenous and exogenous [18]. In general, the amount absorbed is nearly equal to the amount ingested if the substance is soluble and is readily absorbed, e.g., NaCl, $NaHCO_3$, Na citrate, K citrate, NH_4Cl. An insoluble substance such as $CaCO_3$ must react with gastric acid to become soluble and absorbable. In some cases, only the anionic part is absorbable; in other cases, the situation is reversed. The overall effects of these substances on the body's acid-base balance are complex, and are often unrelated to the acidity or alkalinity of the substance. The following describes 5 different categories of chemical substances and the effects of their ingestion on acid-base balance.

Ingestion of Nonabsorbable or Poorly Absorbable Cation Accompanied by Absorbable but Nonmetabolizable Anion. Examples: $CaCl_2$, $MgSO_4$

$CaCl_2$ given intravenously is neutral, but when taken by mouth it is an acidifying agent. When it is ingested, virtually all of the Cl but only a fraction of the Ca is absorbed [19]. The excess Ca remaining in the gut combines with CO_3, organic anions, or phosphate and is excreted in the stool; excretion with CO_3 or organic anions represents loss of alkali. Ingested $CaCl_2$ enters the duodenum and reacts with $NaHCO_3$: $CaCl_2 + 2\ NaHCO_3 \rightarrow 2\ NaCl + CaCO_3$. NaCl is absorbed and $CaCO_3$ is excreted in stool. The loss of alkali is then the result of what is essentially an exchange between Cl and CO_3 or organic anions. When Cl exchanges with another nonmetabolizable anion, HPO_4, there is no net gain or loss of alkali: $CaCl_2 + Na_2HPO_4 \rightarrow 2\ NaCl + CaHPO_4$. The gain of acid after ingestion of $CaCl_2$ can be calculated as: chloride absorbed – (calcium absorbed + phosphate excreted in stool in exchange for chloride[2]).

[2] For the estimation of net acid gain, phosphate excreted in stool will be expressed as mmol of phosphate excreted $\times$ 1.8, while all other electrolytes will be expressed as mEq.

The effect of oral ingestion of $MgSO_4$ has not been studied carefully. However, since SO_4 is more readily absorbed than Mg, the overall effect of $MgSO_4$ ingestion on the whole body acid-base balance would be similar to that of ingestion of $CaCl_2$. On the other hand, Mg is better absorbed than Ca, but is a less effective binder of phosphate than Ca [20]. Similarly, it can be predicted that $FeSO_4$ would result in gain of acid since more SO_4 than Fe is likely to be absorbed.

Ingestion of Salts Consisting of Nonabsorbable or Poorly Absorbable Cation Accompanied by Absorbable and Metabolizable Anion. Examples: $CaCO_3$, $Al(OH)_3$, Ca Acetate, Ca Citrate

The amount of alkali gained when these salts are ingested depends on absorption of anions, either with the accompanying cations, or in exchange for nonmetabolizable anion such as phosphate [19]. Administration of $CaCO_3$ results in gain of alkali by the body only when Ca is absorbed or when CO_3 is exchanged for HPO_4. Both processes require that $CaCO_3$ react with HCl in the stomach because $CaCO_3$ is nearly insoluble and will be mostly excreted in stool in the absence of gastric acid. The reaction in the stomach will be: $CaCO_3 + 2\,HCl \rightarrow CaCl_2 + H_2CO_3$. The fate of $CaCl_2$ thus formed will be the same as that of $CaCl_2$ ingested as such: most of it, after reacting with $NaHCO_3$ in the duodenum, is converted back to $CaCO_3$ and is excreted in stool, some is absorbed as $CaCl_2$, and some is excreted as $CaHPO_4$ after reacting with sodium phosphate.

Ingestion of organic salts of calcium such as calcium citrate, calcium lactate, calcium gluconate, and calcium acetate also results in gain of alkali. As with calcium carbonate, alkali gain occurs when these organic anions are absorbed either with calcium or in exchange with phosphate. Unlike calcium carbonate, however, organic salts of calcium do not need reactions with HCl in the stomach for their absorption or binding of calcium with phosphate.

The overall alkalinizing effect of various calcium antacids depends less on the absorption of calcium than on the ability of calcium to bind phosphate, because the amount of calcium absorbed is usually much less than the amount excreted in combination with phosphate. The phosphate-binding ability depends in part on availability of soluble calcium. The amount of phosphate bound by calcium when calcium is ingested as $CaCl_2$ is comparable to that bound by aluminum ingested as $Al(OH)_3$. For each 100 mEq of antacid ingested, phosphate excreted was 11.9 mEq with $CaCl_2$ and 11.4 mEq with $Al(OH)_3$. However, when calcium was ingested as the carbonate salt, only half as much phosphate (6 mEq) was removed for the same amount of calcium. The likely explanation is the poor solubility of $CaCO_3$ and the need for it to react with HCl in the stomach. If conversion of $CaCO_3$ to $CaCl_2$ in the stomach were complete, the reactions in the duodenum would be the same as those following ingestion of $CaCl_2$. However, the fact that phosphate excretion in stool after ingestion of $CaCO_3$ is only half that excreted after ingestion of $CaCl_2$ [19], suggests that conversion to $CaCl_2$ in the stomach is incomplete. Calcium acetate binds phosphate better than $CaCO_3$, most likely because the former is more soluble than the latter. Calcium acetate is much more effective binder of phosphate than calcium citrate, because it is more soluble than citrate, and also because citrate competes with phosphate for binding with calcium more effectively than acetate [20].

In the case of $Al(OH)_3$ or aluminum carbonate, because aluminum is virtually unabsorbable, hydroxide or carbonate absorption is possible only in exchange for phosphate. However, net gain of alkali is much greater with $Al(OH)_3$ than with $CaCO_3$, because Al is a more effective binder of phosphate than most calcium salts. The binding of phosphate by $Al(OH)_3$ or aluminum carbonate occurs only when it is administered in a colloidal form, probably because the noncolloidal $Al(OH)_3$ does not have a big enough contact area for chemical reactions [19, 20].

Ingestion of Anion Exchange Resins

Anion exchange resins are made of nonabsorbable and nonmetabolizable polycations (usually quarternary ammonium groups attached to polystyrene skeleton) balanced by exchangeable anions [19]. They have either an alkalinizing or acidifying effect depending on whether the accompanying anion is metabolizable or nonmetabolizable. When the counterion is nonmetabolizable, for example chloride, its absorption in exchange for organic anions, carbonate, or bicarbonate represents loss of alkali. Some chloride is exchanged for phosphate, with neither a net gain nor loss of alkali. When the counterion is metabolizable, for example acetate, a net gain of alkali occurs as the metabolizable anion is absorbed in exchange for chloride or phosphate. Exchange of acetate with other organic anions, carbonate, or bicarbonate is a neutral process for the acid-base balance.

Ingestion of Cation Exchange Resins.
Example: Sodium Polystyrene Sulfonate (Kayexalate)

The unabsorbable and nonmetabolizable anion in a cation exchange resin is usually sulfonate attached to polystyrene skeleton. The anion is balanced by exchangeable and absorbable cation. In the case of sodium polystyrene sulfonate (Kayexalate), the exchangeable cation is sodium. Although the specific purpose of administering sodium polystyrene sulfonate is to exchange Na^+ for K^+ for treatment of hyperkalemia, the Na^+ counterion exchanges with virtually all other cations including Ca^{2+}, Mg^{2+}, NH_4^+, and H^+. Exchange of Na^+ for NH_4^+ or H^+ would result in gain of alkali; exchange of NH_4^+ for K^+ is neutral. One might predict that exchange of Na^+ for Ca^{2+} would lead to neither acid nor alkali gain, but the net effect of such exchange is gain of alkali. Normally, the bulk of calcium excreted in stool is in the form of insoluble calcium carbonate, which represents loss of alkali. When Ca^{2+} in calcium carbonate is exchanged for Na^+ in sodium polystyrene sulfonate, Na_2CO_3 is formed and then absorbed, and calcium polystyrene sulfonate is excreted in stool. Administration of sodium polystyrene sulfonate has been shown to cause a substantial decrease in net acid excretion, up to 40 mEq/day. This effect is apparent primarily when the resin is administered with calcium salts [21]. The urinary excretion of phosphate decreases when the resin is administered with calcium salts, probably because some Ca^{2+} that would otherwise be excreted as insoluble calcium phosphate is exchanged for Na^+ in sodium polystyrene sulfonate, resulting in formation of sodium phosphate, which is soluble and absorbable.

The alkalinizing effect of $Al(OH)_3$ or aluminum carbonate is ordinarily limited to exchange of hydroxide or carbonate with phosphate, since Al is nearly entirely unabsorbable; most of the $Al(OH)_3$ ingested is excreted in stool as such. However, when $Al(OH)_3$ or aluminum carbonate is ingested along with Na polystyrene sulfonate, more alkali is gained because some Al is exchanged for Na, and the resulting NaOH or Na_2CO_3 (after its conversion to $NaHCO_3$) is readily absorbable [22, 23]. The reaction will be: Na polystyrene sulfonate + Aluminum carbonate $\rightarrow$ 3 Na_2CO_3 + Al-polystyrene sulfonate. Similarly, the ingestion of $Mg(OH)_2$ concomitantly with Na polystyrene sulfonate has an alkalinizing effect, because the exchange of Mg in poorly absorbable $Mg(OH)_2$ with Na in sodium polystyrene sulfonate leads to production of NaOH, which will be converted quickly to $NaHCO_3$ and absorbed [24].

Calculation of Net Alkali Content in a Complex Mixture of Substances

The magnitude of the various chemical reactions described above cannot be predicted with accuracy, because many endogenous and exogenous factors are involved, such as the rate of HCl secretion and the nature of simultaneously ingested food or drugs. For example, $CaCO_3$, nearly insoluble in the absence of HCl secretion, would become soluble if it were administered with orange juice, which contains citric acid: $CaCO_3$ + citric acid → Ca citrate + carbonic acid. Neutral sodium or potassium phosphate ingested alone is nearly completely absorbed, and has no effect on acid-base balance. However, when it is ingested with calcium salts, a substantial amount of phosphate is lost in the stool as calcium phosphate, and consequently alkali is gained. It is clear that the overall effect of an ingested substance on acid-base balance depends not only on its alkalinity or acidity but the pattern of its absorption along the GI tract.

The pattern of absorption of even a single chemical is difficult to predict. When ingested material is as complex as food which consists of literally thousands of different chemical substances, the prediction of the effect of food ingestion on acid-base balance on the basis of the absorption pattern of the component chemicals in food would be nearly impossible. Measurements are made instead by analyzing net alkali content of food and feces. The difference between these two is considered to be net alkali absorption.

Net alkali content of food and feces is estimated by the electrolyte balance technique, i.e., the sum of noncombustible cations (Na + K + Ca + Mg) minus the sum of noncombustible anions (Cl + 1.8 P). The assumption in the use of this formula is that when nonmetabolizable cations are accompanied by metabolizable anions, their subsequent absorption and metabolism of their anions would result in gain of alkali, and that ingestion of nonmetabolizable anions accompanied by metabolizable cations would lead to a gain of acid [2]. Hence, the difference between the two represents net gain of alkali. The concentrations of all of the electrolytes are expressed in mEq, except phosphate, which is expressed in mmol and then multiplied by the factor 1.8. The molar ratio of HPO_4/H_2PO_4 at pH 7.4 is 4/1, which gives an average valence of phosphate of 1.8.

The actual pH of food or the pH of feces is not a determinant of the average valency of phosphate for the calculation of net alkali content, since that pH is relevant only in reference to the blood pH. Take the following example. Suppose that a given quantity of food contains 30 mmol of sodium phosphate and that the pH is 7.4. Since the pK of H_2PO_4 is 6.8, the content of Na_2HPO_4 will be 24 mmol and the content of NaH_2PO_4 6 mmol at pH 7.4, with a ratio of HPO_4/H_2PO_4 of 4/1. Complete absorption of such a mixture will not affect acid-base balance. The electrolyte balance technique also predicts neutrality. The total content of noncombustible cation, i.e. Na, is 54 mEq (24 × 2 + 6 = 54) and the total content of noncombustible anion, i.e., phosphate, is also 54 mEq (30 mmol multiplied by 1.8, the average valance of phosphate). If 9 mEq of acetic acid are added to this mixture, acetic acid will react with Na_2HPO_4, forming Na acetate and NaH_2PO_4. The mixture will contain 9 mEq of Na acetate, 15 mmol of Na_2HPO_4, and 15 mmol of NaH_2PO_4. Upon absorption, 9 mEq of Na acetate will generate 9 mEq of alkali, while absorption of 15 mmol of Na_2HPO_4 and 15 mmol of NaH_2PO_4 into blood that has a pH of 7.4 will lead to consumption of exactly 9 mEq of alkali, with the result that net gain of acid is zero. Net alkali content of food after addition of acetic acid, calculated with the use of the electrolyte balance technique, is also zero since addition of acetic acid does not alter the total content of noncombustible cations and anions. Addition of 9 mEq of HCl instead of acetic acid will result in the same overall changes in the ratio of HPO_4/H_2PO_4 as with the addition of acetic acid, but the net alkali content of the food mixture will contain −9 mEq. It will contain 9 mEq of NaCl, and 15 mmol each of Na_2HPO_4 and NaH_2PO_4. Absorption of NaCl will have no effect on acid-base balance, but absorption of 15 mmol each of Na_2HPO_4 and NaH_2PO_4 into an environment with pH 7.4 will lead to consumption of 9 mEq of alkali, as 9 mmol of NaH_2PO_4 is converted to 9 mmol of Na_2HPO_4 in order to reestablish the ratio of HPO_4/H_2PO_4 at 4/1. The answer is the same when the electrolyte balance technique is used; noncombustible cations − noncombustible anions = Na − (Cl + P × 1.8) = 54 − 9 + (30 × 1.8) = −9. For the same reasons, the actual pH of stool is not important in estimating the average valence of phosphate in order to determine net alkali content of stool by the electrolyte balance technique.

For the calculation of net alkali content of food and feces, the only nonmetabolizable ions presumed to be present are 4 cations, Na, K, Ca, and Mg, and 2 anions, Cl and P. The reason for this assumption is simply that no other ions are present in food in any significant amount. Obviously other nonmetabolizable cations such as lithium, iron, aluminum, the counterion of an anion exchange resin may be ingested in significant amounts, but only as pharmacological agents. The major nonmetabolizable anions include sulfate, nitrate, and bromide, and among these, only sulfate has quantitative importance. However, sulfate is not ingested as such; sulfate in the body originates almost exclusively from sulfur-containing amino acids [1]. Hence urinary excretion of sulfate is separately measured in order to estimate its contribution to the whole body acid-base balance. On a rare occasion, ingestion of sulfate would lead to the overestimation of net GI absorption of alkali using the electrolyte balance technique by the amount of sulfate ingested. However, because the measurement of urinary sulfate would lead to overestimation of acid gain (sulfuric acid production) by exactly the same magnitude, the two errors would cancel each other out.

Some organic anions found in food are metabolized very poorly or not at all by the human body. Tartarate, present in may fruits and used in baking powder, is such an anion. However, it has been shown that the bulk of the metabolism of ingested tartarate in humans is attributable to the action of bacteria in the colon, and only about 14% of that which is ingested appears in urine unchanged [25]. Other organic cations, e.g. choline, may be metabolized only by the colonic bacteria [26], but in that event their effects on acid-base balance would be the same as if they were metabolized after intestinal absorption.

Even if an organic anion in food is absorbed and excreted in urine unmetabolized, the quantification of acid-base balance would not be affected so long as net alkali absorption is measured by the electrolyte balance technique, and endogenous acid production by the excretion of urinary organic acid as estimated by the Van Slyke and Palmer method. The presence of nonmetabolizable organic anions in food would result in a falsely increased estimation of net GI alkali absorption. On the other hand, all organic anions excreted in urine estimated by the Van Slyke and Palmer method are assumed to originate from organic acids and hence to have consumed body alkali. A nonmetabolizable organic anion absorbed and excreted as such would therefore result in a falsely increased measurement of acid production by the same extent. The two errors, alkali acquired and alkali consumed, would cancel each other out.

The measurement of GI absorption of alkali by the analysis of food and feces is difficult and prone to inaccuracies. It assumes that the difference between dietary intake and fecal excretion represents what is absorbed. However, some conditions must be met before we can accept such an assumption. First, the subjects must be adapted to their diet for a period sufficiently long to insure that the average fecal composition reflects the new fixed dietary conditions, and the time period of fecal collection should also be sufficiently long. Second, in order to assure complete absorption of the fixed cations and anions there must be no intestinal disease. Furthermore, technical difficulties associated with prolonged stool collection and accurate measurements of electrolytes in such inhomogeneous samples as stool and food require that subjects under study must be admitted to a clinical research center and kept on a specially prepared diet. No study has measured net GI absorption of alkali on a normal diet in an outpatient setting, because the needed measurements would be virtually impossible to obtain in an outpatient setting.

Recently, a simplified technique for measurement of net alkali absorption from the GI tract has been proposed [27]. In this technique, net GI absorption of alkali is estimated as urinary noncombustible cations minus urinary noncombustible anions. The method is based on the assumption that the difference between the amounts of noncombustible cations and noncombustible anions absorbed from the gut equals the difference between the amounts excreted in the urine under stable conditions. The validity of this assumption is supported by empirical as well as theoretical evidence. Empirically, a comparison of measurements of net GI absorption of alkali by analysis of urinary electrolytes with those obtained by analysis of food and fecal electrolytes is shown in figure 1. Highly significant correlation ($r = 0.99$; $p < 0.0001$) between the two sets of results was demonstrated.

The theoretical basis for the use of urinary electrolytes is provided by the following arguments. Since the total extracellular content of divalent ions is small (about 45 mEq of Ca, 30 mEq of Mg, and 30 mEq of P), in the absence of substantial net flux from or into bone, the amounts of these ions excreted in the urine must be virtually equal to the amounts absorbed from the GI tract. Net flux of calcium, magnesium, and phosphate into or out of bone in a normal adult without active bone formation or bone resorption is insignificant. Even a substantial net formation or resorption of bone would have a trivial effect on net GI absorption of alkali estimated by the urinary electrolyte technique. For example, net deposition of 100 mg Ca/day (as hydroxyapatite) would cause a loss of 2.3 mEq of alkali from the extracellular fluid. Conversely, negative calcium balance of 100 mg/day with dissolution of bone would result in overestimation of net GI absorption by 2.3 mEq/day.

In the steady state, the amounts of Na, K, and Cl excreted in the urine should be nearly equal to the amounts absorbed from the GI tract. However, because the extracellular volume tends to fluctuate, especially with varying salt intake, urinary Na and Cl content in a given 24-hour period may be different from the amounts absorbed from the GI tract during the same period. However, daily fluctuations in Na and Cl excretion would not affect the calculation of net GI absorption of alkali as long as these ions are

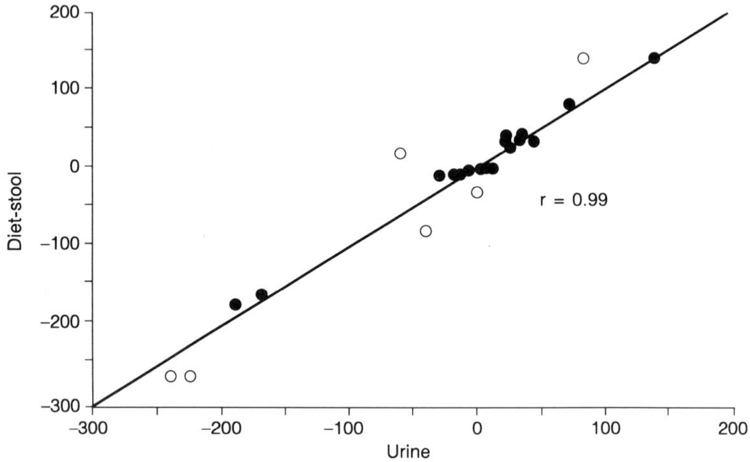

Fig. 1. Correlation of net GI absorption of alkali measured by the analysis of food and feces compared with that estimated by analysis of urinary electrolytes. Both values are expressed as mEq/day. The open circles represent values obtained during unstable periods of acid-base balance [from 27].

retained or lost together; the effect of increased or reduced Na excretion in urine would be exactly cancelled by parallel alterations in excretion of Cl. For the same reasons, extrarenal loss of salt, e.g. in sweating, would have negligible effect on the calculation of net alkali absorption from the GI tract. To the extent that sodium or potassium is balanced by HCO_3 or lactate in sweat, alkali will be lost, but such loss is negligible.

Even if urinary excretion of noncombustible ion did not precisely quantify the pattern of the GI absorption in individual cases, a mean value of net GI absorption of alkali in a large population would be accurate, even in an outpatient setting. Thus, it is now possible to measure the net GI absorption of alkali in a large population ingesting their usual diet in an outpatient setting. Such measurement has been made, and the amount of alkali absorbed was found to be rather substantial, 29.6 mEq/day [Oh and Carroll, unpubl. data].

References

1 Lemann J Jr, Relman AS: The relation of sulfur metabolism to acid-base balance and electrolyte excretion: The effects of *DL*-methionine in normal man. J Clin Invest 1959;38:2215–2223.

2 Relman AS, Lennon EJ, Lemann J Jr: Endogenous production of fixed acid and the measurement of the net balance of acid in normal subjects. J Clin Invest 1961;40: 1621–1630.

3 Lemann J Jr, Litzow JR, Lennon EJ: The effect of chronic acid loads in normal man: Further evidence for the participation of bone mineral in the defense against chronic metabolic acidosis. J Clin Invest 1966;45:1608–1614.

4 Goodman AD, Lemann J Jr, Lennon EJ, Relman AS: Production, excretion, and net balance of fixed acid in patients with renal acidosis. J Clin Invest 1965;44:495–506.

5 Lemann J Jr, Lennon EJ, Goodman AD, Litzow JR, Relman AS: The net balance of acid in subjects given large loads of acid or alkali. J Clin Invest 1965;44:507–517.

6 Lennon EJ, Lemann J Jr, Relman AS: The effect of phosphoproteins on acid balance in normal subjects. J Clin Invest 1962;41:637–645.

7 Lennon EJ, Lemann J Jr, Litzow JR: The effects of diet and stool composition on the net external acid balance of normal subjects. J Clin Invest 1966;45:1601–1607.

8 Lemann J Jr, Lennon EJ: A potential error in the measurement of urinary titratable acid. J Lab Clin Med 1966;67:906–913.

9 Lentner C: Urine volume rate and physiochemical data; in Geigy Scientific Tables. Basel, Ciba-Geigy Ltd, 1981.

10 Scriber CR, Rosenberg LE: Distribution of amino acids in body fluids; in Amino Acid Metabolism and Its Disorders. Philadelphia, Saunders, 1973, p 39.

11 Van Slyke DD, Palmer WW: Studies of acidosis. XVI. Titration of organic acids in urine. J Biol Chem 1920;41:567.

12 Dawson J, Dempsey E, Bartter F, Leaf A, Albright F: Evidence for the presence of an amphoteric electrolyte in the urine of patients with 'renal tubular acidosis'. J Clin Invest 1954;33:225–237.

13 Kaufman AM, Brodmiller C, Kahn T: Role of citrate excretion in acid-base balance in diuretic-induced alkalosis in the rat. Am J Physiol 1985;248:F796.

14 Brown JC, Packer RK, Knepper MA: Role of organic anions in renal response to dietary acid and base loads. Am J Physiol 1989;257:F170–F176.

15 Kesner L, Muntlwyler E: Determinants of non-nitrogenous organic acids in the urine in conditions of altered acid-base balance. J Lab Clin Med 1963;61:605.

16 Lawson AM, Chalmers RA, Watts RWE: Urinary organic acids in man. I. Normal patterns. Clin Chem 1976;22:1283.

17 Goldberg G, Paul W, Gonick H: A titrimetric method for determining organic acids and bases in urine. Clin Chem 1966;12:830.

18 de Strihou C van Y: Importance of endogenous acid production in the regulation of acid-base equilibrium: the role of the digestive tract. Adv Nephrol 1980;9:367–385.

19 Hurst PE, Morrison RBI, Timoner J, Metcalfe-Gibson A, Wrong O: The effect of oral anion exchange resins on faecal anions. Comparisons with calcium salts and aluminium hydroxide. Clin Sci 1963;24:187.

20 Sheikh MS, Maguire JA, Emmett M, Santa Ana CA, Nicar MJ, Schiller LR, Fordtran JS: Reduction of dietary phosphorus absorption by phosphorus binders. J Clin Invest 1989;83:66–73.
21 de Strihou C van Y, Dieu JP: Acid-base and metabolic consequences of the gastrointestinal chelation of cations. Eur J Clin Invest 1978;9:A333.
22 Baluarte HJ, Presbis J, Goldberg M, Gruskin AB: Metabolic alkalosis in an anephric child caused by the combined use of Kayexalate and Basaljel. J Pediatr 1978;92: 237.
23 Schroeder ET: Alkalosis resulting from combined administration of a 'nonsystemic' antacid and a cation-exchange resin. Gastroenterology 1969;56:868.
24 Fernandez PC, Kovnat PJ: Metabolic acidosis reversed by the combination of magnesium hydroxide and a cation-exchange resin. N Engl J Med 1972;286:23.
25 Chadwick VS, Vince A, Killingley M, Wrong OM: The metabolism of tartarate in man and the rat. Clin Sci Mol Biol 1978;54:273–281.
26 De la Huerga J, Gyorgy P, Waldstein B, Katz R, Popper H: Effects of anti-microbial agents upon choline degradation in intestinal tracts. J Clin Invest 1953;32:1117.
27 Oh MS, Carroll HJ: A new method for the measurement of net GI absorption of alkali. Kidney Int 1989;36:915–917.

Man S. Oh, MD, Department of Medicine, SUNY Health Science Center at Brooklyn, 450 Clarkson Avenue, Brooklyn, NY 11203-2098 (USA)

Berlyne GM (ed): The Kidney Today. Selected Topics in Renal Science.
Contrib Nephrol. Basel, Karger, 1992, vol 100, pp 105–117

Perspectives in Acid-Base Balance in Advanced Chronic Renal Failure

G.M. Berlyne[a], *A.J. Adler*[a], *R.H. Barth*[a], *D. Burke*[b], *C.E. Palant*[a,1]

[a]Nephrology Service, Medical Service and [b]Dietitic Service, Brooklyn Veterans Administration Medical Center, Brooklyn, N.Y., USA

The production of hydrogen ion was investigated some 25–30 years ago in both normal individuals and in the uremic patient suffering from chronic renal failure. In normal man, the hydrogen ion appearing in the urine had been measured using the formula:

$$\Sigma H^+ = NH_4^+ + TA - HCO_3^-.$$

The possibility that the urinary ΣH^+ did not include all the hydrogen ion excreted from the body was at first not appreciated, although by the late 1960s and early 1970s, Wrong and his group had published a series of articles on the acid base parameters of colonic fecal water which indicated that hydrogen ion could be excreted in the feces, in various forms, although this was not the main thrust of their research [1]. After this series of publications the role of the gut in hydrogen ion homeostasis was initially ignored, and most of the research effort went towards solving the problem arising from the obvious inability of the acidotic patient with advanced renal failure to excrete hydrogen ion in the urine and to determine if the bone crystal was a buffer; and to characterize the buffering capacity of the bone in its carbonate and hydroxyapatite moieties [2–6]. It should be added that Lemann and his group were well aware of the importance of the bowel in the acid base balance in uremia, but in those days the techniques of measurement of some of the parameters, such as organic acids, were

[1] We thank Drs M.S. Oh, M. Halperin, D.E. Atkinson and E. Bourke for stimulating discussion about this subject.

indirect, unlike the more sophisticated methods of direct measurement developed since then, e.g., HPLC, and now readily available in the majority of research laboratories.

We are therefore, now in 1992, in a far better position to look at the earlier papers critically. Some of the arguments raised by Oh [7] are particularly convincing and formidable and it was the stimulation of his work and that of Halperin and Jungas [8] that prompted this review of the literature and some conclusions which I believe indicate the need for further clinical research using modern techniques. The challenges are:

(1) To quantitate the production of hydrogen ion in health.

(2) To quantitate the influence of various animal and vegetarian protein intakes on the hydrogen ion production in health.

(3) To determine 1 and 2 above in various stages of chronic renal failure.

Meanwhile, until the data are available, in this article we will critically appraise what is so far known about acid base balance in advanced chronic renal failure.

Amount of Hydrogen Ion Generated

The amount of hydrogen ion generated in health on a regular protein intake is about 1 mEq/kg of body weight, as is reflected in the total hydrogen ion excretion in the urine calculated from the formula $\Sigma H^+ = NH_4^+ + TA - HCO_3^-$, where ΣH^+ = total hydrogen ion excretion (mEq/24 h), NH_4^+ = urinary 24 h ammonium excretion in mEq/24 h urine, TA = titratable acidity (mEq/day), and HCO_3^- = bicarbonate excreted in mEq/day.

Titratable acidity is principally a measure of phosphorus excretion in the urine in normal health. Halperin and Jungas [8] point out that the current formula used for ΣH^+ in urine has at least two problems: (1) NH_4^+ and (2) TA, where P excreted probably overstated H^+ buffering. Their arguments are cogent and clearly stated and the reader is advised to read their review article. Urinary bicarbonate is negligible if urine pH is <6.2.

Lemann and his colleagues, in a series of papers [2–6], defined the net hydrogen ion production in healthy man and in uremics by measuring diet and stool and assaying Na, K, Mg and Ca to calculate inorganic cations; they measured Cl and $1.8 \times P$ in an effort to include inorganic anions. The difference between inorganic anions and cations was ascribed to anions of organic acids, and, in addition, sulfate from metabolized methionine and cysteine. The difference between 'organic anion' in diet and that in the feces was equivalent to the amount of organic anions absorbed from the gastrointestinal tract. These organic anions give rise to hydroxyl ion or bicarbonate during metabolism. It is necessary to subtract the urinary sulfate moiety from the total organic anions. The figure of $1.8 \times P$ was reached in an effort to express the mean valency of inorganic phosphates at a urine pH of 7.4.

In 1983, Halperin and Jungas [8] reformulated this concept with a detailed version of hydrogen ion buffering and hydrogen ion generation in a healthy man eating approximately 120 g of beef protein a day, assuming that he was in metabolic equilibrium. They looked at four sources of H^+ or OH^-:

(1) Methionine and cysteine in protein yielding 72 mEq of hydrogen ion per day.

(2) Cationic amino acids give rise, with histidine, to 138 mEq of hydrogen ion per day.

(3) Anionic amino acids which give rise to bicarbonate or hydroxyl ion, i.e., hydrogen ion buffer of 100 mEq/day.

(4) Anions of organic acids, e.g., lactate, which give rise to the generation of 60 mEq/day of bicarbonate or hydroxyl ion.

They used the following formula to indicate the overall acid base balance equation as a result of the ingestion of about 120 g beef protein per day:

$$\begin{aligned}\Sigma H^+ &= (\text{sulfur containing amino acids} + \text{cationic amino acids})\\ &\quad - (\text{anionic amino acids} + \text{organic acid anions})\\ &= (79 + 138) - (100 + 60) = 50 \text{ mEq/day.}\end{aligned}$$

This figure of 50 mEq/day is considerably less than is found as total hydrogen ion excreted in the urine, usually at least 70 mEq/day on a regular protein intake. Thus, Giovannetti's group [9] found about 70 mEq/day in their normal persons taking an animal protein diet of about 70 g/day. Using the meat intake of Halperin and Jungas' [8] example, and assuming that hydrogen ion excretion in the urine is related to protein catabolism, it would be expected that the excretion of hydrogen in the urine should be $120/70 \times 70 \times 70 = 120$ mEq/day. This is an approximation but it is likely that Halperin's figures are not quite comparable because he has used a diet where the protein is mainly of beef. 120 g of animal protein is rarely taken even in USA. There is of course animal protein which is not in the form of beef in the US normal diet as described by the latest (1987–88) survey of the US Department of Agriculture. It includes 14.5 g of nonbeef protein (e.g., in bread, milk, milk products, and eggs), and 65.6 g of beef protein in a 110 g protein diet, and in addition there is 28.8 g vegetable protein, i.e., potatoes, grains, vegetables [10].

The figures of Halperin and Jungas [8] for hydrogen ion produced by diet can be examined more meaningfully using the standard average US diet with the figures from the Human Nutrition Survey of the US Department of Agriculture for 1987–88 [10]. The average diet of males in the age group of >20 years is 80.1 g animal origin protein, 28.8 g vegetable origin proteins. On this diet, using Halperin and Jungas' [8] sources of hydrogen ion, the following would be found:

S-containing amino acids: If all were to be broken down to inorganic sulfate, which is *not* likely to be the case, this would give at the most 10.5 m*M* of methionine and 2.66 mmol of cysteine. This would yield 31.6 mEq of hydrogen ion. However, only half the urinary sulfur is in the form of inorganic sulfate, so the yield of H^+ is about 16 mEq.

Cationic amino acids would produce at most 40.2 mEq of hydrogen ion: if they yield any hydrogen ion is arguable [11]. This figures histidine as a *source* of hydrogen ion of 50% of its molar concentration; it is possibly totally a hydrogen ion *buffer*. However, using Halperin and Jungas' [8] formula, total hydrogen ion production for a normal US diet in men >20 years old is 40.2 mEq H^+/day from cationic amino acids.

Anionic amino acids produce 87.94 mEq of HCO_3^- a day from 31.9 m*M* of aspartate and 56 mmol of glutamate in the diet.

Organic acids: Fruits and vegetables are rich in organic anions but few tables contain the required data other than beef, as a source for beef organic anions (lactate, succinate and glycolate), and various vegetables and fruits including potatoes, kohlrabi and carrots. Using only the former two, there are 16.55 m*M* of lactate, 1.26 m*M* of succinate and 5.21 m*M* of glycolate in the 65.6 g of beef found in the normal US diet, and in 86 g of potatoes there are 2.33 m*M* of citric acid and 0.58 m*M* of malic acid, with smaller amounts of succinic, gluconic and salicylic acid. (The dietary details of the US survey do not give enough details about most of the vegetables and fruits to calculate organic acid content.) There is a total of 26 m*M* of bicarbonate generated by these three organic acids in beef and potatoes ignoring all the content of organic acids in the remainder of the food. This would make a normal diet give rise to a minimum of:

$$31.6 + 40.2 - 87.9 - 26 \text{ mEq of } H^+ = -42.1 \text{ mEq, i.e., } 42.1 \text{ mEq of } HCO_3^-.$$

This does not take into account most of the vegetable/fruit origin organic anions and is therefore less than if these organic anions were fully included.

What does this imply? It suggests that even meat diets are alkalinizing, i.e., produce more HCO_3^- than H^+. It would seem that vegetarian or vegan diets would produce even less hydrogen ion because of the smaller quantities of S-containing amino acids, so they should be even more alkalinizing.

Why then is the urine normally acid? It may be – and this is an unproven hypothesis – that the gut excretes base, i.e., bicarbonate and carbonate in an insoluble form which is *not in dynamic equilibrium with fecal water.* The kidney then would have a role in health of excreting just enough protons to neutralize the base which would appear to be about 1 mEq/g of ingested protein. This would imply the excretion in the stool in normal health of an equivalent amount of bicarbonate/carbonate, i.e., 70 mEq/day in a man taking a 70 g protein intake. This is still unproven, as is the possibility that in kidney damage the gut loss of base continues but as the ability of the kidney to excrete acid falls, the patient becomes acidotic because of the continuous loss of base via the gut without a corresponding loss of hydrogen ion in the urine. However, this is speculation, but it certainly is possible that hydrogen ion excretion in the urine is a mirror of the quantity of bicarbonate generated by the diet and excreted in the gut. Moreover, Atkinson [pers. commun., 1991] has pointed out that when protein is digested, 1,000 m*M* of bicarbonate are liberated from peptide bonds for each 100 g of digested protein, being absorbed and metabolized.

Nevertheless, the figures for animal protein capabilities of hydrogen ion production are comparable insofar as S-containing amino acids, cationic amino acids and anionic amino acids are concerned. The question of organic acid anions is one which is in limbo in the absence of adequate data on food organic acids in the major data dietary table sources at the time of writing.

A further problem is revealed if we take the example of a person eating a low protein diet, e.g., a 40 g protein diet given in renal failure. In the example of Halperin and Jungas [8], 120 g protein intake yielded 50 mEq of urinary hydrogen ion. Therefore, a 40 g intake should yield 17 mEq/day of urinary hydrogen ion. In practice we have shown a mean urinary H^+ excretion of 27 mEq per 24 h in patients with a GFR of 6 ml/min [11]. These patients were eating a supplemental 18 g protein diet, i.e., certainly less than a 40 g diet, yet were excreting in their urine 1.6 times more H^+ ion that would be expected using the data of Halperin and Jungas [8]. In addition, our group of patients had severe metabolic acidosis, so they were presumably producing more hydrogen ion than they were capable of excreting in their urine, or were excreting more alkali in the gut than their kidneys were capable of metabolically compensating for by excreting H^+ in the urine. How can this apparent paradox be resolved?

(1) If some of the hydrogen ion is buffered in the gut. There are some data on this (see below).

(2) Phosphoproteins give rise to metabolic OH^-, and there are phosphoproteins in beef, whereas phospholipids are likely to generate H^+ in metabolism.

(3) The food tables are incorrect. This is highly unlikely.

(4) The entire concept is wrong. The protein-containing diets are producing a net excess of HCO_3^- metabolically, much of which is excreted in the feces, and the loss of H^+ in the urine is modulated to maintain acid base homeostasis. This is unproven but needs looking at in more detail.

Difference between Vegan and Animal Protein Diets in Hydrogen Ion/HCO_3^- Generation

Vegan diets are, by definition, strict vegetarian diets in which there is neither milk nor eggs. Giovannetti and his group [9] have been instrumental in exploring the use of vegan diets, supplemented by keto acids (KAA) and essential amino acids (EAA), in advanced chronic renal failure. They have shown that in subjects with normal renal function a 70 g vegan diet produces just over half the amount of urinary hydrogen ion that is excreted after a 70 g animal protein diet. This is partly due to the lower amounts of S-containing amino acids (methionine and cysteine) in vegetable proteins where compared to animal proteins.

In advanced renal failure, patients on restricted animal protein diets have acidosis, whereas those patients taking vegan diets of equal protein content are generally free of acidosis [9]. This may be of importance in the therapy of chronic renal failure, and is likely to be a reflection of the fact that 14 g vegetable protein diets supplemented with KAA and EAA contain less methionine and cysteine, and thus generate less hydrogen ion from sulfate than corresponding amounts of animal protein, or, it may be, are producing more HCO_3^- than a diet containing animal proteins.

It has also been established that P absorption from the gastrointestinal tract is less on vegan diets in normal persons, so that urinary titratable acidity is reduced and this is a major factor for total urinary hydrogen ion excretion in persons on vegan diets being close to half of that in persons eating similar amounts of animal protein diets. In summary, vegan diets generate about 0.5 mEq of urinary H^+ per gram of protein a day while animal protein diets generate 1 mEq of urinary hydrogen ion per gram of protein a day. The higher P excreted in the feces may be an expression of the increased excretion of alkali in the stools on vegan diets.

The lower absorbed P load makes hyperphosphatemia less common in advanced chronic renal failure patients on restricted protein vegan KAA-EAA diets; hyperparathyroidism, as a result, is a rarity in contrast to its high frequency in chronic renal failure patients on animal protein diets.

If the theory that all diets produce alkali residues were true, it is not difficult to understand why acidosis is less or absent in patients on vegan diets in quite advanced renal failure. If all this alkali were excreted in the stools then the vegan diet might at first be thought to be likely to produce even more acidosis. This is not the case. It is feasible that only some of the diet generated alkali is excreted in the feces, or that less hydrogen ion from S-containing amino acids remains in the body when vegan diets are ingested. Alternatively, the entire theory may be incorrect and vegan diets may be the nails in the coffin of the theory.

Bone Buffering

We have observed that acidosis is frequent in end-stage renal failure patients maintained on low protein diets of animal origin but is unusual in patients given a vegan low protein diet [11]. The duration of life on diet commencing after the onset of severe renal failure and before death (or dialysis/transplantation) is shown in table 1. It is seen that with these extremely severe restricted diets, on the average an extra year of life is given to the patients in which they can survive without the need for extraordinary measures such as dialysis or transplantation. This is a year extra of life, but the duration is not uniform. Thus, polycystic kidney, interstitial nephritis, etc., survive longer on diet than do patients with malignant hypertension.

Lemann's [2–6] and Pellegrino's groups [12–14] have produced a series of papers in which they concluded that the bone salts lost in chronic

Table 1. Survival on diet of 104 patients

Disease	Patients		Mean survival, months	
	dead	alive	dead	alive
Malignant hypertension	24	6	4.9	5.0±
Chronic glomerulonephritis	9	10	7.8	15.0±
Chronic pyelonephritis	9	11	12.1	7.3±
Polycystic kidneys	10	7	11.3	6.1±
Obstructive uropathy	2	4	5.5	8±
Hypoplastic kidneys	4	1	–	2±
Fanconi syndrome	0	1	–	25.0
Lupus nephritis	1	0	17.2	0
Gouty nephritis	1	0	12.0	0
Cortical necrosis	1	0	42.5	0
Familial nephritis (Alport's)	1	2	4.0	8.5
Total	62	42	13.5±11.3	>9.6±7.2

renal failure, both hydroxyapatite and calcium carbonate, buffer hydrogen ion either directly on the bone crystal or in their excretion from the body. Pellegrino's group [12–14] concluded that up to one-third of the bone carbonate could be lost in the course of 3 years. However, only a relatively small percentage of patients with renal disease have such prolonged acidosis in severe renal failure with 3-year survival, and this implies either a major tubular abnormality predominantly such as distal RTA, or, more commonly, a mixed form with relative glomerular sparing. The glomerular disease has to be severe to cause a reduction in hydrogen ion excretion in the urine unless there is a predominantly tubular lesion such as occurs in interstitial nephritis or hydronephrosis. There is, as we have shown in a population of 104 end-stage renal failure patients treated conservatively (with the 18 g protein Manchester-Berlyne diet), on the average a year's survival before death (or dialysis/transplantation to avoid death); similar figures result with Giovannetti's 14 g vegan protein diet supplemented with KAA and EAA [15]. Thus, survival for 3 years on conservative therapy is unusual and is restricted to a minority of patients. It should not be the major problem in calculations aimed at determining whether bone crystal buffering is possible as the major buffer of excess hydrogen ion in chronic renal failure. Thus, Oh [7] has pointed out that in less than half a year, skeletal stores of carbonate would be depleted if 19 mEq of H^+ per day had to be buffered because they were produced metabolically but could not be excreted in the urine due to advanced renal disease. Oh's arguments should be carefully examined because they shed new light on what was accepted hitherto. Oh [16] has also pointed out that in the steady state, the urinary excretion of excess of inorganic cations over inorganic anions (Cl + 1.8 $\times$ P) will equal the organic acids. Unfortunately, phosphorus is not simply a hydrogen ion donor: depending on its source it can be a hydrogen ion donor or a hydroxyl/bicarbonate ion donor. This is discussed in the section on phosphorus in this article. Moreover, when Lemann's group described this method of measuring organic acids in 1966, this was the only acceptable way to measure them. There are now (1992) straightforward HPLC methods to measure the organic acids individually in both food and feces, so that formal balances of organic acids can now be performed directly. So far, I am unaware of anybody carrying out such studies which are now commencing in my laboratory. An additional error, well recognized by Lemann's group, is that the organic acid anion both in the feces and in Oh's urine method [16] includes also sulfate which should be separately measured and deducted from all indirect organic acid anion measurements.

Gut

If the bones are able to account for only some of the retained hydrion in chronic renal failure, according to Oh's classical paper, then where is the remainder of the hydrogen ion buffered? This assumes to a certain extent that we have precise knowledge of the amount of hydrogen ion generated in renal failure, which we do not yet possess. A possible answer is the buffering capacity of the gut. 30% of the P in the diet is in the form of food additives [17]. These are principally in the form of monohydrogen calcium or sodium phosphate, dibasic sodium phosphate or calcium phosphate (table 2). Because stool water pH in renal failure is 7.45, then there is a certainty that most of the food P additives which are not absorbed will be excreted together with hydrogen ion which they buffer, *if they are soluble.* In addition, the gut will excrete hydrogen ion with the calcium salts appearing in the bowel derived from endogenous hydroxyapatite liberated from bone breakdown. Also, patients in advanced renal failure are often

Table 2. Some typical additives in common US foods (found in one visit to VA PX)

Product	Additive
Cremora	K_2HPO_4
Ritz crackers	$Ca_3(PO_4)_2$
Graham crackers	$Ca_3(PO_4)_2$
Pepperidge Farm biscuits	Na_2HPO_4
Stella D'Oro biscuits	$NaAlPO_4$, calcium propionate
Lance chocolate chip cookies	$CaHPO_4$
Lance fig bar	$CaHPO_4$

given aluminum hydroxide, aluminum carbonate, calcium carbonate, magnesium carbonate and magnesium oxides as phosphorus binders. All these salts may act as OH^-/HCO_3^- donors and are reportedly capable of overall buffering of hydrogen ion in the gastrointestinal tract, although we found that $CaCO_3$ did not have any effect on acid base balance presumably because of its insolubility.

Thus, the gut is capable of binding H^+ ion by the following:

(1) P as dietary additives becoming HPO_4^{2-} and $H_2PO_4^-$, depending on solubility of the initial phosphate.

(2) $Ca_3(PO_4)_2$ being excreted as HPO_4^{2-} and $H_2PO_4^-$ salts of calcium in the stools. This is predicated on solubility and is doubtful.

(3) P from phosphoproteins liberated in digestion and acting as hydrogen ion buffers.

(4) Carbonate and hydroxides of calcium, magnesium and aluminum used for P binding, buffer H^+ in colon. This is predicated on solubility, which is low.

(5) Any natural $Ca_3(PO_4)_2$, NaH_2PO_4 or Na_3PO_4 in food will not all be absorbed but may contribute to H^+ buffering in the gastrointestinal tract. Again, this depends on solubility.

We have seen that in 4 out of 5 of the above, solubility has been mentioned. It may be that the insoluble calcium phosphates, calcium carbonate, and magnesium carbonate salts in the feces are *not* in dynamic equilibrium with the fecal water, in which case they may not be buffering anything approaching the quantities of protons we are presuming in this analysis. Supporting this is Wrong's group's [1] observation that only 5% of fecal P is dialyzable. Thus, the gut may not be overall buffering H^+ ion but may be excreting alkaline-insoluble salts, an overall loss of alkali. This is so far inadequately investigated in health, let alone in renal disease.

Phosphorus

Clearly what is taken in the diet is of prime importance from the point of view of the hydrogen ion generated, i.e., from proteins and the hydroxyl or bicarbonate ion generated from proteins and from various phosphates whether absorbed in the gastrointestinal tract or excreted in the feces. The medications being taken will also be of importance in the final degree of gastrointestinal buffering. An example is noteworthy: In patients on 18 g of protein intake, GFR < 10 ml/min, 290 mg of P were excreted in the urine and 710 mg were excreted in the stool; a total of 1,000 mg P were in the diet. If 50% of this is in the form of Na_3PO_4 or $Ca_3(PO_4)_2$ or then we are capable of buffering 17.7 mmol of P – up to 35 mEq of H^+ if the P is soluble or in dynamic equilibrium with fecal water, which is not certain. This figure of 35 mEq of H^+ is, of course, variable and may be less if the P is in the form of monohydrogen phosphates, but this is an indication of the *potential* buffering capacity of the gastrointestinal tract.

Lemann's group in 1966 showed that some of the organic P in the diet (phosphoesters) could give rise to H^+ and the salts of these could act as OH^- or HCO_3^- donors. Phospholipids tend to be H^+ producers and phosphoproteins are OH^- producers [Halperin, pers. commun., 1991].

Thus the presence of inorganic phosphorus in the urine is no guarantee that it represents hydrogen ion *production* during the course of protein metabolism. On the contrary, some of it certainly represents the *negative* hydrogen ion production, i.e., HCO_3^- or OH^- generation during metabolism of the original organic P compound. The presence of P binders, usually oxides, hydroxides or carbonates of Group 2A and 3B metals, adds to the body's capacity to buffer hydrogen ion. This too will vary depending on which drug is given and its water solubility and dosage.

In conclusion, phosphorus may be a hydrogen ion donor or receptor, and significant amounts of phosphoproteins and phosphate additives may contribute to bicarbonate or hydroxyl generation and so buffer hydrogen ion in the gastrointestinal tract. This is also dependent on the pH of fecal water and this varies widely in uremia but has a mean value of pH 7.45, about 0.05 pH units higher than that in normals [1]. It is unlikely that all the fecal inorganic phosphorus is soluble and will buffer H^+ ion in the gastrointestinal tract by combination with hydrogen ion in the proportions dictated by the Henderson-Hasselbalch formula.

Conclusions

The amount of net hydrogen ion generated in the body in health on various protein intakes is unknown. What reliable data are available relate solely to urinary excretion of hydrogen ion in both health and renal failure. No reliable data are yet available on the production, absorption and neutralization of hydrogen ion or bicarbonate in the gastrointestinal tract in health and disease on differing protein intakes, including animal protein diets and vegetable protein diets. Apart from Lemann's group, the literature on classical balance techniques in uremia is remarkably deficient, possibly due to the reluctance to process feces due to the position of laboratories in the US modern hospitals in the core of the tower type buildings with difficulty in preventing the spread of malodorous fumes to patient areas. It would appear that bone crystal buffering of excess hydrogen ion is not the sole buffer available, and that the gastrointestinal tract plays a key function in hydrogen ion balance. Phosphate-binding medications and various food additives containing phosphorus may have a pivotal effect on hydrogen ion balance in uremia. It may be that all diets taken by man produce an alkaline residue metabolically, and that this is excreted in the feces, whereas an important renal tract function is to ensure acid/base homeostasis by urinary excretion of the necessary amount of acid. In advanced renal failure, renal ability to excrete acid to match this proposed gastrointestinal loss may be the reason for the development of metabolic acidosis.

References

1 Wrong OM, Mercalfe-Gibson A: The electrolyte content of feces. Proc R Soc Med 1965;58:1007–1009.
2 Relman AS, Lennon EJ, Lemann J: Endogenous production of fixed acid and the measurement of the net balance of acid in normal subjects. J Clin Invest 1961;40:1621–1630.
3 Lemann J, Lennon EJ: Role of diet, gastrointestinal tract and bone in acid-base homeostasis. Kidney Int 1972;1:275–279.

4 Lemann J, Litzow JR, Lennon EJ: The effects of chronic acid loads in normal man: Further evidence for the participation of bone minerals in the defense against chronic metabolic acidosis. J Clin Invest 1966;45:1608–1615.
5 Litzow JR, Lemann J, Lennon EJ: The effect of treatment of acidosis on calcium balance in patients with chronic azotemic renal disease. J Clin Invest 1967;46:280–286.
6 Goodman AD, Lemann J, Lennon EJ, Relman AS: Production, excretion and net balance of fixed acid in patients with renal acidosis. J Clin Invest 1965;44:495–506.
7 Oh MS: Irrelevance of bone buffering to acid-base homeostasis in chronic metabolic acidosis. Nephron 1991;59:7–10.
8 Halperin ML, Jungas RL: Metabolic production and renal disposal of hydrogen ions. Kidney Int 1983;24:709–713.
9 Giovannetti S: The effect of nutritional therapy on blood chemical abnormalities; in Giovannetti S (ed): Nutritional Treatment of Chronic Renal Failure. Boston, Kluwer, 1989, pp 233–234.
10 Human Nutrition Survey, 1987–1988. Washington, US Department of Agriculture, 1988.
11 Berlyne GM, Mallick MP, Gaan D: Dietary treatment of chronic renal failure; in Wrong OM (ed): Fourth Symposium on Advanced Medicine. Aulander (NC), Pittman, 1968, pp 99–114.
12 Pellegrino ED, Biltz RM: The composition of human bone in uremia. Medicine 1965;44:397–418.
13 Pellegrino ED, Biltz RM: Bone carbonate and the Ca to P molar ratio. Nature 1968; 219:1261–1262.
14 Pellegrino ED, Biltz RM, Letteri JM: Interrelationships of carbonate, phosphate, monohydrogen phosphate, calcium, magnesium and sodium in uraemic bone: Comparison of dialyzed and non-dialyzed patients. Clin Sci Mol Med 1977;53:307–316.
15 D'Amico G: Economic implications of nutritional treatment of chronic renal failure; in Giovannetti S (ed): Nutritional Treatment of Chronic Renal Failure. Boston, Kluwer, 1989, pp 275–278.
16 Oh MS: A new method for estimating gastrointestinal absorption of alkali (technical note). Kidney Int 1989;36:915–916.
17 Aviolo LV: Recommended Dietary Allowances, ed 10. Washington, National Academy Press, 1989, pp 184–185.

Prof. Dr. G.M. Berlyne, MD, Chief of Renal Disease, Brooklyn V.A. Hospital, 800 Poly Place, Brooklyn, NY 11209 (USA)

Berlyne GM (ed): The Kidney Today. Selected Topics in Renal Science.
Contrib Nephrol. Basel, Karger, 1992, vol 100, pp 118–126

The Effect of Aluminum on Na-K-ATPase Activity in vitro

Andrew J. Adler, Carmine Caruso, Geoffrey M. Berlyne

Department of Medicine (111), Brooklyn Veterans Administration Hospital, Brooklyn, N.Y., USA

Since Berlyne et al. [1] first suggested that aluminum posed a significant threat to patients with renal disease, a number of aluminum-associated disorders have been recognized [2–11]. Yet, despite continued investigation, the mechanisms for its toxicity remain poorly understood. Among the mechanisms that have been considered is an impairment of sodium transport systems, in particular Na-K-ATPase. In 1980, Lai et al. [12] reported that aluminum inhibits Na-K-ATPase activity in rat brain synaptosomes. However, the 50% inhibitory concentration of 8.3 m*M* was well above any physiologically relevant range. Staurnes et al. [13] found that at pH 5, aluminum in micromolar concentrations is toxic to Na-K-ATPase in the gills of salmonids, but the relevance of these findings to mammalian physiology is uncertain. Recently, Bittar et al. [14], in an elegant study of the effect of aluminum on sodium efflux from barnacle muscle, demonstrated significant inhibition of sodium transport by aluminum. They suggested that the effect is, at least in part, due to indirect inhibition of Na-K-ATPase. This conclusion, however, was based entirely on sodium transport studies and not on direct estimates of Na-K-ATPase activity. Consequently, we studied the effect of aluminum at micromolar concentration and at physiological pH on Na-K-ATPase activity in vitro.

Materials and Methods

Materials. All chemicals used were of analytical grade and purchased from Fisher Scientific (Springfield, N.J.). Magnesium ATP, Tris-HCl, ouabain, and Na-K-ATPase, extracted from canine kidney and rat brain, were purchased from Sigma Chemical Co. (St. Louis, Mo.).

Enzyme Activity. Assays of Na-K-ATPase activity were performed according to the method of Bonting [15]. 0.1 unit of purified Na-K-ATPase was dissolved in 1 ml of 6 m*M* Tris-HCl buffer, pH 7.4, containing 60 m*M* NaCl and 5 m*M* KCl. Aluminum, when added, was in the form of $AlCl_3$. The mixture was preincubated for 10 min at 37 °C in a Dubnoff shaking water bath, followed by the addition of Mg-ATP to a final concentration of 1 m*M*. The mixture was then incubated an additional 10 min at 37 °C, whereupon the reaction was terminated with the addition of 2 ml ice-cold 10% (w/w) trichloroacetic acid. Phosphate determinations were made using the method described by Baginski et al. [16]. Briefly, 50 µl of the reaction mixture are added to 1 ml of ascorbic sulfuric acid and 250 µl of Teepol 610 dispersant solution (diluted 2:1 with deionized water) and vortexed immediately. 500 µl of molybdate reagent were then added, followed by 1 ml arsenite citrate solution.

The optical density of the resultant mixture was read at 660 nm after incubation for 10 min at room temperature.

Statistical Analysis. Significance of difference between two groups was assessed by Student's t test. One-way analysis of variance was used to determine significance of differences among more than two groups. All results are reported as means ± SD.

Results

The effect of aluminum on the activity of canine kidney and rat brain Na-K-ATPase is shown in figure 1. The results demonstrate that at aluminum concentrations of up to 20 µ*M* there is no discernible effect on either kidney or brain Na-K-ATPase. However, at concentrations of 40 µ*M* and above, aluminum significantly inhibits kidney Na-K-ATPase activity in a concentration-dependent manner. At 120 µ*M* aluminum, kidney Na-K-ATPase activity is reduced nearly 35% compared to control levels ($p < 0.0005$). The inhibitory effect on brain Na-K-ATPase, although less marked, becomes statistically significant compared to control values at 60 µ*M* aluminum ($p < 0.02$) reaching 10% at 120 µ*M* aluminum ($p < 0.0005$). The differences in activity between kidney and brain Na-K-ATPase are statistically significant at all aluminum levels from 60 µ*M* and above.

The effect of aluminum on the ouabain-insensitive component of renal Na-K-ATPase activity is shown in figure 2. Renal Na-K-ATPase

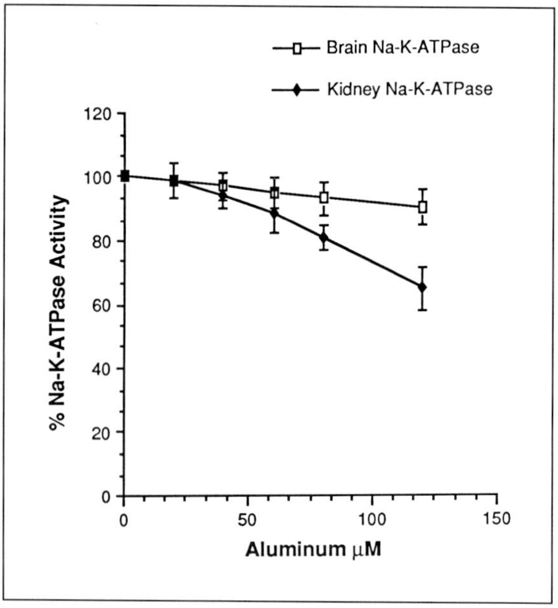

Fig. 1. The effect of aluminum on Na-K-ATPase activity. The reaction mixture consisted of 0.1 unit Na-K-ATPase, 6 m*M* Tris-HCl, pH 7.4, 60 m*M* NaCl, 5 m*M* KCl and 1 m*M* Mg ATP. Aluminum was added in concentrations shown above. The mixture was incubated for 10 min at 37 °C in a Dubnoff shaking water bath and terminated by the addition of 2 ml 10% trichloroacetic acid. Closed symbols, rat brain Na-K-ATPase; open symbols, canine kidney Na-K-ATPase. n = 8 for each group except at 120 μ*M* aluminum where n = 7 for kidney and n = 6 for brain. Results are expressed as mean percentages ± SD of the activity in the aluminum-free controls. Brain enzyme activity became significantly different from control at 60 μ*M* aluminum ($p < 0.02$) reaching a p value of <0.0005 at 120 μ*M* aluminum. Kidney enzyme activity became significantly different from control at 40 μ*M* aluminum ($p < 0.05$) reaching a p value of <0.0005 at 120 μ*M* aluminum. Differences between brain and kidney became significant at 60 μ*M* aluminum reaching a p value of <0.0005 at 120 μ*M* aluminum.

activity was assayed following incubation with 40 and 80 μ*M* aluminum in the presence of 100 μ*M* ouabain. The results indicate that aluminum has no effect on the ouabain-insensitive component of Na-K-ATPase activity and that inhibition by aluminum is, therefore, exclusively on ouabain-sensitive activity which constitutes 85% of the total activity.

As ATP has an appreciably higher affinity for aluminum than magnesium, we determined whether the observed inhibition of Na-K-ATPase

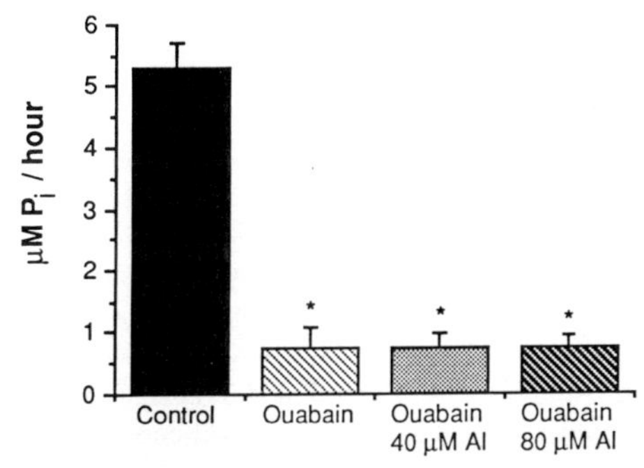

Fig. 2. The effect of aluminum on ouabain-insensitive, kidney Na-K-ATPase activity. The reaction mixture consisted of 0.1 unit Na-K-ATPase, 6 m*M* Tris-HCl, pH 7.4, 60 m*M* NaCl, 5 m*M* KCl and 1 m*M* Mg ATP. Ouabain, where indicated, was at 100 μ*M*. Aluminum was added in concentrations shown above. The mixture was incubated for 10 min at 37 °C and terminated by the addition of 2 ml 10% trichloroacetic acid. n = 5 for each group. Results are expressed as means ± SD. * $p < 0.0005$ compared to ouabain-free control.

activity could be due to a reduction in effective ATP levels brought about through the displacement of magnesium from ATP by aluminum. Na-K-ATPase activity was determined in the absence of aluminum, but with ATP concentrations reduced to levels equivalent to the amount that would remain available should all the administered aluminum be bound by ATP on a mole-to-mole basis. As shown in table 1, Na-K-ATPase activity at an ATP concentration of 0.89 m*M* was 90% of control values for both kidney and brain Na-K-ATPase, indicating that aluminum binding to ATP is not responsible for the reduction in ATPase activity.

We also investigated the possibility that the inhibitory effect of aluminum could be entirely artifactual due to binding of liberated P_i by aluminum and rendering it undetectable by the phosphate assay we employed. Modified reaction mixtures consisting of micromolar concentrations of $NaHPO_4$, but without Na-K-ATPase, were incubated for 30 min in the presence or absence of $AlCl_3$. As shown in table 2, aluminum, at concentrations up to 150 μ*M*, did not alter the detectable levels of P_i. These findings exclude the possibility that the decline in Na-K-ATPase activity is due to artifact.

Table 1. Effect of ATP levels on Na-K-ATPase activity

ATP concentration m*M*	Na-K-ATPase activity, μmol P_i/h	
	brain	kidney
1	6.81 ± 0.03	5.43 ± 0.13
0.95	6.75 ± 0.17	5.07 ± 0.28
0.89	6.14 ± 0.28	4.81 ± 0.14
0.85	5.77 ± 0.06	4.61 ± 0.05

Reaction mixture consisted of 0.1 unit Na-K-ATPase, 6 m*M* Tris-HCl, pH 7.4, and Mg ATP at the concentrations indicated above. The mixture was incubated for 10 min at 37 °C in a Dubnoff shaking water bath followed by the addition of 2 ml 10% trichloroacetic acid. n = 4 for each group. Results are reported as means ± SD.

Table 2. Effect of aluminum on inorganic phosphate determinations

	Phosphate recovered, μmol	
	20 μmol added P_i	30 μmol added P_i
0 μ*M* Al	17.3	30.6
50 μ*M* Al	17.2	31.6
125 μ*M* Al	17.7	31.9

Reaction mixture consisted of 6 m*M* Tris-HCl, pH 7.4, 1 m*M* Mg ATP, and $AlCl_3$ and $NaHPO_4$ at the concentrations indicated above. The mixture was incubated for 20 min at 37 °C in a Dubnoff shaking water bath followed by the addition of 2 ml 10% trichloroacetic acid.

Discussion

These results demonstrate that, at physiological pH, aluminum has a significant inhibitory effect on both canine kidney and rat brain Na-K-ATPase activity in vitro. Both forms of the enzyme display a similar trend, in that inhibition is first observed at a concentration of 40 μ*M* aluminum and increases progressively with increasing concentrations of aluminum. The quantitative effect, as determined by the slope of the inhibition curves, is 2.7 times greater on kidney Na-K-ATPase than on brain Na-K-ATPase. Indeed, inhibition is significantly greater for renal than brain Na-K-ATPase at every level of aluminum studied. The reason for this difference is not apparent, but may be related either to differences in species or to tissue-specific isoforms of the enzyme. The latter may well be a significant factor, in that structural diversity has been established for Na-K-ATPase and distinct differences have been described between the enzymes localized to the brain and kidney [17, 18].

The difference in the responses to aluminum by brain versus kidney Na-K-ATPase also supports the conclusion that the inhibitory effect is direct and not due to a reduction of effective ATP levels. Although aluminum is known to have a strong affinity to ATP nearly 10^7 times greater than Mg [19], when ATP levels are reduced in the reaction mixture it results in similar reductions in Na-K-ATPase activity for brain and kidney. On the other hand, equal concentrations of aluminum have distinctly different quantitative effects on the respective enzyme activities. Moreover, in the case of *renal* Na-K-ATPase, the extent to which activity declines with a reduction of ATP is not sufficient to account for the observed decline in the presence of aluminum. Finally, aluminum has no effect on the ouabain-insensitive component of Na-K-ATPase and therefore could not be acting through a reduction in effective ATP levels since this should have similar effects on ouabain-sensitive and ouabain-insensitive activity.

Previous studies investigating the effect of aluminum on Na-K-ATPase have produced conflicting results. Lai et al. [12] first reported that aluminum inhibits Na-K-ATPase in rat brain synaptosomes and demonstrated a 50% inhibitory concentration of 8.3 m*M*. This concentration, however, is so far above any conceivable physiological level as to be clinically irrelevant and considering the solubility characteristics of aluminum, raises serious questions concerning the methods employed. Staurnes et al. [13], reported 30% inhibition of Na-K-ATPase in the gills of salmonids at concentrations of aluminum below 10 μM. Although generally consistent with our findings, the sensitivity of salmonid Na-K-ATPase to aluminum was at least one order of magnitude greater than we observed. The reason for this discrepancy is likely to be the result of differences in experimental conditions. Besides the obvious difference in the species studied, the most probable explanation is the effect of pH. Staurnes et al. [13] carried out their studies at pH 5, a condition in which aluminum is nearly 10^6 times more soluble than at pH 7.4. The *effective* aluminum concentration attained at 10 μM was, therefore, considerably greater than achieved in our studies at 120 μM. Finally, their experimental design does not appear to adequately differentiate the effect of aluminum from that of pH on Na-K-ATPase activity. In a study investigating the effect of AlF_4 on the activity of porcine kidney Na-K-ATPase, Missiaen et al. [20] found that although AlF_4 significantly inhibits Na-K-ATPase activity, aluminum in the absence of fluoride has no inhibitory effect. These studies, however, were carried out at a single concentration of 10 μM aluminum and, in actuality, demonstrated a 10% inhibition of Na-K-ATPase activity which due to the small number of determinations did not achieve statistical significance. Bittar et al. [14], studying the effect of aluminum on Na transport in barnacle muscle, demonstrated marked inhibition following intracellular microinjection of aluminum chloride. Although a direct effect on Na-K-ATPase was not entirely excluded, they suggested that inhibition is indirect and mediated through an increase in intracellular free calcium. They proposed that free calcium, a recognized inhibitor of Na-K-ATPase [21], is increased through its displacement by aluminum from binding sites within the cell. This mechanism may, indeed, occur in vivo, but it cannot account for the findings of our studies, since the in vitro conditions under which they were carried out do not allow for such a possibility. This is not to imply that the indirect mechanism proposed by Bittar et al. [14] is not a factor in vivo, but merely that a direct effect by aluminum on Na-K-ATPase has also been demonstrated and may be physiologically relevant in vitro as well. It would seem quite reasonable to expect that, in vivo, aluminum may interfere with cellular Na efflux by several different mechanisms which could include indirect inhibition through the mechanism proposed by Bittar et al. [14], a reduction of available substrate by binding to ATP, and direct inhibition of Na-K-ATPase as shown in this study. A recent study by Caspers et al. [22], demonstrating that micromolar concentrations of aluminum increases the binding of [^{3}H]-ouabain to Na-K-ATPase, suggests yet another intriguing mechanism. Aluminum could possibly increase the binding of naturally occurring digoxin-like substances to Na-K-ATPase and thereby further suppress sodium transport in vivo.

As this study indicates that aluminum has a direct inhibitory effect on Na-K-ATPase activity in vitro, this mechanism may, under appropriate conditions, serve to impair sodium transport in vivo. Considering the concentrations at which aluminum exerts this inhibitory effect and the differential effect on Na-K-ATPase from different anatomical sites, aluminum's action in vivo may be limited to specific susceptible sites such as the kidney or gastrointestinal tract.

References

1 Berlyne GM, Yagil R, Ben Ari J, Weinberg G, Knopf E, Danovitch GM: Aluminum toxicity in rats. Lancet 1972;i:564–568.

2 Drueke T: Dialysis osteomalacia and aluminum intoxication. Nephron 1980;26: 207–210.

3 Flendrig JA, Krvis H, Das HA: Aluminium and dialysis dementia. Lancet 1976;i: 1235.

4 Alfrey AC, LeGendre GR, Kaehny WD: The dialysis encephalopathy syndrome: Possible aluminum intoxication. N Engl J Med 1976;294:184–188.

5 Short ASIK, Winney RJ, Robson JS: Reversible hypochromic microcytic anemia in dialysis patients due to aluminum intoxication. Proc Eur Dial Transplant Assoc 1980;17:226–233.

6 Berlyne GM, Yagil R: Aluminum-induced porphyria in rats. Lancet 1973;ii:1501–1502.

7 Netter P, Kessler M, Burnel D, Hutin MF, Delones S, Benoit J, Gaucher A: Aluminum in the joint tissue of chronic renal failure patients treated with regular hemodialysis and aluminum compounds. J Rheumatol 1984;11:66–70.

8 Ebina S, Okada S, Hanazaki S, Midorikawa O: Liver, kidney and central nervous system toxicity of aluminum given intraperitoneally to rats. A multiple-dose subchronic study using aluminum nitrilotriacetate. Toxicol Appl Pharmacol 1984;75: 211–218.

9 Perl DP, Brody AR: Alzheimer's disease: X-ray spectrophotometric evidence of aluminum accumulation in neurofibrillary tangle bearing neurones. Science 1980;208: 297–299.

10 Martyn CN, Barker JP, Osmond C, Harris EC, Edwardson JA, Lacey RF: Geographical relationship between Alzheimer's disease and aluminum in drinking water. Lancet 1989;i:59–62.

11 Joshi JG: Neurochemical hypothesis: Participation by aluminum in producing critical mass of colocalized errors in brain leads to neurological disease. Comp Biochem Physiol 1991;100C:103–105.

12 Lai JCK, Guest JF, Leung TKC, Lim L, Davison AN: The effects of cadmium, manganese and aluminium on sodium-potassium-activated and magnesium-activated adenosine triphosphatase activity and choline uptake in rat brain synaptosomes. Biochem Pharmacol 1980;29:141–146.

13 Staurnes M, Sigholt T, Reite OB: Reduced carbonic anhydrase and Na-K-ATPase activity in gills of salmonids exposed to aluminium-containing acid water. Experientia 1984;40:226–227.

14 Bittar EE, Nwoga J, Huang YP: Sensitivity of the sodium efflux in single barnacle muscle fibers to the microinjection of aluminum. Toxicol Appl Pharmacol 1990; 102:174–185.

15 Bonting SL: in Bittar EE (ed): Membranes and Ion Transport. London, Wiley Interscience, 1970, vol 1, p 257.

16 Baginski ES, Epstein E, Zak B: Review of phosphate methodologies. Ann Clin Lab Sci 1975;5:399–416.

17 Martin-Vasallo P, Dackowski W, Emmanuel JR, Levenson R, Nadal-Ginard B: Identification of a putative isoform of Na,K-ATPase beta subunit. Primary structure and tissue-specific expression. J Biol Chem 1989;264:4613–4618.

18 Shyjan AW, Gottardi C, Levenson R: The Na-K-ATPase β_2 subunits expressed in rat brain and copurifies with Na-K-ATPase activity. J Biol Chem 1990;265:5166–5169.
19 Martin BR: The chemistry of aluminum as related to biology and medicine. Clin Chem 1986;32:1797–1806.
20 Missiaen L, Wuytack F, DeSmedt H, Vrolix M, Casteels R: AlF_4 reversibly inhibits 'P'-type cation-transport ATPases, possibly by interacting with the phosphate-binding site of the ATPase. Biochem J 1988;253:827–833.
21 Yingst DR: Modulation of the Na^+ + K^+-ATPase by Ca and intracellular proteins. Annu Rev Physiol 1988;50:291–303.
22 Caspers ML, Kwaiser TM, Grammas P: Control of [^{3}H]ouabain binding to cerebromicrovascular (Na^+ + K^+)-ATPase by metal ions and proteins. Biochem Pharmacol 1990;39:1891–1895.

Andrew J. Adler, MD, Department of Medicine (111),
Brooklyn Veterans Administration Hospital, 800 Poly Place,
Brooklyn, NY 11209 (USA)

Berlyne GM (ed): The Kidney Today. Selected Topics in Renal Science.
Contrib Nephrol. Basel, Karger, 1992, vol 100, pp 127–138

Adverse Events of Subcutaneous Recombinant Human Erythropoietin Therapy

Horst Klinkmann[a], *Reinhard Schmidt*[a], *Lothar Wieczorek*[b], *Paul Scigalla*[b]

[a]Department of Internal Medicine, University of Rostock, Rostock, and
[b]Boehringer Mannheim, Mannheim, FRG

The history of erythropoietin is an ideal example for biotechnical and clinical progress ever since the basic observation of Jourdanet [1] in 1863 until its molecular cloning and first clinical application in the late 1980s [2–4].

In less than 5 years the clinical use of recombinant human erythropoietin (rhEPO) has already substantially improved the quality of life of thousands of patients with anemia due to renal insufficiency. Realistic expectations are now that a broad range of diseases and a countless number of patients will soon benefit from the therapeutic application of rhEPO as well as from its use in autologous blood storage [5–9].

At this stage of development, two basic questions have high priority: (1) the efficacy of the different routes of application and (2) the safety profile of rhEPO. We are conducting a large European Multicenter Trial over 2 years, assessing the efficacy of subcutaneously administered rhEPO as well as its clinical tolerance and influence on morbidity and safety profile. This report summarizes the safety issues related to the subcutaneous administration of rhEPO in this trial.

Patients and Methods

362 hemodialyzed patients (162 males, 200 females) from 16 European dialysis centers (fig. 1) entered the study after giving informed consent to participate in a controlled trial with half of the patients receiving rhEPO therapy at the beginning of the trial

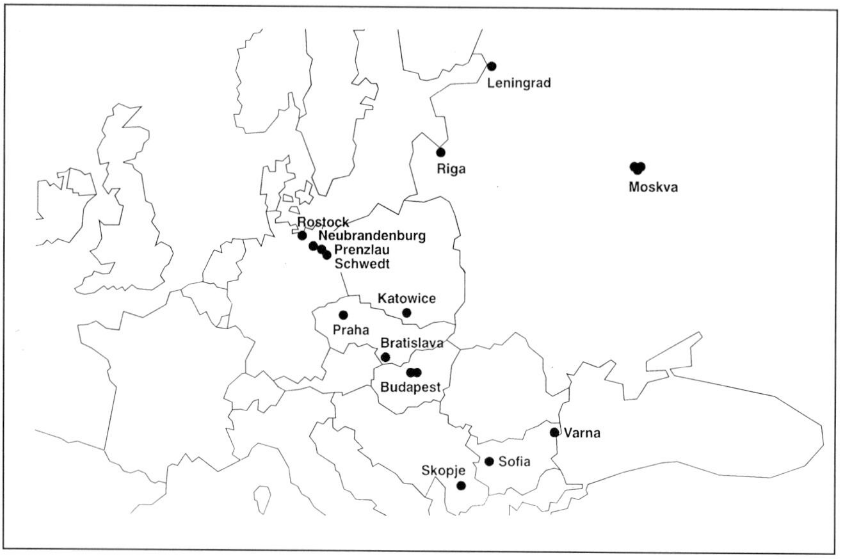

Fig. 1. Participating centers in the European Multicenter Trial (see text for details).

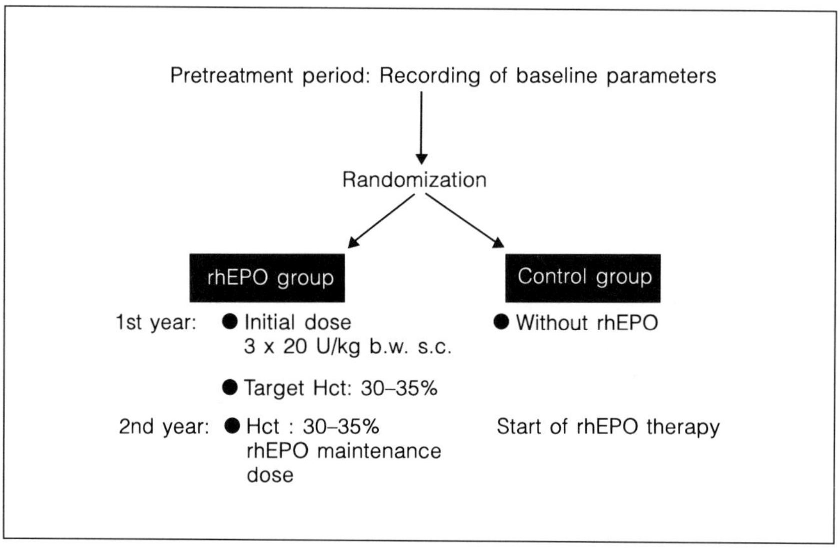

Fig. 2. Study protocol.

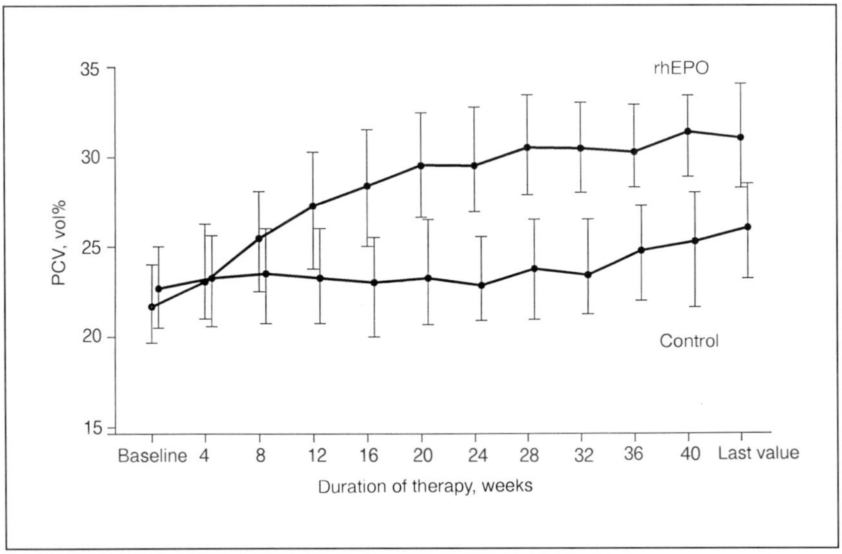

Fig. 3. Course of hematocrit in ESRD patients with and without rhEPO treatment (median, interquartile ranges).

and the other half serving as control group in the first year and receiving rhEPO therapy thereafter in the second year (fig. 2). The trial was initiated with a 2-week pretreatment period for adjustment prior to randomization. The initial rhEPO dose of Reconorm (Boehringer Mannheim) in the therapy group was 3 × 20 U/kg body weight per week subcutaneously until the target hematocrit of 30–35 vol% was reached. When the weekly increase in hematocrit was 0.5 vol% the initial dose was increased by 20 U/kg body weight per single dose every 4 weeks until the target hematocrit was reached. After reaching the target hematocrit, the last individually administered rhEPO dose was then reduced by 50% and adjusted subsequently to the development of the hematocrit (fig. 3).

Safety was evaluated in all patients based on serial histories, repeated physical examinations, blood chemistries, hematological values, chest x-rays, electrocardiograms and determination of rhEPO antibodies. The observation time reported in this paper comprises an average of 266 days. During the entire observation period all adverse events (AE) were documented irrespective of a possible causality with rhEPO therapy. The analysis is based on the AE reported to the principal clinical investigator. AE are described as any undesired noxious or pathological change in a patient as indicated by signs, symptoms and for laboratory changes. An AE is classified as a serious one if any of the following occurs: (1) the event was fatal or life-threatening (WHO grade IV); (2) hospitalization was required or prolonged; (3) the patient was permanently disabled; (4) the event was a congenital anomaly or cancer [10].

Table 1. Summary of death causes occurring during the study

rhEPO group (n = 14)	n	Control group (n = 18)	n
Progressive heart failure	2	Cardiorespiratory failure	7
Apoplexy	1	Myocardial infarction	2
Cerebral hemorrhage	1	Subarachnoidal hemorrhage	1
Intracranial bleeding	1	Intracranial bleeding	1
Pulmonary edema	1	Hemorrhagic cerebral infarction	1
Brain edema	1	Gastrointestinal bleeding	1
Hyperhydration and hyperkalemia	3	Pulmonary embolism	1
Sepsis	2	Pneumonia	1
Pneumonia	1	Peritonitis	1
Spondylitis purulenta	1	Tuberculosis septica	1
		Diabetic coma	1

Analysis of the outcome parameters was centrally performed on patients only, who comprehend all study parameters required. A total of 73 patients dropped out of the study, 34 from the therapy group, 39 from the control group. A summary of the withdrawals are listed in figure 4. Renal transplantation was the cause for discontinuation in 21 patients. 32 patients died, 14 from the therapy group, 18 from the control group. The causes of death are shown in table 1. Among the remaining 20 withdrawals, 8 showed a noncompliance to the protocol requirements, 1 was a protocol violator and in 3 patients essential data for analysis were missing. Among the 8 serious AE (6 in the therapy group, 2 in the control group) that caused the discontinuation of the study in these patients, hypertension or hypertensive-related episodes are prevailing.

Results

During the observation period a total of 837 AE occurred among the 362 patients included in this trial. 277 of these AE were classified as serious and 560 as nonserious (fig. 5). The relation between the two group studies for the total AE showed 468 = 55.9% in the therapy group and 369 = 44.1% in the control group. The number of serious AE was slightly higher in the control group (n = 149) compared to the rhEPO-treated group (n = 128). A summary of the documented serious AE during the observation period is given in table 2. In this report only the serious AE are analyzed and reported.

Death (table 1). A total of 32 deaths (14 from the therapy group, 18 from the control group) have been reported for the study population. Per-

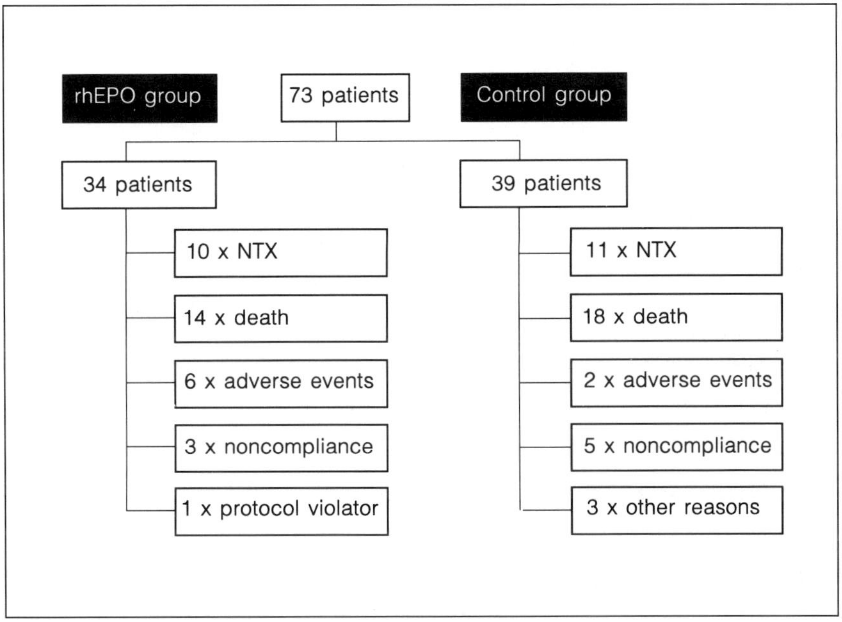

Fig. 4. Summary of withdrawals during the study.

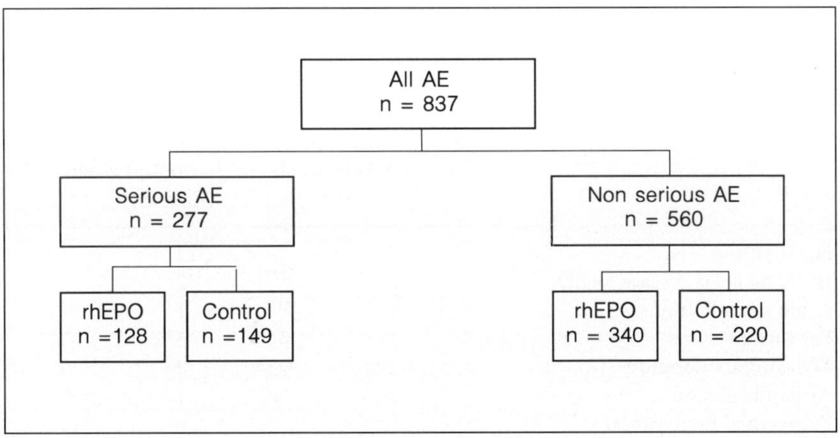

Fig. 5. AE in controlled clinical study on the influence of subcutaneous rhEPO therapy on morbidity (see text for details).

Table 2. Serious AE in controlled clinical study on the influence of subcutaneous rhEPO therapy on morbidity in hemodialyzed patients

Serious AE	rhEPO group	Control group	Total
Cardiovascular system	14	25	39
Hypertension	19	7	26
Hypotension	–	3	3
Peripheral circulatory disturbances	2	1	3
Shunt occlusions	23	17	40
Shunt infections	3	2	5
Other shunt problems	3	8	11
Respiratory system	12	19	31
Gastrointestinal tract	4	8	12
Liver, pancreas, gallbladder	9	7	16
Kidneys	2	2	4
Hyperkalemia	4	3	7
Hyperhydration	5	4	9
Exsiccosis	–	1	1
CNS	9	7	16
Peripheral nervous system	–	2	2
Sepsis	3	4	7
Others	14	27	41

Table 3. Serious AE in cardiovascular system

	rhEPO group (n = 30)	Control group (n = 31)
Heart failure	5	12
Ischemic heart disease/angina	2	2
Acute heart failure	1	1
Pericarditis	1	5
Myocardial infarction	2	3
Atrial fibrillation	2	1
Paroxysmal bradycardia	1	–
Hypertension/encephalopathy	9	2
Uncontrolled hypertension	7	5

centagewise this reflects 9% of the patients entering the study and mirrors the disease state of this patient population. Death due to heart disease is significantly higher in the control group (n = 9) than in the therapy group (n = 1). In the rhEPO-treated group, we observed a higher incidence of death caused by hyperhydration in conjunction with cerebral and pulmonary edema and hyperkalemia (n = 5) compared to the control group (n = 0). Death cases based on bleeding and/or vascular lesions are equally distributed among the therapy group (n = 3) and the control group (n = 3). The same applies to death cases caused by infection (3:3).

Cardiovascular System (table 3). It is noteworthy that the number of patients with serious symptoms from the cardiovascular system including hypertension is almost as equally high in the therapy group (n = 30) as in the control group (n = 31). If, however, hypertension is excluded, the overall picture changes completely. 14 serious AE are in the therapy group but 24 in the control group. While 9 acute hypertensive episodes with encephalopathy-like syndromes occurred among the therapy group, only 2 were observed in the control group. The development and/or aggravation of hypertension was registered 7 times in the therapy group and 5 times in the control group. Taking the total number of patients into consideration, these data also show convincing evidence that the overall incidence of hypertensive events is quite modest and differs only by 4.9% between the therapy group and control group.

Vascular Access. The observed total number of 40 shunt occlusions in 27 patients was almost equally divided between the two study groups (table 2). The number of patients with a so-called fistula at risk was with 5 patients in each group also equally distributed in both groups.

Gastrointestinal System (table 4). Whereas the total number of serious AE originating from the gastrointestinal system is comparably low in numbers (n = 12), the distribution among the two groups is clearly in favor of the therapy group (n = 4) compared to the control group (n = 8).

Respiratory System (table 5) *and Infection.* Altogether, 31 serious AE are reported as being related to the respiratory system: 12 occurred in the therapy group and 19 in the control group. A distinct difference between the two groups was observed in the incidence rate of pneumonia (8 in the therapy group, 13 in the control group) and in pleuritis (0:3). As reported

Table 4. Serious AE in gastrointestinal system

	rhEPO group	Control group
Erosive gastritis	1	1
Chronic gastritis	–	1
Erosive duodenitis	1	–
Gastric bleeding	–	1
Gastrointestinal bleeding	2	3
Peritonitis	–	1
Acute abdomen	–	1

Table 5. Serious AE in respiratory system

	rhEPO group	Control group
Pneumonia	8	13 (12 pts)
Bronchitis	2	2
Upper respiratory tract infection	2	–
Pleuritis	–	3
Acute lung embolism	–	1

in previous studies with intravenous application [5, 10–13], the respiratory tract is the main site of infection. Among the reported 42 serious AE due to infection, 28 were located in the respiratory tract, followed by gallbladder (n = 7). The remaining 85 serious AE, which do not belong to the above described main groups, are rather equally distributed among the therapy group and control group, They reflect the disease status of the patient population with regular dialysis treatment. The observed biochemistry profile shows no significant changes that could be attributed to rhEPO therapy. No circulating rhEPO antibodies were detected.

Discussion

The safety profile of intravenous rhEPO therapy in anemic hemodialyzed patients has been documented in several studies [10–12]. Due to different pharmacokinetic courses of intravenously or subcutaneously administered rhEPO [14, 15], it is theoretically conceivable that the occurrence or characteristics of adverse events may be different among the two modes of application. Based upon this assumption, one of the goals of this controlled study on 362 patients was to assess the safety profile of subcutaneous rhEPO therapy. A detailed analysis of the nonserious adverse events will follow later.

The design of this controlled multicenter study has aimed at the comparison between two groups of hemodialyzed, transfusion-dependent patients with an almost identical medical background and the same quality in dialysis treatment. The general problem of the influence of the disease state of the patients on the results that previous studies were also facing, remains unchanged and may be even more important in this particular study involving mainly Eastern European dialysis centers. Patients from this part of Europe enter the dialysis program usually later compared to Western European countries. Therefore most of the dialysis patients in this study had numerous other medical problems apart from anemia. On the one hand, it might be therefore almost impossible to judge which of the serious adverse events originate from other sources than rhEPO therapy; on the other hand, however, the comparatively equal amount of serious adverse events to previously reported studies with intravenous administration in a different patient population indicates the high safety profile of subcutaneous rhEPO therapy.

Analyzing the 32 cases of death, it is obvious that death, resulting from cardiovascular insufficiency including myocardial infarction, is significantly lower in the therapy group (n = 1) than in the control group (n = 9). The correction of renal anemia has without any doubt a protective effect on the high-risk cardiovascular situation of patients with terminal renal failure [16, 17]. The higher incidence of death caused by fluid overload and hyperkalemia may well be an indication for a reduced discipline in these patients, since an improved general well-being leads to an infringement of dietary and liquid intake rules [10, 13, 18, 19]. No difference was found between the therapy group and the control group regarding hypertension as the cause of death. A direct relation between subcutaneous rhEPO therapy and death can be rated unlikely in this study.

The correlation between the serious and nonserious AE in this trial is in the average range of 1:2 (277 serious AE, 560 nonserious AE). A comparison of the two groups taking the total number of AE into account showed a higher number in the therapy group (n = 468) than in the control group (n = 369). It is however remarkable that the relation between the two groups changes if one only compares serious AE (therapy group n = 128, control group n = 149). The lower number of serious AE in the rhEPO-treated group is almost exclusively based on the lower incidence rate of complications arising from the cardiovascular-respiratory system which again supports the opinion of a protective effect of rhEPO therapy on the cardiovascular system.

Hypertension as the most common AE observed in patients with rhEPO treatment remains to be an area of concern also with subcutaneously administered rhEPO [20–22]. If one looks however at the percentage difference between the two groups (8.8%/year in the rhEPO group, 3.9%/year in the control group) the difference of 4.9%/year between the therapy group and the control group is not very large. The different pharmacokinetics as well as the lower dose of subcutaneously administered rhEPO as well as our better understanding of the clinical situation may well be the reason for this beneficial development in controlling the problem of hypertension in rhEPO treatment.

Fistula problems are almost equal in the therapy group and in the control group and are very well comparable to fistula complications in the general hemodialyzed patient population. It reflects in our opinion that a hematocrit between 30 and 35% does not introduce significant changes in blood viscosity and rheology [23]. The comparatively rather low number of serious AE based on infections is most likely an indication for a better oxygen supply and oxygen transport as well as an overall increase in resistance to infection. Only 12 of the 30 reported respiratory tract infections occurred in the therapy group.

In summary, subcutaneous rhEPO treatment has demonstrated a reliable safety profile in this study which might even be better than that in rhEPO treatment with intravenous application.

References

1 Jourdanet D: De l'anémie des altitudes et de l'anémie en général dans ses rapports avec la pression de l'atmosphère. Paris, Ballière, 1863, p 44.

2 Jacobs K, Shoemaker C, Rudersdorf R, Neill SD, Kaufman RJ, Mufson A, Seehra J, Jones SS, Hewick R, Fritsch EF, Kawakita M, Shimizu T, Miyake T: Isolation and characterization of genomic and cDNA clones of human erythropoietin. Nature 1985;313:806.

3 Lin F, Suggs S, Lin C, Browne JK, Smalling R, Egrie JC, Chen KK, Fox GM, Martin F, Stabinsky Z, Badrawi SM, Lai P, Goldwasser E: Cloning and expression of the human erythropoietin gene. Proc Natl Acad Sci USA 1985;82:7580.

4 Eschbach JW, Kelly MR, Haley NR, Abels RI, Adamson JW: Treatment of the anemia of progressive renal failure with recombinant human erythropoietin. N Engl J Med 1989;321:158–163.
5 Scigalla P, Wieczorek L, Bicker U: Treatment of renal anaemia with recombinant human erythropoietin: European experience; in Garnick MB (ed): Erythropoietin in Clinical Applications. New York, Dekker, 1990, pp 141–182.
6 Means RT, Olsen NJ, Krantz SB, Graber SE, Dessypris EN , Graber SE, Stohe WJ, O'Neil VO, Pincus T: Treatment of the anemia of rheumatoid arthritis with recombinant human erythropoietin: Clinical and in vitro studies. Arthritis Rheum 1989; 32:638–642.
7 Levine EA, Rosen AL, Gould SA, Sehgal LR, Egrie JC, Browne JK, Sehgal HL, Moss GS: Reconmbinant human erythropoietin and autologous blood donation. Surgery 1988;104:365–369.
8 Toy PTCY, Strauss RG, Stehling LC, Sears R, Price TH, Rossi EC, Collins ML, Crowley JP, Eisenstaedt RS, Goodnough LT, Greenwalt TJ, Johnston MFM, Kennedy MS, Lenes BA, Lusher JM, Mintz PD, Patten ED, Simon TL, Westphal RG: Predeposited autologous blood for elective surgery: A national multicenter study. N Engl J Med 1987;316:517–520.
9 Sherwood JB, Goldwasser G, Chilcote R, Carmichael LD, Nagel RL: Sickle cell anemia patients have low erythropoietin levels for their degree of anemia. Blood 1986;67:46–49.
10 Samtleben W, Ehmer B, Lutz-Knoechenhauer I, Hagmann C, Scigalla P, Gurland HJ: Side effects during recombinant human erythropoietin therapy in 2,000 ESRD patients. Contrib Nephrol, Basel, Karger, 1991, vol 88, pp 107–116.
11 Eschbach JW, Egrie JC, Downing MR, Browne JK, Adamson JW: The safety of epoetin-alpha: Results of clinical trials in the United States. Contrib Nephrol, Basel, Karger, 1991, vol 88, pp 72–80.
12 Bahlmann J, Schöter K-H, Scigalla P, Gurland HJ, Hilfenhaus M, Koch KM, Muthny FA, Neumayer HH, Pommer W, Quellhorst E, Sieberth HG, Weber U: Morbidity and mortality in hemodialysis patients with and without erythropoietin treatment: A controlled study. Contrib. Nephrol, Basel, Karger, 1991, vol 88, pp 90–106.
13 Wong KC, Li PKT, Lui SF, Nicholls MG, Lai KN: The adverse effects of recombinant human erythropoietin therapy; in Adverse Drug Reaction. Acute Poisoning Review. Oxford, Oxford University Press, 1990, vol 9, pp 183–206.
14 MacDougall IC, Roberts DE, Coles GA, Williams JD: Clinical pharmacokinetics of Epoetin (recombinant human erythropoietin). Clin Pharmacokinet 1991;20:99–113.
15 Bommer J, Samtleben W, Koch KM, Baldamus CA, Grützmacher P, Scigalla P: Variations of recombinant human erythropoietin application in hemodialysis patients. Contrib Nephrol, Basel, Karger, 1989, vol 76, pp 149–158.
16 Akizawa T, Koshikawa S, Takaku F, et al: Clinical effect of recombinant human erythropoietin on anaemia associated with chronic renal failure. A multi-institutional study in Japan. Int J Artif Organs 1988;11:343–350.
17 Onoyama K, Kumagai H, Takeda K, Shimamatsu K, Fujishima M: Effects of human recombinant erythropoietin on anaemia, systemic haemodynamics and renal function in predialysis renal failure patients. Nephrol Dial Transplant 1989;4:966–970.

18 Delano BG: Improvements in quality of life following treatment with rHuEPO in anemic hemodialysis patients. Am J Kidney Dis 1989;14(suppl 1):14–18.
19 Zehnder C, Gluck Z, Descocudres C, Uehlinger E, Blumberg A: Human recombinant erythropoietin in anaemic patients on maintenance haemodialysis. Secondary effects of the increase of haemoglobin. Nephrol Dial Transplant 1988;3:657–660.
20 Brunkhorst R, Nonnast-Daniel B, Koch KM, Frei U: Hypertension as a possible complication of recombinant human erythropoietin therapy. Contrib Nephrol, Basel, Karger, 1991, vol 88, pp 118–125.
21 Bommer J, Barth HP, Zeier M, Mandelbaum A, Bommer G, Ritz E, Reichel H, Novack R: Efficacy comparison of intravenous and subcutaneous recombinant human erythropoietin administration in hemodialysis patients. Contrib Nephrol, Basel, Karger, 1991, vol 88, pp 136–143.
22 Buckner FS, Eschbach JW, Haley NR, Davidson RR, Adamson JW: Correction of the anemia in hemodialysis patients with recombinant human erythropoietin (rHuEpo): Hemodynamic changes and risks for hypertension. Kidney Int 1989;35: 190A.
23 Lerche D, Schmidt R, Zoellner K, Meier W, Paulitschke M, Distler B, Klinkmann H: Rheology in whole blood and in red blood cells under recombinant human erythropoietin therapy. Contrib Nephrol, Basel, Karger, 1989, vol 76, pp 299–305.

Prof. Dr. Drs. Horst Klinkmann, Klinik für Innere Medizin, Universität Rostock, Ernst-Heydemann-Strasse, D-O–2500 Rostock (FRG)

Berlyne GM (ed): The Kidney Today. Selected Topics in Renal Science.
Contrib Nephrol. Basel, Karger, 1992, vol 100, pp 139–154

Recombinant Human Growth Hormone in Pubertal Patients with Chronic Renal Disease

Ora Yadin[a], *Elaine S. Kamil*[a], *Kimberly Pyke-Grimm*[a], *Pauline A. Nelson*[a], *Inez M. Boechat*[c], *Barbara M. Lippe*[b], *Richard N. Fine*[d]

UCLA School of Medicine, Department of Pediatrics, Divisions of [a]Pediatric Nephrology and [b]Pediatric Endocrinology, and [c]Department of Radiology, Los Angeles, Calif.; [d]Department of Pediatrics, State University of New York at Stony Brook, N.Y., USA

Puberty is characterized by an increase in linear growth velocity (GV) and the appearance of secondary sexual characteristics [1]. The first sign of pubertal development in boys is the increase of testicular size [2]; the first pubertal change in girls is breast development [3]. Subsequently, pubic hair appears; 1½ years later in boys than in girls. The peak GV is reached approximately 2 years later in boys than in girls [4].

In children with chronic renal failure (CRF) sexual development and pubertal growth spurt are usually delayed, and are often incomplete [5]. Despite major advances in the clinical management of children with preterminal CRF and end-stage renal disease (ESRD) retarded growth, delayed pubertal development and an insufficient growth spurt remain unresolved clinical problems. Many factors have been implicated in the etiology of delayed pubertal development and growth retardation [6–11]. Uremia may affect the endocrine function by altering synthesis, release, protein binding, degradation, elimination, receptor binding and postreceptor activity of various hormones [12].

Endocrine function has been studied extensively in adults with CRF and a disturbance at the gonadal level has been postulated [13–15]. In prepubertal and pubertal boys with CRF, low or low-normal plasma testosterone levels have been reported [16], which have been attributed to damage of testicular tissue as a result of the uremic milieu occurring either before or during puberty [17].

The level of spontaneous circulating immunoreactive growth hormone (GH) is either normal or elevated in uremia [18, 19], and the response of GH production to known stimuli is increased in uremic children [20]. The pathogenetic mechanisms for these pertubations of GH levels in uremia could be related to either decreased excretion or impaired metabolism [21].

The magnitude of the pubertal growth spurt is also suppressed in patients with CRF. When compared to a late maturing control group, the pubertal height gain in patients with CRF was only 58% for boys and 48% for girls [22]. This impaired pubertal growth spurt significantly adversely affects the final height of children with CRF [22] despite successful renal transplantation either prior to or following puberty [22].

We have previously demonstrated a salutary affect of treatment with supraphysiological doses of recombinant human GH (rhGH) on GV in growth-retarded children with CRF and ESRD [23–25]. This report will describe the effect of rhGH treatment on the growth rate and pubertal development of growth-retarded pubertal patients with preterminal CRF, those undergoing peritoneal dialysis (PD) and those post-renal transplantation (PT).

Table 1. Patient data

Pat. No.	Sex	Primary renal disease	Category	Tanner stage	Age at initiation years	Bone age years	SDS for height at initiation
1	m	renal dysplasia	CRF	III	16.3	12.5	–5.93
2	m	oligomegalonephronia	CCPD	I	10.8	8.0	–1.30
3	m	obstructive uropathy	CCPD	II	15.0	12.5	–4.41
4	m	anti-GBM	CCPD	I	17.7	11.5	–5.90
5	m	FSGS	transplant	I	17.3	11.5	–7.10
6	m	Alport's syndrome	transplant	I	14.0	12.5	–2.10
7	m	MCD	transplant	III	14.9	11.0	–4.60
8	m	renal dysplasia	transplant	II	12.9	11.5	–2.83
9	m	renal dysplasia	transplant	III	12.5	12.5	–1.90
10	m	obstructive uropathy	transplant	II	14.3	10.0	–3.03
11	m	Goodpasture's syndrome	transplant	III	17.8	12.5	–5.53

anti-GBM = Antiglomerular basement membrane; FSGS = focal segmental glomerular sclerosis; MCD = medullary cystic disease; CCPD = continuous cycling peritoneal dialysis.

Materials and Methods

Ten growth-retarded (standard deviation score (SDS) –2.1 to –7.1) patients and 2 patients with a reduced GV during at least 1 year of observation (SDS –1.3 and –1.9), who were either pubertal or became pubertal during the course of treatment (Tanner stage II–III), with either preterminal CRF, undergoing PD or PT were enrolled into the study. One patient undergoing PD discontinued rhGH after 12 months and a PT patient elected to discontinue GH after 30 months of treatment. Neither patient demonstrated a marked increase in GV following rhGH treatment, and noncompliance was suspected in both cases.

The age, sex, bone age, Tanner stage, SDS for height at baseline, primary renal disease, treatment modality (CRF, PD or PT) and duration of rhGH treatment are shown in table 1. All patients were males, aged 10.8–17.8 years (mean ± SD = 15.2 ± 1.9), with a bone age of 8–13 years (mean ± SD = 11.45 ± 1.46) at initiation of the study. One patient had preterminal CRF, 3 were undergoing PD, and 7 were PT. The mean age at transplantation for the 7 PT patients was 10.5 years (range 6.6–13 years). The dose of prednisone was between 0.11 and 0.43 mg/kg/day (table 2), and the glomerular filtration rate (GFR) was between 23.5 and 104 ml/min/1.73 m^2 at entrance into the study. The 3 patients undergoing PD were aged 10.8, 15 and 17.7 years, and had a bone age of 8, 12.5 and 11.5 years, respectively, at initiation of the study. Currently, 2 PD patients have completed 24 months of rhGH treatment, and the third died 10 months after discontinuing 1 year of rhGH treatment from an arrhythmia during an episode of peritonitis. The patient with CRF was 16.3 years old at initiation of the study, with a bone age of 13 years and a Tanner stage of III (patient 1 in table 1). All patients were significantly growth retarded at the initiation of the study, except for patients 2 and 9 (SDS –1.30 and –1.9), as indicated by an SDS for height of –2.1 to –5.93 (table 1).

Treatment Protocol

The treatment protocol was approved by the UCLA Human Subjects Protection Committee and written informed consent was obtained from each parent(s) and/or patient prior to initiation of the study. Caretakers and/or patients were trained to administer the rhGH subcutaneously, or intraperitoneally (IP). Genentech Inc. provided the

Table 2. Prednisone dose at initiation of study in PT group

Pt. No.	Age at Tx	Prednisone dose, mg/kg
5	12	0.27
6	7.5	0.19
7	12	0.12
8	11.6	0.42
9	10.8	0.11
10	6.6	0.13
11	13	0.17

Table 3. Mean hormone levels prior to and following rhGH treatment

Time on study months	Testosterone ng/dl	LH mIU/ml	FSH mIU/ml	TSH mU/ml	T_4 µg/dl	Prolactin ng/ml
Baseline	147.5±200.3	6.3±4.1	3±1.9	2.5±2.3	8.9±1.9	5.8±4.5
12	199.4±188.6	10.3±7.4*	4.3±3.1***	2.16±1.9	9.4±2.3	9.0±6.0
24	324±156	17.6±10.5**	7.2±7	2.2±0.6	7.7±1.5	9.0±4.8
36	443.5±98.3	12.5±0.7	2±0	1.75±1.3	7.6±2.3	6.0±0.8

* p = 0.016; ** p = 0.02; *** p = 0.045.

Table 4. Stimulated GH levels at 60 and 90 min

Pt. No.	GH stimulation, ng/ml	
	60 min	90 min
1	18.4	10.1
2	0.3	7.9
3	10.6	10.6
4	24 h	GH sampling
5	1.0	0.7
6	13.3	8.7
7	3.1	5.2
8	50.0	50.0
9	13.1	2.8
10	8.4	4.3
11	1.0	1.0

Table 5. Mean fasting glucose and insulin levels at baseline and at 6-month intervals following rhGH treatment

		Baseline	12 months	24 months	36 months
Glucose	fasting	92±11	97±11	98±11	76±8
mg/dl	postprandial	112±28	112±24	116±25	100±19
Insulin	fasting	32±44	16±9	19±11	22±10
µU/ml	postprandial	89±77	291±98	75±38	80±52

rhGH (Protropin®, met-hGH) in 5-mg vials which was reconstituted with 1 ml of bacteriostatic water for the subcutaneous route and 5 ml for the IP route. The initial dosage of rhGH was either 0.125 mg/kg three times weekly or 0.053 mg/kg daily, and this dose was doubled if after a 6-month period the GV had not increased by at least 50% above that of the previous year.

Endocrine Studies

Testosterone, luteinizing hormone (LH), follicle-stimulating hormone (FSH), T_4, thyroid-stimulating hormone (TSH) and prolactin blood levels were determined prior to initiation of the study, and at 6-month intervals thereafter (table 3). Plasma insulin-like factor 1 (IGF-1) determinations were performed following acid chromatography [26] by Dr. Raymond Hintz of Stanford University School of Medicine, Palo Alto, Calif., at baseline and at 6-month intervals thereafter. Ten of the patients underwent an *L*-dopa/propranolol GH stimulation test as described by Fass et al. [27], while 1 patient had a 24-hour serial sampling of spontaneous GH secretion with samples being obtained every 30 min. The results of the stimulation tests are shown in table 4. Each patient had a fasting and 2-hour postprandial blood glucose and insulin determination prior to initiating the study and at 6-month intervals thereafter (table 5).

Routine Laboratory Studies

At the initiation of the study and at 2, 4, 6, 9, and 12 months following rhGH treatment and then at 3-month intervals thereafter, the following studies were performed: complete blood count, fasting SMAC biochemical panel, and serum calcium level.

Radiologic Studies

Bone age was evaluated by one of us (M.I.B.) on the radiograph of the left wrist using the standard Greulich-Pyle method [28] at baseline and at 6-month intervals thereafter.

Pubertal Status

Tanner stage [29] was evaluated at baseline and at 6-month intervals thereafter.

Clinical Follow-Up and Anthropometric Measurements

Follow-up examinations were performed at 2, 4, 6, 9, and 12 months following the initiation of the rhGH treatment, and every 3 months thereafter. One of us (P.A.N.) obtained the anthropometric measurements, without reference to previous determinations. Height was measured using a fixed, wall-mounted stadiometer, according to previously published methods [30]. Measurements were repeated until three consecutive observations agreed with each other within a range of 0.2 cm. Weight was measured on the same beam scale each time with the patient wearing light clothing. Data for normal children were obtained from the Hanes I study [31], using the 50th percentile of children of the same chronological age and sex for calculation of the height and weight SDS. Triceps skinfold (TSF) was measured using standard calipers (Lange Skinfold Calipers, Cambridge Scientific Industries Inc.). Measurements were repeated until three consecutive observations agreed with each other within a range of 2 mm. Midarm circumference (MAC) was obtained from the nondominant arm, using a nonstretch tape, and midarm muscle circumference (MAMC) was derived mathematically using a previously published formula [32], at baseline and at 6-month intervals thereafter.

Table 6. GV (in cm) for the previous year and at 12-month intervals following rhGH treatment

Pt. No.	Baseline	12 months	24 months	36 months
1	3.0	6.2	4.2	
2	4.0	7.3	6.5	
3	4.0	3.7	4.3	
4	0.1	0.4		
5	0.1	2.2	2.3	
6	0.2	4.0	3.5	5.1
7	1.9	7.0	7.9	8.4
8	2.3	4.4	3.7	
9	5.8	9.2		
10	2.9	10.0		
11	0.6	4.3		

Statistical Analysis

Results are expressed as mean ± SD. The significance of differences was assessed by the paired two-tailed Student's t test, and a $p < 0.05$ was accepted as significant.

Results

Growth

The GV and SDS for the 12 months prior to initiation of the study and for 12-month periods following rhGH treatment are shown in table 6. The mean GV increased from 2.26 ± 1.9 to 5.34 ± 2.91 cm/year ($p = 0.0006$), 4.63 ± 1.92 ($p = 0.01$) and 5.2 ± 3.15 cm/year (NS) following 12, 24 and 36 months of rhGH treatment, respectively. The mean weight gain for the 12 months prior to treatment was 2.58 ± 2.77 kg. It was 3.26 ± 2.58, 2.69 ± 1.5 and 0.92 ± 0.78 kg following 12, 24 and 36 months of rhGH treatment, respectively. The weight gain with treatment was not statistically different from the year prior to rhGH treatment. Although TSF decreased from a mean of 13 ± 6.7 mm at baseline to 10 ± 4.2 mm after 12 months of rhGH treatment, to 7.9 ± 3.1 mm after 24 months and to 7.75 ± 1.0 mm after 36 months of treatment, the changes were not statistically significantly different from baseline. MAC was not changed at any time interval during rhGH treatment. MAMC increased from a mean of 19.3 ±

1.5 cm at baseline to a mean of 20.1 ± 2.0 cm at 6 months of rhGH treatment ($p = 0.05$), but there was no further significant increase at other treatment intervals.

Bone Age

The mean bone age at baseline was 11.45 ± 1.4 years. At 12, 24 and 36 months of rhGH treatment the mean bone age was 12.3 ± 1.6 ($p = 0.015$), 12.17 ± 1.8 and 13.25 ± 1.0 years, respectively. During the 12- to 36-month period of rhGH treatment the bone age did not increase more than the increase in chronological age in any of the patients, despite the acceleration in GV. The mean Δheight age/Δbone age was 1.14 ± 0.92, indicating that GV was accelerated more than the increase in bone age.

Renal Function

The creatinine clearance (C_{cr}) of the 1 patient with CRF and the 7 renal allograft recipients at the initiation of the study and at 6-month intervals thereafter were calculated. Mean C_{cr} was 57.3 ± 24.8 ml/min/ 1.73 m^2 prior to initiation of rhGH, and 54.7 ± 32.4, 42.9 ± 23.4, and 46.2 ± 31.6 ml/min/1.73 m^2 at 12, 24 and 36 months, respectively. There was no significant decline in the mean C_{cr} during the 36 months of treatment. Patient 9 whose C_{cr} decreased from 63 ml/min/1.73 m^2 at initiation of the study to 28 ml/min/1.73 m^2 after 18 months of rhGH treatment had two episodes of acute rejection secondary to admitted noncompliance with immunosuppressive therapy.

Endocrine Studies

The Tanner stage, LH, FSH, TSH, T_4 and prolactin levels at initiation of the study and at 6-month intervals thereafter are shown in table 2. Four of the 11 patients entered the study at Tanner stage I, 3 were at Tanner stage II and 4 were at Tanner stage III. Six patients progressed by 1 Tanner stage during 12–18 months on rhGH; 2 patients progressed from Tanner stage I to IV during 30 and 36 months of rhGH treatment and 1 patient progressed from Tanner stage III to V during 36 months of rhGH treatment. Three of the patients remained at their baseline Tanner stage after 12 months of rhGH treatment. The mean testosterone levels were 14.8 ± 8.2 ng/ml for Tanner stage I, 55.6 ± 67.3 ng/ml for Tanner stage II, 348.9 ± 96.3 ng/ml for Tanner stage III, 412 ± 157 ng/ml for stage IV and 359.5 ± 60.1 ng/ml for Tanner stage V.

GH Studies

The GH stimulation studies are shown in table 3. Of the 10 patients undergoing the *L*-dopa/propranolol stimulation test, 6 had GH levels at either 60 or 90 min that exceeded 10 ng/ml. One PT patient (No. 8) had very high levels – 50 ng/ml both at 60 and 90 min. This patient was on a high mean prednisone dose of 0.42 mg/kg and had a mean C_{cr} of 50.8 ± 6.4 ml/min. Patient 7 (PT) had a level > 5 ng/ml at 90 min. Two additional PT patients failed to demonstrate a stimulated GH level > 5 ng/ml both at 60 and 90 min. Patient 5 had levels of 1 and 0.7 ng/ml at 60 and 90 min, respectively, and patient 11 had a level of 1 ng/ml at both time periods. These 2 patients were receiving mean doses of 0.27 and 0.17 mg/kg of prednisone, respectively (table 2), and had a mean C_{cr} of 23.6 ± 2.4 and 63.6 ± 5.5 ml/min/1.73 m^2, respectively. Patient 2, who is undergoing PD, had a low GH level of 0.3 ng/ml at 60 min, but a higher level of 7.9 ng/ml at 90 min. Patient 4 who underwent the 24-hour spontaneous GH sampling had one peak of > 10 ng/ml at night during sleep.

IGF-1 Levels

IGF-1 levels at baseline and at 6-month intervals following initiation of rhGH treatment were determined by acid chromatography. The mean IGF-1 levels increased from 266 ± 97 ng/ml at baseline to 440 ± 299, 424 ± 158, 490 ± 174 and 638 ± 471 ng/ml at 6, 12, 24 and 30 months, respectively, following initiation of rhGH treatment. IGF-1 levels were higher after initiation of the rhGH in all patients, but the increment did not reach statistical significance.

Glucose and Insulin Levels

The mean fasting and 2-hour postprandial blood glucose levels tested at 6-month intervals up to 36 months of rhGH treatment were not different from baseline values in 10 of the patients (table 4). In 1 patient (No. 6), the postprandial glucose and insulin levels became abnormal at 36 months of rhGH treatment. A glucose tolerance test (GTT) demonstrated a diabetic pattern. rhGH was discontinued in this patient following 36 months of treatment.

Biochemical Data

There was no significant increase in the mean serum triglyceride, cholesterol, calcium or phosphorus levels following 6, 12, 18, 24, 30 or 36 months of treatment (data not shown). However, the mean serum alkaline phosphatase level did increase significantly at 24 months following initiation of treatment ($p = 0.012$).

Discussion

Retarded growth and delayed sexual maturation are a frequent complication of CRF [7]. Neither hemodialysis nor PD alter the phenomena producing the delayed sexual maturation [7, 16, 33]. Successful renal transplantation may restore normal sexual maturation provided the maintenance dosage of corticosteroids is low and renal allograft function is optimal [16, 34]. Rees et al. [35] demonstrated a significant increase in GV after transplantation in pubertal children; however, GV declined at the expected age of the normal pubertal growth spurt and appearance of secondary sexual characteristics was delayed. These impaired growth and development patterns were thought likely to be due to the long-term corticosteroid treatment posttransplantation. In a more recent publication, Rees et al. [36] demonstrated significantly improved GV in 6 prepubertal patients with CRF, 6 prepubertal PT patients and in 6 pubertal PT patients treated with rhGH for a year. The prepubertal PT group had the least obvious response to rhGH with no correlation with the corticosteroid dosage or severity of suppression of spontaneous GH secretion. The authors note that the pubertal growth spurt could have a confounding effect on the improved GV with rhGH treatment in the pubertal PT patients, despite their previous data indicating depressed pubertal growth in patients receiving corticosteroids.

David-Nato et al. [37] studied the spontaneous secretion of GH in 15 PT pediatric patients after conversion of their immunosuppressive regimen from azathioprine and prednisone to azathioprine and ciclosporin. In 6 of the 15 patients the spontaneous GH secretion and GV increased significantly following discontinuation of the prednisone. Schaefer et al. [38] who studied pulsatile spontaneous GH secretion in 40 pubertal PT patients, observed a strong inverse relationship between the GH peak amplitude and the corticosteroid dosage. The authors concluded that pubertal growth retardation despite successful renal transplantation appears to be due to corticosteroid-induced GH hyposecretion.

Retardation of bone age has been considered a critical factor in predicting the magnitude of GV after transplantation. The child whose bone age is retarded in proportion to his height age retains the potential for growth restoration after transplantation [39–41]. However, at puberty there may be a rapid increase in bone age in such patients unaccompanied by similar increments in linear growth [16]. Endocrine function has been studied extensively in adults with CRF. A disturbance at the gonadal level has been postulated in adult males based upon the observation that plasma LH and less commonly FSH levels are elevated in the presence of low levels of circulating androgens and impaired spermatogenesis [13, 42]. In adult females, similar abnormalities have been observed [13, 14]. Plasma estrogen and progesterone levels are decreased, and secondary amenorrhea and/or anovulatory cycles occur frequently [14, 43]. Several studies have shown that during the age range associated with normal puberty, there is an increase in pulsatility of gonadotrophins [35, 44, 45]. Rees et al. [35] demonstrated considerable blunting of gonadotrophin secretion in pubertal patients with renal insufficiency. Prolactin, which at high levels has been reported to decrease LH pulsatility in men [46], was within normal limits in that study [35]. In prepubertal and pubertal boys with CRF, low or low-normal plasma testosterone levels have been reported [16]. Total plasma testosterone and dihydrotestosterone concentrations have been found to be decreased, while the ratio of testosterone to dihydrotestosterone and the percentage of free testosterone in plasma have been normal [12], indicating impaired Leydig cell function; 5α-reductase activity and plasma protein binding of testosterone were not disturbed. Testosterone response to human chorionic gonadotropin (HCG) stimulation was decreased in prepubertal and pubertal boys with CRF [17]. Therefore, damage to testicular tissue may occur either before or during puberty due to the uremic milieu. In our study, testosterone levels increased appropriately with advancement of Tanner stages in the pubertal patients receiving rhGH, despite the fact that there was no parallel increase in the gonadotrophin levels. There was a statistically significant increase in mean LH levels between baseline and 12 months and 12 and 24 months of treatment with rhGH, and between mean FSH levels at baseline and 12 months of treatment.

GH levels are normal or elevated in children with CRF [12, 18, 19]. In a recent study, stimulated GH levels were lower than normal in PT patients receiving more than 5 mg of prednisone daily [47]. In our study, prior to administration of rhGH, 1 of the PT patients receiving a high daily prednisone dose had high GH levels after stimulation, but 4 other PT patients did not have stimulated levels >10 ng/ml, while 1 patient had a level > 10 ng/ml at 60 min but < 10 ng/ml at 90 min. Somatomedin (IGF-1) by bioassay is significantly diminished in CRF patients, but Saenger et al. [41] demonstrated restoration of somatomedin levels, measured by bioassay, to normal after renal transplantation. In our study, IGF-1 levels by RIA following acid chromatography were normal at initiation of GH, and increased steadily during all treatment periods. Tejani et al. [47] have also demonstrated accelerated growth in 3 of 4 pubescent PT children with growth retardation, requiring maintenance prednisone, who received short-term rhGH treatment.

We have previously reported the efficacy of rhGH treatment in growth-retarded children with preterminal CRF, those undergoing PD and following kidney transplantation [23–25, 48]. Patients aged 2.5–17.7 years were included in those reports, with no differentiation between the different pubertal stages. Many studies have shown that pubertal growth in children with ESRD, those undergoing dialysis, and even those after successful kidney transplantation was inferior to that of normal children [33, 39, 49, 50]. The European Dialysis and Transplantation Association (EDTA) pediatric registry reported that children with CRF begin their pubertal growth spurt at a time when that of healthy children is already declining, and that appreciable late growth continues at an age when normal individuals already have attained their final adult height [5, 51]. In contrast, Schaefer et al. [22], in a recent study, indicate that the duration of the pubertal period, and the duration of the pubertal growth spurt in children with CRF, were shorter than in control populations, concluding that the diminished magnitude and duration of the growth spurt both contribute to the diminished height gain during puberty. In the present study, we have shown that pubertal boys with CRF, those undergoing dialysis and following transplantation with growth retardation had a significant improvement in GV with rhGH treatment. During the study period, the increase in bone age did not exceed the increase in chronological age. Testosterone and gonadotrophin levels increased during the gradual advancement in Tanner stage in all patients.

Since rhGH may cause hyperglycemia, each patient had fasting and 2-hour postprandial glucose and insulin levels prior to initiation of the rhGH, and at 6-month intervals thereafter; only 1 patient had a significant decrease in glucose tolerance and treatment was discontinued.

Both endogenous [52] and exogenous [53] GH have been shown to increase GFR and renal plasma flow (RPF). Short-term rhGH administration results in an increase in GFR in normal adults [53–57] but it is not

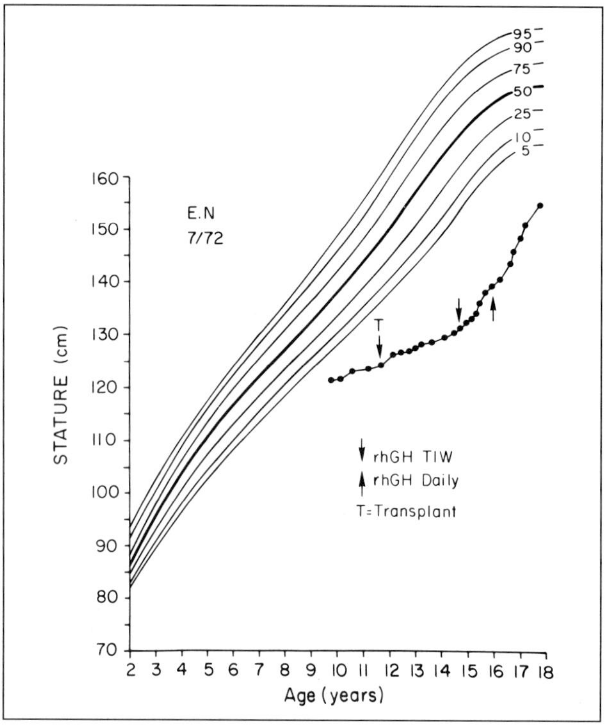

Fig. 1. Growth curve of patient 7. PT patient (7): rhGH initiated at age 14.9 years, Tanner stage III, BA 11 years, SDS for height −4.60. Following 36 months of treatment: Tanner stage V, SDS for height −3.36, GV 7.0 cm after 12 months, 7.9 cm after 24 months and 8.4 cm after 36 months of rhGH treatment.

known whether long-term treatment could produce hyperfiltration with resultant glomerulosclerosis and an accelerated decline in renal function. To assess whether rhGH deleteriously affected renal function in the patient with CRF and the transplant recipients in our study, C_{cr} was followed serially. No significant decline in the mean calculated C_{cr} during up to 36 months of treatment was noted. Only 1 patient, who had two episodes of acute rejection due to noncompliance, had a substantial decline in C_{cr}. Unfortunately, more sophisticated methods were not used to assess GFR serially in this study.

Figure 1 shows the growth curve of patient 7, demonstrating the salutary affect of rhGH treatment in a pubescent boy. He received a maternal kidney transplant at the age of 12 years, but despite a very good allograft function, his GV was suboptimal, reaching an SDS for height of −4.61 at the age of 14.9 years, with a Tanner stage III. rhGH treatment was initiated and the patient's GV improved from a pretreatment GV of 1.9 cm/year to 7.0 cm/year at 12 months of treatment, 7.9 cm/year at 24 months and 8.4 cm/year at 36 months of treatment.

In summary, these data demonstrate that the long-term use of rhGH in growth-retarded pubertal children with CRF, those undergoing dialysis and renal allograft recipients, has the potential to improve the GV of these children at a time when their healthy peers are entering the normal pubertal growth spurt, and when, as shown in previous studies without rhGH, their growth spurt may be markedly diminished [22]. Although the ultimate benefit of this therapeutic intervention on final adult height requires longer follow-up, it is anticipated to have a salutary effect.

References

1 Oertel PJ, Lichtwald K, Häfner S, Rauh W, Schönberg D, Schärer K: Hypothalamo-pituitary-gonadal axis in children with chronic renal failure. Kidney Int 1983; 24(suppl 15):S34–S39.

2 Zachmann M, Prader A, Kind HP, Häfliger H, Budliger H: Testicular volume during adolescence. Cross-sectional and longitudinal studies. Helv Paediatr Acta 1974;29: 61–72.

3 Kulin HE: Normal pubertal development; in Rudolph AM, Hoffman JIE (eds): Pediatrics, ed 17. East Norwalk, Appleton-Century-Crofts, 1982, p 1558.

4 Marshall WA, Tanner JM: Variations in the pattern of pubertal change in boys. Arch Dis Child 1970;45:13–23.

5 Schärer K, Chantler C, Brunner FP, et al: Combined report on regular dialysis and transplantation of children in Europe, 1975. Proc Eur Dial Transplant Assoc 1976; 13:59–105.

6 Chantler C, Holliday MA: Growth in children with renal disease with particular reference to the effects of calorie malnutrition: A review. Clin Nephrol 1973;1:230–242.

7 Lewy JE, New MI: Growth in children with renal failure. Am J Med 1975;58:65–68.

8 Stickler GB: Growth failure in renal disease. Pediatr Clin North Am 1976;23:885–894.

9 Mehls O, Ritz E, Gilli G, Kreusser W: Growth in renal failure. Nephron 1978;21: 237–247.

10 Schärer K: Growth in children with chronic renal failure. Kidney Int 1978;13(suppl 8):S68–S71.

11 Broyer M: Growth in children with renal insufficiency. Pediatr Clin North Am 1982; 29:991–1003.

12 Rauh W, Oertel PJ: Endocrine function in children with ESRD; in Fine RN, Grushkin CM (eds): End-Stage Renal Disease in Children. Philadelphia, Saunders, 1984, pp 296–306.

13 Lim VS, Auletta F, Kathpalia S: Gonadal dysfunction in chronic renal failure. Dial Transplant 1978;7:896–907.

14 Swamy AP, Woolf PD, Cestero RVM: Hypothalamic-pituitary-ovarian axis in uremic women. J Lab Clin Med 1979;93:1066–1072.

15 Drüeke T: Endocrine disorders in chronic hemodialysis patients (with the exclusion of hyperparathyroidism); in Hamburger J, Crosnier J, Grünfeld JP, Maxwell MH (eds): Advances in Nephrology. Chicago, Year Book Medical Publisher, 1981, pp 351–382.

16 Ferraris J, Saenger P, Levine L, et al: Delayed puberty in males with chronic renal failure. Kidney Int 1980;18:344–350.

17 Schärer K, Broyer M, Vecsei P, Roger M, Arnold-Schwender E, Usberti J: Damage to testicular function in chronic renal failure of children. Proc Eur Dial Transplant Assoc 1980;17:725–729.

18 Wright AD, Lowy D, Fraser TR: Serum growth hormone and glucose intolerance in renal failure. Lancet 1968;ii:798–801.

19 Samaan NA, Freeman RM: Growth hormone levels in severe renal failure. Metabolism 1970;19:102–113.

20 Ijaiya K: Pattern of growth hormone response to insulin, arginine and hemodialysis in uremic children. Eur J Pediatr 1979;131:185–198.

21 Cameron DP, Burger HG, Catt KJ, et al: Metabolic clearance rate of human growth hormone in patients with hepatic and renal failure, and in the isolated perfused pig liver. Metabolism 1972;21:895–904.

22 Schaefer F, Seidel C, Binding A, et al: Pubertal growth in chronic renal failure. Pediatr Res 1990;28:5–10.

23 Fine RN, Pyke-Grimm K, Nelson PA, et al: Recombinant human growth hormone (rhGH) treatment of children with chronic renal failure (CRF): long-term (one to three years) outcome. Pediatr Nephrol, in press.

24 Koch VH, Nelson PA, Boechat MI, et al: Accelerated growth after recombinant human growth hormone treatment of children with chronic renal failure. J Pediatr 1989;115:365–371.

25 Kamil ES, Yadin O, Koch VH, et al: Intraperitoneal (IP) recombinant human growth hormone (met-hGH) treatment (Rx) of children undergoing peritoneal dialysis. 32nd Annu Meet American Society of Nephrology, Washington, D.C., Dec 1989.

26 Powell DR, Rosenfeld RG, Baker BK, Liu F, Hintz RL: Serum somatomedin levels in adults with chronic renal failure; the importance of measuring insulin-like growth factors (IGF-1) and IGF-2 in acid chromatographed uremic serum. J Clin Endocrinol Metab 1986;63:1186–1192.

27 Fass B, Lippe BM, Kaplan SA: Relative usefulness of three growth hormone stimulating tests. Am J Dis Child 1979;133:931–933.

28 Greulich WW, Pyle SI: Radiographic Atlas of Skeletal Development of the Hand and Wrist, ed 2. Stanford, Stanford University Press, 1950.

29 Tanner JM: Growth at adolescence with a general consideration of the effects of

hereditary and environmental factors upon growth and maturation from birth to maturity, ed 2. Oxford, Blackwell Scientific Publications, 1962, pp 28–39.

30 Roche AF: Growth assessment in abnormal children. Kidney Int 1978;14:369–377.

31 Hamil PVV: Growth curves for children; in Hamil PVV (eds): National Center for Health Statistics, 1977; Vital and health statistics; series 2; DHEW Publ No (PHS)78-1650 National Health Survey No 165, Washington, 1977.

32 Gurney J, Jelliffe D: Arm anthropometry in nutritional assessment; nomogram for rapid calculation of muscle circumference and cross-sectional muscle and fat areas. Am J Clin Nutr 1973;26:912.

33 Kleinknecht C, Broyer M, Gagnadoux M, et al: Growth in children treated with long-term dialysis: a study of 76 patients; in Hamburger J, Crosnier J, Grunfeld J, Maxwell MH (eds): Advances in Nephrology. Chicago, Year Book Medical Publishers, 1980, vol 9, pp 133–163.

34 Najarian JS, Simmons RL, Tallent MB, et al: Renal transplantation in infants and children. Ann Surg 1971;174:583–600.

35 Rees L, Greene SA, Adlard P, et al: Growth and endocrine function after renal transplantation. Arch Dis Child 1988;63:1326–1332.

36 Rees L, Rigden SPA, Ward G, Preece MA: Treatment of short stature in renal disease with recombinant human growth hormone. Arch Dis Child 1990;65:856–860.

37 David-Neto E, Vilares S, Lando V, et al: Conversion from azathioprine/prednisone to azathioprine/cyclosporin promotes catch-up growth in pediatric renal allograft recipients. Clin Transplant 1990;4:229–234.

38 Schaefer F, Stanhope R, Preece MA, Schärer K: Pulsatile growth hormone secretion in peripubertal patients with chronic renal failure (in press).

39 Grushkin CM, Fine RN: Growth in children following renal transplantation. Am J Dis Child 1973;125:514–516.

40 Boyer M, Kleinknecht C, Loirat C, Marti-Henneberg C, Roy MP: Growth in children treated with long-term hemodialysis. J Pediatr 1974;84:642–649.

41 Saenger R, Wiedeman E, Schwartz E, et al: Somatomedin and growth after renal transplantation. Pediatr Res 1974;8:163–169.

42 Mies R, Von Baeyer H, Figge H, Finke K, Winkelmann W: Investigation on pituitary and Leydig cell function in chronic hemodialysis and after renal transplantation. Klin Wochenschr 1975;53:611–615.

43 Wass VJ, Wass JAH, Rees L, Edwards CRW, Ogg CS: Sex hormone changes underlying menstrual disturbances on hemodialysis. Proc Eur Dial Transplant Assoc 1978; 15:178–186.

44 Lee PA, Plotnick LP, Steele RE, et al: Integrated concentrations of luteinising hormone and puberty. J Clin Endocrinol Metab 1976;43:168–172.

45 Lee PA, Plotnick LP, Migcon CJ, et al: Integrated concentrations of follicle stimulating hormone and puberty. J Clin Endocrinol Metab 1978;46:488–490.

46 Winters SJ, Bardin CW: Altered pulsatile secretion of luteinising hormone in hypogonadal men with hyperprolactinemia. Clin Endocrinol (Oxf) 1984;21:257–263.

47 Tejani A, Butt KMH, Rajpoot D, et al: Strategies for optimizing growth in children with kidney transplants. Transplantation 1989;47:229–233.

48 Fine RN, Yadin O, Nelson PA, et al: Recombinant human growth hormone treatment of children following renal transplantation. Pediatr Nephrol 1991;5:147–151.
49 Boineau FG, Lewy JE: Sexual maturation in children with renal insufficiency: response to dialysis and transplantation; in Fine RN, Grushkin CM (eds): End-Stage Renal Disease in Children. Philadelphia, Saunders, 1984, pp 291–295.
50 Schärer K, Gilli G: Growth in children with chronic renal insufficiency; in Fine RN, Grushkin CM (eds): End-Stage Renal Disease in Children. Philadelphia, Saunders, 1984, pp 271–290.
51 Donckerwolcke RA, Chantler C, Brunner FP, et al: Combined report on regular dialysis and transplantation of children in Europe, 1977. Proc Eur Dial Transplant Assoc 1978;15:77–114.
52 Ikkos D, Ljunggren H, Luft R: Glomerular filtration rate and renal plasma flow in acromegaly. Acta Endocrinol 1956;21:226–236.
53 Christiansen JS, Gammelgaard J, Orskov H, Andersen AR, Telmer S, Parving HH: Kidney function and size in normal subjects before and during growth hormone administration for one week. Eur J Clin Invest 1981;11:487–490.
54 Hirschberg RR, Kopple JD: Increase in renal plasma flow and glomerular filtration rate during growth hormone treatment may be mediated by insulin-like growth factor 1. Am J Nephrol 1988;8:249–253.
55 Hirschberg R, Rabb H, Bergamo R, Kopple JD: The delayed effect of growth hormone on renal function in humans. Kidney Int 1989;35:865–870.
56 Haffner D, Ritz E, Mehls O, et al: Growth hormone induced rise in glomerular filtration rate is not obliterated by angiotensin-converting enzyme inhibitors. Nephron 1990;55:63–68.
57 Haffner D, Zacharewicz S, Mehls O, Heinrich U, Ritz E: The acute effect of growth hormone on GFR is obliterated in chronic renal failure. Clin Nephrol 1989;32:266–269.

Richard N. Fine, MD, Children's Medical Center at Stony Brook,
Department of Pediatrics, State University of New York at Stony Brook,
Stony Brook, NY 11794-8111 (USA)

Berlyne GM (ed): The Kidney Today. Selected Topics in Renal Science.
Contrib Nephrol. Basel, Karger, 1992, vol 100, pp 155–171

Blood-Dialyzer Interactions: Their Role in Long-Term Complications of Hemodialysis

Gerhard Lonnemann[a], *Jürgen Floege*[a], *Stanley Shaldon*[b], *Karl Martin Koch*[a]

[a]Department of Nephrology, Medizinische Hochschule, Hannover, FRG;
[b]Department of Nephrology, University Hospital, Nîmes, France

Hemodialysis therapy has been improved over the past 20 years due to advances in technology of dialysis machines, dialysate preparation and distribution systems (including the reintroduction of bicarbonate dialysate, and the improvement in the production of cellulosic membranes as well as the introduction of synthetic dialyzer membranes. However, there is no doubt that none of the treatment modalities of terminal renal failure including hemodialysis, hemodiafiltration, hemofiltration or peritoneal dialysis, is capable of replacing renal function completely. Therefore, with intermittent dialysis therapy, parts of the uremic syndrome persist and may contribute to some of the long-term complications seen in end-stage renal disease (ESRD) patients. In addition, circulation of blood through extracorporeal devices may induce biological reactions, which may cause long-term complications as was postulated in 1983 with the interleukin hypothesis [1]. A major problem in the discussion of such biological reactions is to distinguish between causative factors inherent to the extracorporeal procedure and others related to uremia, bacterial infections or an underlying inflammatory kidney disease. With this restriction in mind, inflammatory processes induced during hemodialysis are the main subject of this review. In the first part, current knowledge about cellular activation as well as indicators of impaired cellular functions in patients on hemodialysis therapy is discussed. In the second part, a critical review of the existing evidence for a link between blood-dialyzer interactions and long-term complications of hemodialysis is given.

Cellular Responses due to Blood-Dialyzer Interactions

All types of circulating leukocytes including granulocytes, monocytes, and lymphocytes as well as platelets are activated during infections or injury. In addition to these classical causes of cellular activation, cells may be stimulated by adherence to foreign materials such as glass, plastic or the extracorporeal hemodialysis circuit. Indeed, lysosomal enzymes released from granulocytes as well as inflammatory mediators such as cytokines produced and secreted from stimulated mononuclear cells have been studied as indicators of incompatibility between blood and the extracorporeal circuit [2, 3]. The interaction between the extracorporeal circuit and granulocytes, lymphocytes, monocytes, and platelets concerning: (a) mechanisms of hemodialysis-dependent cell activation, (b) type of inflammatory mediators released, and (c) further evidence for functional changes of these cells will now be discussed.

Granulocytes

Indirect evidence of granulocyte activation during hemodialysis has been provided by early studies describing a severe leukopenia during the early phase of hemodialysis [4]. Subsequent studies showed that leukopenia was the consequence of the activation of the alternate complement pathway on the dialyzer membrane [5]. This complement activation was found to be most pronounced with certain cellulosic membranes such as cuprophan, and less or almost absent with modified cellulosic membranes (e.g. hemophan, cellulose triacetate) or synthetic membranes. C5a generated by the alternate complement pathway activates granulocytes via specific C5a receptors [6]. Activated granulocytes increase the expression of a surface glycoprotein called M01 which mediates leukocyte adhesion [7]. This is probably the cause of leukocyte aggregation and subsequent sequestration in the lung [5].

Another indicator of granulocyte activation is the release of elastase, which is dependent on the dialyzer membrane used [8]. The increases in plasma levels of granulocyte elastase during hemodialysis were significant with cellulose hydrate, cuprophan and polymethylmethacrylate but not with ethylene-vinyl alcohol polymer and polyacrylonitrile [8], or polysulfone [9]. As polymethylmethacrylate stimulates granulocyte elastase release in the absence of C5a generation, other mechanisms than complement activation, e.g. sterilization procedures, have been discussed to be involved in granulocyte activation during hemodialysis [8]. Granulocyte activation is also indicated by decreased intracellular levels of elastase and lactoferrin in polymorphonuclear cells from ESRD patients, especially diabetic, on hemodialysis with cuprophan or polysulfone membranes [10]. A further indication of granulocyte activation during hemodialysis was the increased production of reactive oxygen species in granulocytes harvested during hemodialysis [11].

In addition to parameters indicating granulocyte activation, there is evidence for an impaired function of these cells in hemodialysis patients. The phagocytic ability of polymorphonuclear neutrophils is reduced [12, 13], and neutrophil chemotaxis is decreased [14]. The reason for the impaired function of polymorphonuclear cells could in part be due to a recently characterized polypeptide found in the ultrafiltrate of ESRD patients on high-flux hemodialysis [15]. This protein has been shown in vitro to inhibit chemotaxis, oxidative metabolism and intracellular bacterial killing by granulocytes.

Lymphocytes

Although it is still controversial whether ESRD patients on hemodialysis therapy have decreased or normal numbers of circulating lymphocytes, there is evidence for the presence of preactivated T cells [16] as indicated by an increased interleukin-2 (IL-2) receptor expression on T cells of these patients compared to healthy controls. In vitro studies showed functional changes of T cells such as a reduced IL-2 production and a diminished proliferative response following stimulation with mitogens [17]. A similar pattern of T-cell abnormalities, although less profound, was observed in uremic patients not yet on hemodialysis therapy [18]. The impaired response to hepatitis B vaccination often seen in hemodialysis patients and the persistence of an antigen-positive carrier state has been related to an altered monocyte-T-cell interaction resulting in a diminished IL-2 production by T cells. IL-2 production is critically dependent on a monocyte-derived signal which has not yet been identified [19]. IL-2 in small dose has been used to induce antibody formation against HBsAg in previously nonresponding dialysis patients [20].

Impaired lymphocyte function is also indicated by the reduced cytotoxic activity of natural killer cells in vitro [21] and in vivo dialysis as well as in predialysis patients [22]. Together these studies suggest that uremia per se is responsible for impaired lymphocyte functions and that hemodialysis may have an additional negative effect.

Mononuclear Cells, Monocytes, Macrophages

Activated mononuclear cells produce proinflammatory cytokines such as interleukin-1 (IL-1) and tumor necrosis factor-α (TNFα) [23] which are important mediators of the acute-phase response. Several in vitro dialysis experiments have identified possible hemodialysis-associated cytokine-inducing mechanisms. Cytokine-inducing bacterial substances including lipopolysaccharides and derived fragments have been shown to penetrate intact dialysis membranes [24–26]. In addition, low molecular weight LAL-negative products of Pseudomonas species grown from dialysate which are likely to represent breakdown products of Pseudomonas exotoxin A have been identified and shown to have cytokine-inducing activity [27]. The permeability to cytokine-inducing bacterial substances depends upon the dialyzer membrane. If the membrane activates complement, an enhancement effect for bacterial substances may be seen. However, the major factors is the adsorptive capacity for bacterial substances. Regenerated cellulose, cellulose triacetate and polyacrylonitrile (AN69) appear to be more permeable than polysulfone and polyamide dialyzer membranes [28]. An alternative mechanism has demonstratd that when mononuclear cells adhere to the dialyzer membrane, in the absence of plasma proteins, IL-1 production can be induced [29, 30].

Finally, in vitro hemodialysis recirculation using diluted donor whole blood and pyrogen-free dialysate showed that regenerated cellulosic dialyzers stimulate transcription of mRNA coding for IL-1 and TNFα in mononuclear cells more than hemophan or synthetic membranes, without translation and subsequent synthesis of the respective cytokines [31]. However, these normal donor mononuclear cells containing cytokine mRNA (so-called primed cells) produced more IL-1 and TNFα protein in response to a second stimulus such as endotoxin than 'unprimed' cells [31].

To demonstrate hemodialysis-dependent cytokine induction in vivo, two approaches were made, namely the measurement of cytokine levels in plasma and the separation of mononuclear cells from patient's blood with the subsequent ex vivo investigation of mononuclear cell activation.

Using bioassays to detect IL-1 activity in the plasma, early studies suggested that circulating IL-1 levels were elevated in ESRD patients on hemodialysis compared to levels detectable in the plasma of healthy controls [32, 33]. Results of studies using specific immunoassays for IL-1 and TNFα remain controversial for the following reasons. First, the preparation of the plasma sample is crucial. Although it has been described that interfering inhibitory substances need to be extracted before performing immunoassays for IL-1 [34], nonpretreated plasma samples have been measured and undetectable levels have been reported [35]. Furthermore, the considerable inter- and intraindividual variability in cytokine production [36, 37] makes it very difficult to obtain reliable and reproducible results. The principle counter-arguments against the use of plasma measurements for the determination of increased cytokine production are that cytokines act predominantly in the paracrine situation of cell-cell interactions and that extracellular cytokines are quickly inactivated by degradation [23]. Thus, plasma levels may only represent an overflow of cytokines during a short period of time following activation of the mononuclear cell system and therefore may not be reliable indicators of cytokine induction. Indeed, the existence of a specific circulating receptor antagonist to IL-1 would suggest that IL-1 plasma levels do not reflect IL-1 activity at the tissue level in regard to inflammation. This possibility seems less likely to apply for TNFα which may indeed have a true humoral role [23]. Nevertheless, plasma levels of TNFα in stable nonfebrile ESRD patients on dialysis have been reported to be elevated [38] or normal [39]. However, the ESRD patients following exposure to endotoxin shows an acutely elevated plasma TNFα level without change in the IL-1 level [40].

Because of the obstacles inherent to the measurement of cytokine plasma levels, several groups separated mononuclear cells from the blood of ESRD patients and studied cytokine production and release of these cells ex vivo.

A recent in vivo study [41], confirming previous in vitro results [31], showed that hemodialysis in ESRD patients induces transcription of mRNA coding for IL-1β in mononuclear cells. However, the priming effect seen in vitro (i.e. and an increase in cytokine synthesis following a second stimulus) was not demonstrable with these uremic mononuclear cells [41]. Several studies showed that compared to healthy controls, ESRD patients have elevated cell-associated levels of the IL-1 peptide in mononuclear cells [42–45]. These elevated pretreatment levels increased further during hemodialysis [43, 45]. However, uremic patients not yet on chronic hemodialysis therapy were also found to have increased cell-associated levels of IL-1 [45]. Thus, elevated cell-associated cytokine levels in mononuclear cells of ESRD patients may be the consequence of uremia as well as hemodialysis. The most convincing evidence that cytokine activation results from blood membrane interaction is indirect. Normal volunteers were dialyzed with cuprophan dialyzers without dialysate. Their femoral venous blood content of tyrosine, phenylalanine and 3-methylhistidine increased.

This suggests that breakdown of muscle protein ocurred during cuprophan dialysis, but not during dialysis with AN69 or polysulfone membranes [46, 47]. Previously, hemodialysis in ESRD patients had been shown to be catabolic [48]. A possible explanation for these important findings could be the synergistic action between IL-1 and TNFα acutely infused into rats resulting in muscle protein catabolism and negative nitrogen balance [49].

There is evidence, in addition to activation of monocytes, that there is impairment of function of monocyte/macrophage cells in ESRD patients. The uremic monocyte may fail to produce a signal which is important in the induction of IL-2 production by T cells [19], it appears unable to respond to a second stimulus after priming [41], and there is an impaired function of the macrophage Fcγ receptor, which is important in the clearance of antigens and immune complexes [50]. Whether these impairments of monocte/macrophage function are due to uremia per se or to uremia and hemodialysis is uncertain. However, it seems reasonable to speculate that this impairment of function protects the patient against the inflammatory ravages of the extracorporeal procedure.

Platelets

During hemodialysis with cellulosic dialyzer membranes thrombocytopenia and platelet activation as indicated by an increased release of thromboxane-β_2 and platelet factor 4 have been reported [51–53]. The mechanism of platelet activation is not fully understood, both platelet adherence to the dialyzer membrane as well as platelet aggregation caused by activated complement components are likely explanations.

In summary, both indicators of cellular activation and of impaired cellular functions can be observed in patients on hemodialysis therapy. Uremia and the hemodialysis procedure may be causative factors in the genesis of these phenomena.

Long-Term Clinical Consequences of Blood-Dialyzer Interactions

Given the observation that virtually all inflammatory mediator systems may be stimulated and multiple cellular functions may be impaired in ESRD patients, one would expect to see a multitude of clinical consequences in chronic hemodialysis patients. However, in contrast to some long-term complications of hemodialysis such as cardiovascular problems and renal bone disease, clinical consequences of blood-dialyzer interactions remain as yet poorly defined. The following hemodialysis and/or uremia-related pathological mechanisms which may be related to certain specific clinical entities of the dialysis patient will be discussed: (1) Relative immunodeficiency in hemodialysis patients has been suggested to contribute to increased infection rates [54]. (2) Impaired cellular immunity and/or exposure to potentially carcinogenic environmental factors present in the hemodialysis circuit have been proposed to be associated with an increased incidence of malignancies [55]. (3) Repetitive induction of the acute phase response (see above) might result in the development of a serum amyloid A protein-derived amyloidosis (AA amyloidosis, secondary amyloidosis). (4) Retention and/or overproduction and/or defective processing of β_2-microglobulin have been implicated in the development of a β_2-microglobulin-derived amyloidosis [56–58]. (5) Chronic intermittent stimulation of inflammatory events might result in clinical symptoms related to fibrotic or sclerotic changes. (6) Protein malnutrition is a common feature in the long-term dialysis patient and in part may be related to the repeated stimulation of monocytes and production of proinflammatory cytokines [59].

Firstly, one will discuss whether these clinical problems indeed exist in long-term hemodialysis patients. In the second part, one will then examine the evidence available to implicate a role for blood-dialyzer interactions in their pathogenesis.

Infections

Increased infection rates have been recognized as an important cause of morbidity and mortality in patients on chronic hemodialysis [54]. With respect to the possible cause(s) these patients are exposed to a higher infectious risk than the general population. Indeed it has been shown that most septic deaths in hemodialysis patients are due to staphylococcal infections, possibly acquired in association either with an indwelling arteriovenous (AV) access or following needling of AV fistulae [54]. Other risk factors for infections were found to be old age, a low Karnofsky activity score and poor hygiene [60]. Investigations of immune functions in patients with ESRD have revealed an impairment of macrophage Fcg receptors, depression of cell-mediated immunity, lymphocyte abnormalities and depression of phagocytosis and chemotaxis by granulocytes (see above). Iron overload has been suggested to contribute to this immunodeficiency [61], although iron overload may also have a direct growth-promoting effect on bacteria by increasing the availability of iron molecules, which are essential for bacterial growth. Indeed, iron overload in hemodialysis patients has been shown to correlate with the development of septicemia [61]. In contrast, although a multitude of in vitro findings (as discussed above) makes a role of the disturbed immunological defense in the genesis of increased infection rates highly likely, little direct supporting evidence (apart from the persistence of the HBsAg carrier state and the impaired response to HBsAg vaccination [19] is available.

Malignancies

Early studies on small, selected populations of hemodialysis patients have suggested an increased overall cancer frequency in hemodialysis patients [55]. However, in a recent study on 28,049 hemodialysis patients in the United States [62] and in large surveys by the European Dialysis and Transplant Association (EDTA) [63, 64] the overall risk of cancer was not significantly different from that observed in the general population, except that for example neoplasms of the urinary tract developed at an increased rate in patients with analgesic nephropathy [63, 64]. Thus, if certain cancer types develop at an increased frequency in hemodialysis patients, this appears to be related to the underlying renal disease rather than to impaired cellular immunity or presumed carcinogen exposure.

Amyloidosis

Despite evidence for a repetitive induction of the acute phase response (see above), including SAA protein [65], no cases of unexplained secondary amyloidosis (AA amyloidosis) have been published in chronic hemodialysis patients. This suggests that the blood levels of SAA do not permanently reach such a magnitude as to be capable of causing amyloidosis.

In contrast, a specific dialysis-related amyloidosis has recently been recognized in hemodialysis patients. Dialysis-related amyloidosis usually is confined to patients on hemodialysis for more than 5 years [56, 57]. After 15 years of treatment the prevalence reaches 50–100%. Dialysis-related amyloidosis is derived from β_2-microglobulin [66], which accumulates markedly in end-stage renal failure. Several in vitro studies have demonstrated that cellular β_2-microglobulin synthesis can be enhanced by inflammatory cytokines or by contact of the cells with some dialyzer membranes [67, 68]. It has thus been speculated that blood-dialyzer interactions might affect β_2-microglobulin synthetic rates and thus contribute indirectly to the rate of amyloid formation [56]. Subsequent studies have, however, shown that the enhancement of the synthetic rate observed in vitro is of low order of magnitude in vivo and does not appear to exert a significant influence on the whole body synthetic rate of β_2-microglobulin [69]. However, interactions of blood cells with dialyzer membranes may nevertheless be involved in the pathogenesis of dialysis-related amyloidosis, as recent findings suggest that this amyloidosis is not only a consequence of β_2-microglobulin accumulation, but also may involve a limited proteolysis of this protein subsequent to cellular activation. Thus, in addition to intact β_2-microglobulin, cleaved β_2-microglobulin molecules have been detected in solubilized amyloid fibrils [58], suggesting that proteases, which have been activated for example during hemodialysis [8, 10], may contribute to the amyloidogenesis.

Fibrosis/Sclerosis

Proinflammatory cytokines such as IL-1 and TNFα are potent inducers of fibroblast proliferation in vitro [23]. Repetitive induction of these cytokines could therefore be involved in the development of fibrotic or sclerotic changes. One of the most easily accessible sites to investigate such pathological alterations and their clinical correlates is the skin. In earlier studies on the skin of uremic patients the deposition of a substance which stained like amyloid was described [70], but which on subsequent electron microscopic examination lacked characteristic amyloid fibrils [71]. It was speculated that these deposits might represent altered collagen. Indeed it has recently been reported that ESRD may catalyze the formation of a pentose-derived crosslink from collagen [72]. However, the pathogenetic significance of such altered collagens in hemodialysis patients remains unclear since the typical pathological changes in the skin of hemodialysis patients are characterized by microangiopathy and severe actinic elastosis rather than by sclerotic or fibrotic changes [73, 74]. Qualitatively similar skin changes have also been observed in uremic, but not yet dialyzed patients [70], suggesting that uremia but not blood-dialyzer interaction was the main pathogenetic factor.

In contrast to the skin, fibrotic changes in the other organs, including the pancreas, liver, heart and lung, have been reported in hemodialyzed patients [75–78]. The pathogenesis of these changes has remained speculative and besides hemodialysis-associated factors may also include residuals after prior diseases such as pancreatitis, hepatitis or pneumonia, all of which are frequent in hemodialyzed patients.

Protein Malnutrition

Since the start of long-term hemodialysis more than 30 years ago, wasting and protein malnutrition have been signs of inadequate dialysis and associated with death. A recent review [59] has estimated protein loss per dialysis at over 15 g. The mechanism of the loss appears to be catabolism rather than loss of actual protein through the membrane. Chronic metabolic acidosis associated with acetate dialysis was one of the factors held to be responsible. In a recent analysis of over 20,000 patients, a factor clearly prognostic of impending death was a low serum albumin [79]. Hence there seems to be no doubt that protein malnutrition and wasting exist in many dialysis patients today.

Evidence Supporting a Role of Blood-Dialyzer Interactions in Long-Term Problems of Hemodialysis Patients

Data on the role of blood-dialyzer interactions in any of the aforementioned clinical disorders are very scarce. This is likely to be due to a combination of factors including the following: (a) The problem of defining clinical entities, which could represent long-term consequences of blood-dialyzer interactions, and subsequently the inability to establish causative relationships between such clinical long-term events and single treatment-related events. (b) Multiple changes of treatment-related factors over time in retrospective studies. (c) The paucity of adequate control groups of hemodialysis patients when retrospectively comparing the effects of different membrane types or treatment modalities. (d) The lack of controlled, prospective studies.

Thus, studies suggesting or negating long-term clinical consequences of blood-dialyzer interactions should be analyzed with these obstacles in mind. Since no data are available on the influence of various dialyzer membranes on fibrotic organ changes, and since malignancies and secondary amyloidosis do not show an increased frequency in hemodialysis patients, only infections, dialysis-related amyloidosis and protein malnutrition will be reviewed in the light of blood-dialyzer interactions.

Data comparing the incidence of infections in patients treated with different dialyzer membranes have not been published so far. The only study comparing patients on long-term hemodialysis with either cuprophan or polyacrylonitrile (AN69) membranes could not detect a significant difference of the number of hospitalizations when patients were switched from one treatment to the other [80]. Infection rates, however, were not addressed in this study.

The studies available on the influence of dialyzer characteristics on the development of dialysis-related amyloidosis are retrospective data comparing patients on long-term hemodialysis with cuprophan membranes with those on long-term hemodialysis with high-flux polyacrylonitrile (AN69) membranes. Data from a large multicenter study in Belgium and France comparing patients on cuprophan and AN69 hemodialysis have suggested a lesser prevalence of dialysis-related amyloidosis in the high-flux group, as measured by the presence of carpal tunnel syndrome and/or cystic bone radiolucencies [81]. Similar conclusions were also reached in a smaller study in France [82]. In contrast, neither an EDTA survey performed in 1990 [83], nor a recent smaller survey in France [84] could detect significantly different prevalences of amyloid-associated symptomatology in hemodialysis patients treated with either acrylonitrile or cuprophan membranes. However, in the EDTA study, other (poly)acrylonitrile membranes than the AN69 membrane were also accepted. Furthermore, in both the latter studies, the definition of possible amyloid-related symptoms may have been less rigorous than that in the positive study [81]. On the other hand, based on these apparently contradictory findings, it is obvious that prospective studies and more reliable clinical amyloid markers, such as scanning with radiolabelled β_2-microglobulin [85] or amyloid P component [86], will be necessary to evaluate the issue of prevention of dialysis-related amyloidosis by synthetic, high-flux dialyzer membranes.

The evidence of a membrane cause of muscle protein catabolism has been demonstrated only with cellulosic membranes and not with AN69 or polysulfone [46, 47]. This evidence, together with recent animal work showing the necessity of TNFα and IL-1 to produce synergistically muscle protein catabolism in rats [49], adds support to the role of blood membrane interaction in being partially responsible for a severe and not infrequent complication of long-term hemodialysis. In this respect, a recent hypothesis suggesting that less catabolism occurs with AN69 membranes than with cuprophan [87] is of interest. The results of an ongoing prospective clinical study related to this problem may yield more conclusive proof of the role of the dialyzer membrane as one of the causes of protein malnutrition.

In conclusion, uremia and/or chronic renal disease, in addition to well-established long-term complications such as renal osteopathy and cardiovascular problems, are associated with further long-term complications including dialysis-related amyloidosis, an increased infection rate, atrophic skin changes, organ fibrosis and protein malnutrition. However, despite a multitude of speculations, currently (with the possible exception of dialysis-related amyloidosis and protein malnutrition) no definitive evidence is available to link any of these events to blood-dialyzer interactions.

References

1 Henderson LW, Koch KM, Dinarello CA, Shaldon S: Hemodialysis hypotension: The interleukin hypothension: The interleukin hypothesis. Blood Purif 1983;1:3–8.

2 Basile C, Drüeke T: Dialysis membrane biocompatibility. Nephron 1989;52:113–118.

3 Dinarello CA, Koch KM, Shaldon S: Cytokines and their production during hemodialysis. Am J Kidney Dis, in press.

4 Kaplow LS, Goffinet JA: Profound neutropenia during the early phase of hemodialysis. JAMA 1968;203:1135–1137.

5 Craddock PR, Fehr J, Dalmasso AP, Grigham KL: Hemodialysis leucopenia. Pulmonary vascular leukostasis resulting from complement activation by dialyser cellophane membranes. J Clin Invest 1977;59:879–888.

6 Chenoweth DE, Hugli TE: Demonstration of a specific C5a receptor on intact human polymorphonuclear leukocytes. Proc Natl Acad Sci USA 1978;75:3943–3947.

7 Arnaout MA, Hakim RM, Todd RF III, Dana N, Colten HR: Increased expression of an adhesion-promoting surface glycoprotein in the granulocytopenia of hemodialysis. N Engl J Med 1985;312:457–462.

8 Hörl WH, Steinhauer HB, Schollmeyer P: Plasma levels of granulocyte elastase during hemodialysis: Effects of different dialyzer membranes. Kidney Int 1985;28:791–796.

9 Ward RA, Schmidt B, Blumenstein M, Gurland HJ: Evaluation of phagocytic cell function in an ex vivo model of hemodialysis. Kidney Int 1990;37:776–782.

10 Hörl WH, Schaefer RM, Wanner C, Bahlmann J, Reitinger J, Schollmeyer P, Heidland A: Enhanced plasma and intracellular levels of main granulocyte components in diabetes on dialysis. Blood Purif 1989;7:314–323.

11 Himmelfarb J, Lazarus JM, Hakim R: Reactive oxygen species production by monocytes and polymorphonuclear leukocytes during hemodialysis. Am J Kidney Dis 1991;17:271–276.

12 Hallgren R, Fjellstrom KE, Hakansson L, Venge P: Kinetic studies of phagocytosis. II. The serum-independent uptake of IgG-coated particles by polymorphonuclear leukocytes from uremic patients on regular dialysis treatment. J Lab Clin Med 1979;94:277–284.

13 Vanholder R, Ringoir S, Dhondt A, Hakim R: Phagocytosis in uremic and hemodialysis patients: a prospective and cross-sectional study. Kidney Int 1991;39:320–327.

14 Green WH, Casann RS, Mauer M, Quie P: The effect of hemodialysis on neutrophilic chemotactic responsiveness. J Lab Clin Med 1976;88:971–974.
15 Hörl WH, Haag-Weber M, Georgopoulos A, Block LH: Physicochemical characterization of a polypeptide present in uremic serum that inhibits the biological activity of polymorphonuclear cells. Proc Natl Acad Sci USA 1990;87:6353–6357.
16 Chatenoud L, Dugas B, Beaurain G, Touam M, Drueke T, Vasquez A, Galanaud P, Bach JF, Delfraissy JF: Presence of preactivated T cells in hemodialyzed patients: Their possible role in altered immunity. Proc Natl Acad Sci USA 1986;83:7457–7461.
17 Kurz P, Köhler H, Meuer S, Hütteroth T, Meyer zum Büschenfelde KH: Impaired cellular immune responses in chronic renal failure: Evidence for a T cell defect. Kidney Int 1986;29:1209–1215.
18 Beaurain G, Naret C, Marcon L, Grateau G, Drueke T, Urena P, Nelson DL, Bach JF, Chatenoud L: In vivo T cell preactivation in chronic uremic hemodialyzed and non-hemodialyzed patients. Kidney Int 1989;36:636–644.
19 Meuer SC, Hauer M, Kurz P, Meyer zum Büschenfelde KH, Köhler H: Selective blockade of the antigen-receptor mediated pathway of T-cell activation in patients with impaired primary immune responses. J Clin Invest 1987;80:743–749.
20 Meuer SC, Dumann H, Meyer zum Büschenfelde KH, Köhler H: Low-dose interleukin-2 induces systemic immune response against HBsAg in immunodeficient nonresponders to hepatitis B vaccination. Lancet 1989;i:15–18.
21 Kay NE, Raij L: Differential effect of hemodialysis membranes on human lymphocyte natural killer function. Artif Organs 1987;11:165–167.
22 Cala S, Mazuran R, Kordic D: Negative effect of uraemia and cuprophan haemodialysis on natural killer cells. Nephrol Dial Transplant 1990;5:437–440.
23 Dinarello CA: Interleukin-1 and interleukin-1 antagonism. Blood 1991;77:1627–1652.
24 Bingel M, Lonnemann G, Shaldon S, Koch KM, Dinarello CA: Human interleukin-1 production during hemodialysis. Nephron 1986;43:161–163.
25 Lonnemann G, Bingel M, Floege J, Koch KM, Shaldon S, Dinarello CA: Detection of endotoxin-like interleukin-1-inducing activity during in vitro dialysis. Kidney Int 1988;33:29–35.
26 Laude-Sharp M, Caroff M, Simard L, Pusineri C, Kazatchkine MD, Haeffner-Cavaillon N: Induction of IL-1 during hemodialysis: Transmembrane passage of intact endotoxins (LPS). Kidney Int 1990;38:1089–1094.
27 Mahiout A, Lonnemann G, Schulze M, Shaldon S, Koch KM: Mononuclear cells release cytokines after stimulation with LAL-negative small (< 1 kD) bacterial products (abstract). Nephrol Dial Transplant 1990;5:706–707.
28 Lonnemann G, Behme TC, Lenzner B, Floege J, Mahiout A, Colton CK, Koch KM, Shaldon S: The permeability of dialyzer membranes to TNFα-inducing substances derived from water bacteria. Kidney Int (in press).
29 Lonnemann G, Koch KM, Shaldon S, Dinarello CA: Studies on the ability of hemodialysis membranes to induce, bind and clear human interleukin-1. J Lab Clin Med 1988;112:76–86.
30 Betz M, Haensch GM, Rauterberg EW, Bommer J, Ritz E: Cuprammonium membranes stimulate interleukin-1 release and arachidonic acid metabolism in monocytes in the absence of complement. Kidney Int 1988;34:67–73.

31 Schindler R, Lonnemann G, Shaldon S, Koch KM, Dinarello CA: Transcription, not synthesis, of interleukin-1 and tumor necrosis factor by complement. Kidney Int 1990;37:85–93.

32 Lonnemann G, Bingel M, Koch KM, Shaldon S, Dinarello CA: Plasma interleukin-1 activity in humans undergoing hemodialysis with regenerated cellulosic membranes. Lymphokine Res 1987;6:63–70.

33 Herbelin A, Nguyen AT, Zingraff J, Urena P, Deschamps-Latscha B: Influence of uremia and hemodialysis on circulating interleukin-1 and tumor necrosis factor. Kidney Int 1990;37:1–6.

34 Cannon JG, van der Meer JWM, Kwiatkowski D, Endres S, Lonnemann G, Burke JF, Dinarello CA: Interleukin-1β in human plasma: Optimization of blood collection, plasma extraction and radioimmunoassay methods. Lymphokine Res 1988;7: 457–467.

35 Caruana RJ, Lobel SA, Leffell MS, Campbell H, Cheek PL: Tumor necrosis factor, interleukin-1 and beta-2-microglobulin levels in chronic hemodialysis patients. Int J Artif Organs 1990;13:794–798.

36 Endres S, Ghorbani R, Lonnemann G, van der Meer JWM, Dinarello CA: Measurement of immunoreactive interleukin-1β from human mononuclear cells: optimization of recovery, intrasubject consistency, and comparison with interleukin-1α and tumor necrosis factor. Clin Immunol Immunopathol 1988;49:424–438.

37 Endres S, Cannon JG, Ghorbani R, Dempsey RA, Sisson SD, Lonnemann G, van der Meer JWM, Wolff SM, Dinarello CA: In vitro production of IL-1β, IL-1α, TNF and IL-2 in healthy subjects: distribution, effect of cyclooxygenase inhibition and evidence of independent gene regulation. Eur J Immunol 1989;19:2327–2333.

38 Davenport A, Crabtree J, Androjna C, Will EJ, Davison AM: Tumour necrosis factor does not increase during routine cuprophan haemodialysis in healthy well-nourished patients. Nephrol Dial Transplant 1991;6:435–439.

39 Ghysen J, De Plaen JF, van Ypersele de Strihou C: Effects of membrane characteristics on tumour necrosis factor kinetics during haemodialysis. Nephrol Dial Transplant 1990;5:270–274.

40 Lonnemann G, van der Meer JMW, Cannon JG, Dinarello CA, Koch KM, Granolleras C, Deschodt G, Shaldon S: Induction of tumor necrosis factor during extracorporeal blood purification (letter). N Engl J Med 1987;317:963–964.

41 Schindler R, Schulze M, Dinarello CA, Shaldon S, Koch KM: Transcription of interleukin-1β during hemodialysis (submitted).

42 Blumenstein M, Schmidt B, Ward RA, Ziegler-Heitbrock HWL, Gurland HJ: Altered interleukin-1 production in patients undergoing hemodialysis. Nephron 1988;50:277–281.

43 Haeffner-Cavaillon N, Cavaillon JM, Ciancioni C, Bacle F, Delons S, Katatchkine MD: In vivo induction of interleukin-1 during hemodialysis. Kidney Int 1989;35: 1212–1218.

44 Lonnemann G, Haubitz M, Schindler R: Hemodialysis-associated induction of cytokines. Blood Purif 1990;8:214–222.

45 Herbelin A, Urena P, Nguyen AT, Zingraff J, Deschamps-Latscha B: Influence of first and long-term dialysis on uremia-associated increased basal production of interleukin-1 and tumor necrosis factor-α by circulating monocytes. Nephrol Dial Transplant 1991;6:349–357.

46 Guttierrez A, Alvestrand A, Warren J, Bergström J: Effect of in vivo contact between blood and dialysis membranes on protein catabolism in humans. Kidney Int 1990; 38:487–494.

47 Guttierrez A, Bergström J, Alvestrand A: Protein catabolism in sham hemodialysis: the effect of different membranes. Clin Nephrol, in press.

48 Borah MF, Schoenfeld PY, Gotch FA, Sargent JA, Wolfson M, Humphreys MH: Nitrogen balance during intermittent therapy of uremia. Kidney Int 1978;14:491–500.

49 Flores EA, Bistrian BR, Pomposelli JJ, Dinarello CA, Blackburn GL: Infusion of tumour necrosis factor/cachectin promotes muscle catabolism in the rat. A synergistic effect with interleukin-1. J Clin Invest 1989;83:1614–1622.

50 Ruiz P, Gomez F, Schreiber AD: Impaired function of macrophage Fcy receptors in end-stage renal disease. N Engl J Med 1990;322:717–723.

51 Hakim RM, Schafer AI: Hemodialysis-associated platelet activation and thrombocytopenia. AM J Med 1985;78:575–580.

52 Mahiout A, Lajous-Peter AM, Jörres A, Meinhold H, Kessel M: Extracorporeal thromboxane release and the morphological structure of artificial membrane after blood exposure. Kidney Int 1988;33(S24):75–79.

53 Sultan Y, London GM, Goldfarb B, Toulon P, Marchais SJ: Activation of platelets, coagulation and fibrinolysis in patients on long-term haemodialysis: influence of cuprophan and polyacrylonitrile membranes. Nephrol Dial Transplant 1990;5:362–368.

54 Bradley JR, Evans DB, Caine RY: Long-term survival in hemodialysis patients. Lancet 1987;i:295–296.

55 Digenis G, Pierratos A, Orepoulos DG: Cancer and chronic renal failure. Can Med Assoc J 1986;135:14–16.

56 Bardin T, Zingraff J, Kuntz D, Drüecke T: Dialysis-related amyloidosis. Nephrol Dial Transplant 1986;1:151–154.

57 Charra B, Calemard E, Laurent G: Chronic renal failure treatment duration and mode: their relevance to the late dialysis periarticular syndrome. Blood Purif 1988; 6:117–124.

58 Linke RP, Hampl H, Lobeck H, Ritz E, Bommer J, Waldherr R, Eulitz M: Lysine-specific cleavage of β_2-microglobulin in amyloid deposits associated with hemodialysis. Kidney Int 1989;36:675–681.

59 Bergström J, Alvestrand A, Fürst P: Plasma and muscle free amino acids in maintenance hemodialysis patients without protein malnutrition. Kidney Int 1990;38: 108–114.

60 Kaplowitz LG, Comstock JA, Landwehr DM, Dalton HP, Mayhall G: A prospective study of infections in hemodialysis patients: Patient hygiene and other risk factors for infection. Infect Control Hosp Epidemiol 1988;9:534–541.

61 Seifert A, Von Herrath D, Schaefer K: Iron overload, but not treatment with desferrioxamine favours the development of septicemia in patients on maintenance hemodialysis. Q J Med 1987;48:1015–1024.

62 Kantor AF, Hoover RN, Kinien LJ, McMullan MR, Fraument JF: Cancer in patients receiving long-term dialysis treatment. Am J Epidemiol 1987;126:370–376.

63 Wing A, Broyer M, Brunner FP, Brynger H, Challah S, Donckerwolcke RA, Gretz N,

Jacobs C, Kramer P, Selwood NH: Combined report on regular dialysis and transplantation in Europe, 13, 1982. Proc Eur Dial Transplant Assoc 1983;20:2–75.
64 Brunner FP, Broyer M, Brynger H, Challah S, Fassbinder W, Oulès R, Selwood NH, Wing AJ: Combined report on regular dialysis and transplantation in Europe, 15, 1984. Proc Eur Dial Transplant Assoc 1985;22:5–53.
65 Honkanen E, Grönhagen-Riska C, Teppo AM, Maury CPJ, Meri S: Acute-phase proteins during hemodialysis: Corelations with serum interleukin-1β levels and different dialysis membranes. Nephron 1991;57:283–287.
66 Gejyo F, Odani S, Yamada T, Honma N, Saito H, Suzuki Y, Nakagawa Y, Kobayashi H, Maruyama Y, Hirasawa Y, Suzuki M, Arakawa M: β_2-Microglobulin: a new form of amyloid protein associated with chronic hemodialysis. Kidney Int 1986;30: 385–390.
67 Knudsen PJ, Leon J, Ng AK, Shaldon S, Floege J, Koch KM: Hemodialysis-related induction of β_2-microglobulin and interleukin-1 synthesis and release by mononuclear phagocytes. Nephron 1989;53:188–193.
68 Zaoul PM, Stone WJ, Hakim RM: Effects of dialysis membranes on β_2-microglobulin production and cellular expression. Kidney Int 1990;38:962–968.
69 Floege J, Bartsch A, Schuze M, Shaldon S, Koch KM. Smeby LC: Turnover of ^{131}I-β_2-microglobulin in hemodialyzed patients. J Lab Clin Med, in press.
70 Altmeyer P, Kachel HG, Runne U: Microangiopathie, Bindegewebsveränderungen und amyloidartige Ablagerungen bei chronischer Niereninsuffizienz. Hautarzt 1983; 34:277–285.
71 Kachel HG, Altmeyer P, Kühn KW, Koch KM: Deposition of nonamyloid material in connective tissue in uremia. Blood Purif 1984;2:142–148.
72 Seil DR, Monnier VM: End-stage renal disease and diabetes catalyze the formation of a pentose-derived crosslink from aging human collagen. J Clin Invest 1990;85: 380–384.
73 Gupta AK, Gupta MA, Cardella CJ, Haberman HF: Cutaneous associations of chronic renal failure and dialysis. Int J Dermatol 1986;25:498–504.
74 Anderson CD, Gibson IM, Rossi E: Actinic damage in the skin of patients on hemodialysis for chronic renal failure. Photodermatology 1988;5:15–24.
75 Fairshter RD, Vaziri ND, Mirahmadi MK: Lung pathology in chronic hemodialysis patients. Int J Artif Organs 1982;5:97–100.
76 Pahl MV, Vaziri ND, Dure-Smith B, Miller R, Mirahmadi MK: Hepatobiliary pathology in hemodialysis patients: An autopsy study of 78 cases. Am J Gastroenterol 1986;81:783–787.
77 Vaziri ND, Dure-Smith B, Miller R, Mirahmadi MK: Pancreatic pathology in chronic dialysis patients – an autopsy study of 78 cases. Nephron 1987;46:347–349.
78 Mail G, Huther W, Schneider J, Lundin P, Ritz E: Diffuse intermyocardiocytic fibrosis in uremic patients. Nephrol Dial Transplant 1990;5:39–44.
79 Lowrie EG, Lew NL: Death risk in hemodialysis patients: The predictive value of commonly measured variables and an evaluation of death rate differences between facilities. Am J Kidney Dis 1990;15:458–482.
80 Chanard J, Brunols JP, Melin JP, Lavaud S, Toupance O: Long-term results of dialysis therapy with a highly permeable membrane. Artif Organs 1982;6:261–226.

81 Van Ypersele C, Jadoul M, Malghem J, Maldague B, Jamart J, and the Working Party on Dialysis Amyloidosis: Effect of dialysis membrane and patient's age on signs of dialysis-related amyloidosis. Kidney Int 1991;39:1012–1019.
82 Lavaud S, Toupance O, Roujouleh H, Fakir M, Melin JP, Chanard J: Carpal tunnel syndrome in hemodialysis patients: effect of dialysis strategy; in Man, Mion, Henderson (eds): Blood Purification in Perspective: New Insights and Future Trends. Cleveland, ISAO Press, 1987, pp 125–128.
83 Brunner FP, Brynger H, Ehrloh JHH, Fassbinder W, Geerlings W, Rizzoni G, Selwood NH, Tufveson G, Wing AJ: Case control study on dialysis arthropathy: the influence of two different dialysis membranes: data from the EDTA registry. Nephrol Dial Transplant 1990;5:432–436.
84 Kessler M, Netter P, Maheut H, Wolf C, Prenat E, Cao Huu T, Gaucher A: Highly permeable and biocompatible membranes and prevalence of dialysis-associated artropathy (letter). Lancet 1991;337:1092.
85 Floege J, Burchert W, Brandis A, Gielow P, Nonast-Daniel B, Spindler E, Hundeshagen H, Shaldon S, Koch KM: On the imaging of dialysis-related amyloid (AB-amyloid) deposits with ^{131}I-β_2-microglobulin. Kidney Int 1990;38:1169–1176.
86 Hawkins PN, Myers MJ, Lavender JP, Pepys MB: Diagnostic radionuclide imaging of amyloid: biological targeting by circulating human serum amyloid P component. Lancet 1988;i:1413–1418.
87 Lindsay RM, Spanner E: A hypothesis. The protein catabolic rate is dependent upon the type and amount of treatment in dialyzed uremic patients. Am J Kidney Dis 1989;13:382–389.

Dr. S. Shaldon, 86, rue de Grezac, F-34080 Montpellier (France)

Berlyne GM (ed): The Kidney Today. Selected Topics in Renal Science.
Contrib Nephrol. Basel, Karger, 1992, vol 100, pp 172–187

From a Third Factor to a Glucose-6-Phosphate Dehydrogenase Stimulator[1]

H.E. de Wardener

Research Laboratories, Charing Cross & Westminster Medical School, London, UK

This is a retrotelescopic view of certain experiments my colleagues and I have performed and are performing to elucidate the nature of what was first called a third factor, then a natriuretic hormone, a digitalis-like factor and now a cytochemically evident Na,K-ATPase inhibitor/G6PD (glucose-6-phosphate dehydrogenase) stimulator. Our story starts in 1961 when we found that the natriuresis caused by acute blood volume expansion or the intravenous administration of saline was mainly due to a change in the concentration of a circulating natriuretic substance [1]. About 10 years later it was observed that acute volume expansion was also accompanied by an increase in the ability of plasma to inhibit sodium transport [2, 3] and that this was due to inhibition of Na,K-ATPase [4]. These findings led to the false assumption that the increased natriuretic capacity of the plasma and its increased ability to inhibit sodium transport, both of which occurred with volume expansion, was due to a change in the concentration of the same substance which was named the natriuretic hormone.

The atrial natriuretic peptide was discovered in 1981 [5]. To the surprise of many of us, the atrial natriuretic peptide had no effect on Na,K-ATPase. It was obvious, therefore, that the increased natriuretic capacity of the plasma and its increased ability to inhibit Na,K-ATPase were not necessarily due to the same substance. It followed that the structure of the

[1] This review is based on a lecture given to the 2nd Symposium on Endogenous Digitalis-Like Factor in Kobe on August 31, 1991 [59] and an editorial in *Clinical Science* [60, 61].

substance responsible for the volume controlled capacity of the plasma to inhibit Na,K-ATPase still needed to be determined. Furthermore, we began to pay attention to the well-established but not so well-recognized fact that in a normal animal the natriuretic effect of certain well-known exogenous Na,K-ATPase inhibitors, such as ouabain and digoxin, is negligible. It was therefore ironic that at about this time the atrial natriuretic peptide (an unquestionably powerful short-acting natriuretic substance) was being closely scrutinized, and only slowly accepted, as a possible physiologically relevant natriuretic substance, while the title of 'natriuretic hormone' continued to be bestowed on the circulating Na,K-ATPase inhibitor that some of us suspected might not be natriuretic.

Early experiments in 1974 [6] had shown that the phenomenon of acute volume expansion increasing the plasma's capacity to inhibit sodium transport could be reproduced in a persistent way by decreasing the kidney's capacity to excrete sodium by drastically reducing its mass surgically. This phenomenon was confirmed and considerably extended when it was observed that the capacity of plasma to inhibit Na,K-ATPase was raised in acquired and experimental forms of hypertension in animals [7,8]. We studied essential hypertension in man. White cells from normotensive subjects were incubated in serum obtained from patients with essential hypertension [9]. This caused an impairment of active sodium transport in the normal white cells was similar to that found in the white cells of hypertensive patients. The impairment in sodium transport was due to a fall in the ouabain-sensitive component of the total sodium efflux rate constant. We concluded that the serum of patients with essential hypertension contains a substance with ouabain-like properties which is probably responsible for the impairment of sodium transport in the leucocyte of patients with essential hypertension.

The search for the substance responsible for the endogenous Na,K-ATPase inhibitor has been hampered by two principal difficulties. The first is that more than 100 endogenous compounds are known to modify the activity of Na,K-ATPase and that the inhibitory activity of most of these is not digitalis-like. It is therefore not surprising that the concentrating effect involved in purification procedures has already yielded some well-known, endogenous Na,K-ATPase inhibitors, including unsaturated fatty acids and steroids [10–12]. However, the free plasma concentrations of these substances has been considered to be too low to be physiologically important or it has not changed with volume expansion or salt intake, or it has not been raised in hypertension. However, plasma concentrations may be irrelevant. One of these substances may have a relevant physiological role at its site of production. Unsaturated hydroxy fatty acids, for instance, have a wide range of activity locally; for example, thromboxane B_2 and leukotriene B_4.

The lack of a specific or even a highly selective assay with which to distinguish a relevant substance has also been crippling. Researchers have had to make agonizing choices, based on differing assumptions when deciding which of several fractions, all of which inhibit Na,K-ATPase, should be the one to purify. The most widely held assumption has been that the relevant functional substance should closely resemble ouabain or digitalis: in other words, that it should (1) selectively inhibit Na,K-ATPase, (2) inhibit sodium, potassium or rubidium transport, (3) displace ^3H ouabain from surface receptors, (4) be ionotropic, (5) increase vascular tone, (6) increase intracellular calcium, and (7) cross-react with antibodies raised against digoxin. This approach has led to the discovery that human plasma contains ouabain [13] and human urine digoxin [14]. Other investigators, while agreeing that the relevant substance may share some properties with ouabain, have thought it safer to assume that any substance that inhibits Na,K-ATPase, the concentration of which is controlled by volume changes or salt intake and is raised in hypertension, is potentially relevant. We have chosen to use cytochemical bioassays, initially because they are so sensitive, in some instances about 1,000 times greater than radioimmunoassays [15].

Cytochemical Assays

Cytochemical assays are in use for several known hormones such as thyroid-stimulating hormone, parathyroid hormone and gastrin [15]. Such assays are based on the principle that when a hormone binds to its receptor it produces a transient, rapidly reversible effect on an enzyme that can be revealed by means of a coloured end-product at the site of enzymatic activity. The density of the colour is then measured with a scanning and integrating microdensitometer.

Na,K-ATPase

We have used a technique developed by Chayen et al. [16] to measure Na,K-ATPase activity in fresh tissue. We found that ouabain (4×10^{-4} *M*) and normal human plasma (1/100–1/1,000) both induce a rapid reversible

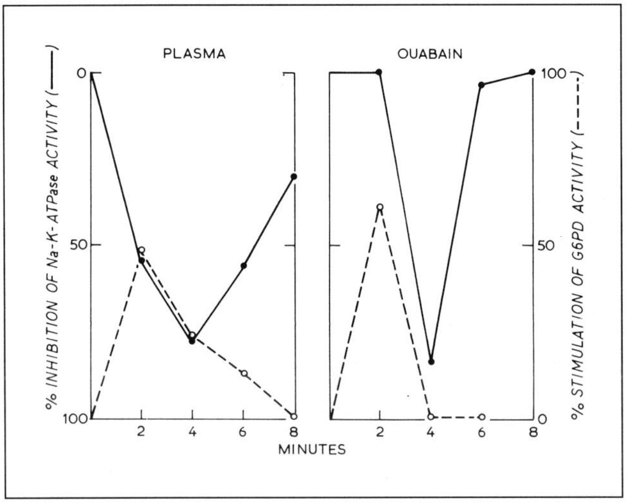

Fig. 1. Percent changes in G6PD activity (●) and Na,K-ATPase activity (○) in proximal tubules of segments of fresh guinea pig kidney exposed to 1:1,000 plasma and 4 × 10^{-4} *M* ouabain for 2–8 min, measured with cytochemical techniques. (Reprinted from Holland et al. [26], with permission.)

inhibition of guinea pig kidney proximal tubule that is maximal at 4–6 min [17] (fig. 1). Using this technique, dilutions of plasma from salt-loaded subjects inhibited Na,K-ATPase at 4 min about 25 times more than plasma from the same persons when on a low salt diet [17].

G6PD Stimulation

The technique with which Na,K-ATPase is measured is semi-quantitative in that though it is possible to obtain a time course curve of inhibition, the small amount of Na,K-ATPase activity remaining in the kidney section has made it difficult to obtain a dose inhibition response curve of sufficient steepness for a reproducible quantitative assay. In a live cell however, inhibition of Na,K-ATPase is associated with stimulation of G6PD activity [18]. The intervening metabolic steps are not known; a change in glutathione or the involvement of certain arachidonic acid metabolites has been suggested [18, 19]. We found that the steepness of the rise in G6PD activity gives a satisfactory dose-response curve. This made it possible to develop and validate a precise quantitative assay that mea-

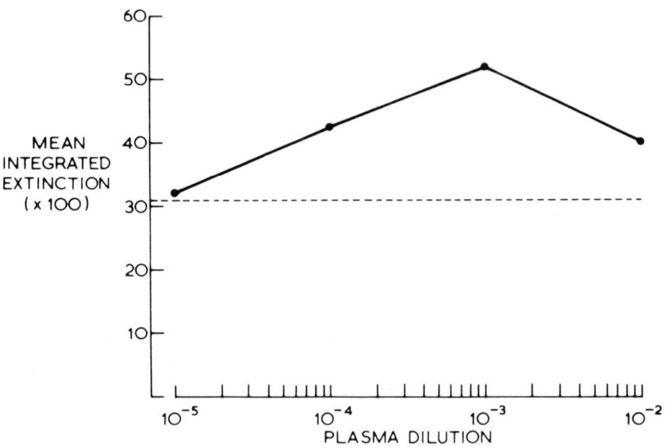

Fig. 2. Changes in G6PD activity as shown by the amount of formazan deposit measured by microdensitometry expressed as mean integrated extinction in the proximal tubules in segments of guinea pig kidney exposed for 2 min to dilutions of a high potency plasma. The broken line illustrates G6PD activity in a segment exposed to culture medium alone. (Reprinted from de Wardener et al. [62], with permission.)

sures the capacity of body fluids or extracts to stimulate G6PD which can be used as an index of their ability to inhibit Na,K-ATPase [20]. Normal human plasma and ouabain both induce a reversible rise in G6PD activity which is maximal at 2 min (fig. 1). A plasma of high potency is used as a standard. Quantitation is derived from a dose-response curve which is biphasic (fig. 2), a phenomenon which is seen in all established cytochemical bioassays for known hormones. It is attributed to the rapid effect of the hormone on enzyme activity, so that at high concentrations the induced effect on enzyme activity (after a 2-min exposure in this instance) is in the recovery phase and thus appears to be less than at a lower concentration. It is the linearity and reproducibility of the G6PD-stimulating activity to zero with increasing dilutions that endows the assay with precision. The amount of G6PD stimulation that produces maximum G6PD stimulation at 2 min is defined as 1 unit of G6PD stimulating activity. To be quantitative, the dose-response curve of the test sample must be parallel with that of the standard.

The plasma from 12 normal subjects (the plasma from 5 of whom had previously been shown to inhibit Na,K-ATPase maximally at 4 min in a manner which was significantly directly related to salt intake) stimulated G6PD maximally at 2 min significantly more when they were on a high sodium diet than when they were on a low sodium diet (39.3 ± 10.6 vs. 1.8 ± 0.5 G6PD-stimulating units/ml, $p < 0.005$) [20].

Possible Site of Production of Cytochemically Detectable Na,K-ATPase Inhibitor/G6PD Stimulator

In order to find the origin of the stubstance responsible for the cytochemically detectable Na,K-ATPase inhibition at 4 min and G6PD stimulation at 2 min, acetone extracts of the following organs were assayed: pituitary, hypothalamus, cerebral cortex, muscle, heart, adrenal, liver, spleen, pancreas, kidney, stomach, duodenum, jejunum and ileum. The only tissue that yielded extracts of high activity was the hypothalamus [21]. The time course of the effect of the hypothalamic extract was the same as that obtained with plasma while the dose-response curve for the hypothalamic material was parallel to that of plasma. The G6PD-stimulating activity from one normal rat hypothalamus was about 10,000 greater than that from 1 ml of plasma, and the G6PD-stimulating activity of hypothalamic extracts from rats which had been on a high sodium intake for 4 weeks was approximately 150 times greater than those obtained from rats which had been on a low sodium diet ($419.6 \pm 110 \times 10^4$ vs. $2.56 \pm 0.59 \times 10^4$ G6PD-stimulating units per hypothalamus, $p < 0.005$) [21]. The G6PD-stimulating activity of the plasma was 6 times more active when it was obtained from a rat on a high sodium diet than when it was deprived of sodium (33.1 ± 5.81 vs. 5.0 ± 0.57 G6PD-stimulating units/ml, $p < 0.005$).

Specificity

Tests of specificity have been performed with a variety of substances including several which are known to inhibit purified preparations of Na,K-ATPase such as fatty acids (linoleic and oleic) dehydroepandrosterone sulphate, lysophosphatidyl choline and sodium vanadate [20, 22, 23].

We have not yet found a known mammalian substance which stimulates guinea pig proximal tubule G6PD activity at 2 min and inhibits Na,K-ATPase activity at 4 min. It would appear therefore that the importance of the cytochemical techniques we are using is as much related to their capacity to act as highly selective markers for a substance in the plasma, and in the hypothalamus, the concentration of which in both is raised by a high salt intake and in hypertension (see below) as in their capacity to be a highly sensitive alternative method for the detection of a Na,K-ATPase inhibitor.

Hypertension

Humans

The ability of plasma to stimulate G6PD maximally at 2 min is significantly greater in hypertensive patients on a normal sodium diet than it is in normotensive subjects (195 ± 52 vs. 22.5 ± 5.8 G6PD-stimulating units/ml, $p < 0.001$). When the two groups are combined, the plasma's ability to stimulate G6PD is significantly correlated with the systolic, mean and diastolic blood pressure ($p < 0.001$) and in the hypertensive patients it is inversely correlated with plasma renin activity ($p < 0.001$) [24].

Rats

The Spontaneously Hypertensive Strain Rat (SHR) and Its Normotensive Control. The ability of the plasma from control normotensive WKY rats to stimulate G6PD at 2 min (12.5 ± 2.6 G6PD-stimulating units/ml) is similar to that of the normal Wistar rat, while the ability of the plasma from SHR rats to stimulate G6PD at 2 min is much greater (772.3 ± 48.1 G6PD-stimulating units/ml, $p < 0.01$).

The content of cytochemical detectable G6PD-stimulating activity in the hypothalamus shows a similar pattern. The mean ability of the hypothalamic extracts from the SHR rat to stimulate G6PD is $2.2 \pm 1.7 \times 10^8$ units/hypothalamus and from the WKY rat it is $4.5 \pm 1.8 \times 10^4$ units/hypothalamus ($p < 0.01$) [25].

Time course curves of the ability of plasma and hypothalamic extracts to inhibit Na,K-ATPase at 2, 4, 6 and 8 min demonstrate that the capacity of the plasma and the hypothalamus of the SHR rat to inhibit Na,K-ATPase is greater than that of the WKY rat.

Milan Hypertensive Rat and Reduced Renal Mass Hypertensive Rat. The plasma concentration of cytochemically detectable G6PD-stimulating/Na,K-ATPase inhibitory activity in both the Milan and the reduced renal mass hypertensive adult rats is significantly greater than in their respective controls ($p < 0.001$). In the Milan hypertensive rat, this difference is detectable at 3 weeks of age, before the onset of hypertension ($p < 0.001$) [26].

In the reduced renal mass hypertensive rat the mean G6PD-stimulating activity of hypothalamic extracts is $2.5 \pm 1.4 \times 10^8$ G6PD-stimulating units per hypothalamus while in the normotensive control it is $6.38 \pm 2.35 \times 10^3$ ($p < 0.01$) [27].

The plasma and hypothalamic concentration of G6PD-stimulating activity in the control normotensive Milan rat and in the normotensive reduced renal mass rat is not significantly different from that of normal Wistar and WKY rats.

Comment

The reduced renal mass rats, both the distilled-water and saline-drinking groups, had equal reductions in renal mass, the only difference between them was the saline intake, therefore the increase in arterial pressure and the increase in the cytochemically assayable G6PD-stimulating activity in the saline-drinking group is due to the increased salt intake. In a normal rat a high salt intake also raises the plasma's and the hypothalamus's ability to stimulate G6PD. Therefore, in a normal rat which has had its ability to excrete sodium impaired by excising most of its renal tissue, which is then given a high intake of salt, it is not surprising to find that the consequent increased ability to stimulate G6PD and to inhibit plasma and hypothalamic Na,K-ATPase is greater than in a normal intact rat given a high intake of sodium.

As raised levels of cytochemically detectable G6PD-stimulating/Na,K-ATPase-inhibiting activity in the plasma and hypothalamus are also found in hereditary forms of hypertension in the rat, it would appear that, in hereditary hypertension, such an increase is unlikely to be due directly to a genetic abnormality. Overall the evidence suggests rather that in the forms of hypertension studied, including the hereditary forms of hypertension and the reduced renal mass rat, the rise of G6PD-stimulating/Na,K-ATPase-inhibiting activity in the plasma and hypothalamus is more likely to be secondary to a renal abnormality which interferes with the urinary excretion of sodium chloride [28].

Physicochemical Characteristics of the Cytochemically Assayable G6PD-Stimulating/Na,K-ATPase-Inhibiting Activity in Plasma and Hypothalamus [29]

Whole Plasma

Temperature has certain effects on the G6PD-stimulating activity of whole plasma which is not evident in extracts of plasma or hypothalamus. The G6PD-stimulating activity of a high potency plasma from the SHR and from hypertensive man is increased by more than 100% by incubating the plasma at 37 °C for 10–15 min. The activity is destroyed after incubation for 45 min. These effects are not influenced by the proteinase inhibitor Trasylol (Bayer UK Ltd.). Incubation of low potency plasma from the normal rat and man for 10–15 min at 37 °C does not consistently increase the plasma's capacity to stimulate G6PD activity, but it is totally destroyed by incubation for 45 min. Freezing and thawing plasma more than once depresses G6PD-stimulating activity.

Plasma and Hypothalamic Extracts

The G6PD-stimulating activity of an acid extract of plasma and hypothalamus remains unchanged if left at room temperature for 3 h but is destroyed by (a) heating under N_2 at 600 °C for 18 h and (b) alkalinization to pH 12 for 3 h at room temperature.

The activity of both extracts is soluble in water, insoluble in chloroform, highly basic, has a molecular weight probably below 500 and appears in the same fraction on silica gel electrophoresis. The activity of the extracts is reduced by prior incubation with an antibody raised against digoxin but the extracts do not inhibit purified preparations on canine renal Na,K-ATPase in vitro nor do they cross-react against antibodies raised against digoxin and they do not appear to be natriuretic [unpubl. observations].

Comment

These similarities between the cytochemically assayable G6PD-stimulating/Na,K-ATPase-inhibiting activity in the plasma and in the hypothalamus and (a) the basic observation that the reversible time course of the cytochemical changes in activity of each is similar, (b) that the dose-response curves are parallel, plus (c) the observation that the activity of both extracts is increased by a high sodium intake, in both man and the rat (see above), (d) that the activity is raised in the plasma and the hypothalamus in the SHR and the reduced renal mass rat (see above) suggests that the cytochemically assayable activity in the plasma and hypothalamus is probably due to the same substance.

Further Observations on the Nature of the Hypothalamic G6PD-Stimulating Substance[2]

As the content of G6PD-stimulating activity is greatest in the SHR hypothalamus, it was decided to use this site as the source material in an attempt to isolate and identify the substance responsible for the cytochemically evident activity.

The active fraction from the silica gel electrophoresis was placed onto reverse-phase HPLC. The G6PD-stimulating activity appears at 16–18 min. UV activity is present throughout the HPLC chromatogram but the fraction containg G6PD-stimulating activity is in a baseline valley of the UV profile, in other words it does not appear to absorb UV light. The active fraction from the HPLC was then placed onto high-voltage electrophoresis. The active fraction from the high-voltage electrophoresis was not stained by ninhydrin or fluorescamine. Calculation of relative mobilities at different pH on high-voltage electrophoresis suggests that the active molecule carries at least three positive charges at pH 2.1 and one less at pH 6.5. The activity of the hypothalamic substance remains after acetylation with acetic anhydride in methanol or esterification in methanol/HCl. It also survives incubation with proteolytic enzymes with endo- and exopeptidases with the exception of leucine amino peptidase which partially inhibits it. It is destroyed by methylation with diazomethane.

Conclusion

The active molecule appears to have a low molecular weight, it is a non-acidic, non-steroidal, non-peptidic substance, possessing a high positive charge to mass ratio; its physico-chemical properties suggest that it

[2] With the collaboration of Prof. H. Morris, Dr. M. Panico and Dr. R. Etienne, Department of Biochemistry, Imperial College, University of London.

may be a choline analogue. Because of these properties and the observations, noted above, that the substance does not inhibit Na,K-ATPase in vitro or cross-react against digoxin antibodies, nor is it natriuretic, it is clear that the cytochemically detectable substance is different from the hypothalamic substances described by Lichtenstein and Samuelov [30]; Akugawa et al. [31]; Morgan et al. [32]; Haber and Haupert [33]; Illescas et al. [34]; Takahashi et al. [35] and Goto et al. [36].

Hypothesis

In both acquired [37–43] and inherited [40–45] hypertension in the rat the hypothalamus manifests certain changes in catecholaminergic, and cholinergic, activity which, although not identical in the two forms of hypertension, appear to be responsible for sympathetic and non-sympathetic hypertensive mechanisms [46–51]. In both forms of hypertension the intraventricular administration of 6-hydroxydopamine, which destroys catecholaminergic neurons, prevents the onset of hypertension but has no effect on established hypertension [52–55]. The administration, by the same route, of hemicholinium-3, which blocks the transfer of choline across the neuron membrane and thus inhibits the synthesis of acetylcholine, prevents the onset of hypertension and reversibly lowers the blood pressure to normal when the hypertension is established [56, 57]. Hemicholinium-3 has no effect on the blood pressure of the normal rat [58]. It appears therefore that the rise in blood pressure in both acquired and inherited hypertension in animals stems from increased hypothalamic cholinergic activity both during and after the initial increase in peripheral sympathetic activity.

The site of the increased cholinergic activity in the SHR appears to be the posterior hypothalamic nucleus where the hemicholinium-3 probably has its blood pressure lowering effect [63, 64; Vargas, H., unpubl. observation]. In contrast, in the dorsomedial nucleus of the hypothalamus, which in normal rats is a cholinergic depressor area [65–69], the neurons are atrophic [70], due possibly to diminished cholinergic activity. Diminished activity in this normally depressor area is likely to habe a pressor effect. It is tentatively suggested that the increase in cytochemically detectable G6PD-stimulating/Na,K-ATPase-inhibiting choline-like substance in the hypothalamus of the SHR rat may either be responsible for, or an indicator of, these disturbances in cholinergic activity.

In summary: Using conventional techniques, initial observations in man and animal showed that the plasma's ability to inhibit Na,K-ATPase is increased after volume expansion and in hypertension. Subsequently a qualitative cytochemical method to measure the plasma's capacity to inhibit Na,K-ATPase and a quantitative assay which measures the plasma's capacity to stimulate G6PD were used to study this phenomenon. (In a living cell, Na,K-ATPase inhibition is reciprocally related to G6PD stimulation.) It has been found that the plasma and hypothalamus of the rat contain cytochemically evident G6PD-stimulating/Na,K-ATPase-inhibiting activity which is directly related to sodium intake, the concentration of which is raised in both acquired and hereditary forms of hypertension. The evidence suggests that the cytochemically evident substance in the plasma and the hypothalamus is the same. The active substance in the hypothalamus does not inhibit canine Na,K-ATPase in vitro or cross-react against digoxin antibodies, nor is it natriuretic, it has a low molecular weight (<500), is non-acidic, non-steroidal, non-peptidic, possessing a high positive charge to mass ratio; its physico-chemical properties suggest that it is a choline analogue. It is proposed that the raised concentration of this choline-like substance in the hypothalamus may either be responsible for, or an indicator of, the cholinergic pressor abnormalities in the hypothalamus of the SHR rat.

References

1 de Wardener HE, Mills IH, Clapham WG, Hayter CJ: Studies on the efferent mechanism of the sodium diuresis which follows the administration of intravenous saline in the dog. Clin Sci 1961;21:249–258.

2 Clarkson EM, Talner LB, de Wardener HE: The effect of plasma from blood volume expanded dogs on sodium, potassium and PAH transport of renal tubule fragments. Clin Sci 1970;38:617–627.

3 Buckalew VM Jr, Martinez J, Green WS: The effect of dialysates and ultrafiltrates of plasma of saline loaded dogs on toad bladder sodium transport. J Clin Invest 1970; 49:926–935.

4 Gonick HC, Kramer HJ, Paul W, Lu E: Circulating inhibitor of sodium-potassium activated adenosine triphosphatase after expansion of extracellular fluid volume in rats. Clin Sci Mol Med 1977;53:329–334.

5 de Bold AJ, Flynn TG: Cardionatrin I: a novel heart peptide with potent diuretic and natriuretic properties. Life Sci 1983;33:297–302.

6 Schmidt RW, Bourgoignie JJ, Bricker NS: On the adaption in sodium excretion in chronic uraemia: the effects of 'proportional reduction' of sodium intake. J Clin Invest 1974;53:1736–1741.

7 Haddy FJ, Overbeck HW: The role of humoral factors in volume expanded hypertension. Life Sci 1976;19:935–948.

8 Haddy FJ, Pamnani MB, Clough DL: Role of a humoral factor in low-renin hypertension; in Lichardus B, Schrier RW, Ponec J (eds): Hormonal Regulation of Sodium Excretion. Amsterdam, Elsevier/North-Holland Biomedical Press, 1980, pp 379–385.

9 Poston L, Sewell RB, Wilkinson SP, Richardson PJ, Williams R, Clarkson EM, MacGregor GA, de Wardener HE: Evidence for a circulating sodium transport inhibitor in essential hypertension. Br Med J 1981;282:847–849.

10 Kelly RA, O'Hara DS, Mitch WE, Smith TW: Identification of Na-K-ATPase inhibitors in human plasma as non-esterified fatty acids and lysophospholipids. J Biol Chem 1985;261:11704–11711.

11 Tamura M, Harris TM, Higashimori K, Sweetman BJ, Blair IA, Inagami T: Lysophosphatidylcholine containing polyunsaturated fatty acids were found as Na-K-ATPase inhibitors in acute volume-expanded dogs. Biochemistry 1987;26:2797–2806.

12 Secombe DW, Pudek MR, Nowaczynski W, Humphries KH: Digoxin-like immunoreactivity, displacement of ouabain and inhibition of Na,K-ATPase by four steroids known to be increased in essential hypertension. Clin Biochem 1989;22:17–21.

13 Ludens JH, Clark MA, DuCharme DW, Harris DW, Lutzke BS, Mandel F, Mathews WR, Sutter DM, Hamlyn JM: Purification of an endogenous digitalis-like factor from human plasma for structural analysis. Hypertension 1991;17:923–929.

14 Goto A, Ishiguro T, Yamada K, Ishii M, Yoshioka M, Eguchi C, Shimora M, Sugimoto T: Isolation of a urinary digitalis-like factor indistinguishable from digoxin. Biochem Biophys Res Commun 1990;173:1093–1101.

15 Chayen J, Bitensky L: Cytochemical Bioassays: Techniques and Clinical Application. New York, Dekker, 1983.

16 Chayen J, Frost GTB, Dodds RA, Bitensky L, Pitchfork J, Bayliss PH, Barnett RJ: The use of a hidden metal-capture reagent for the measurement of Na,K-ATPase activity. A new concept in cytochemistry. Histochemistry 1981;71:533–541.

17 de Wardener HE, Clarkson EM, Bitensky L, MacGregor GA, Alaghband-Zadeh J, Chayen J: Effect of sodium intake on ability of human plasma to inhibit renal Na,K-adenosine triphosphatase in vitro. Lancet 1981;i:411–412.

18 Dikstein S: Stimulability, adenosine triphosphatases and their control by cellular redox processes. Naturwissenschaften 1971;58:439–443.

19 Schwartzman M, Ferrari NR, Carroll MA, Songu-Mize E, McGiff JC: Renal cytochrome P450-related arachidonate metabolite inhibits Na,K-ATPase. Nature 1985; 314:620–622.

20 Fenton S, Clarkson E, MacGregor G, Alaghband-Zadeh J, de Wardener HE: An assay of the capacity of biological fluids to stimulate renal glucose-6-phosphate dehydrogenase activity in vitro as a marker of their ability to inhibit sodium-dependent adenosine triphosphatase activity. J Endocrinol 1982;94:99–110.

21 Alaghband-Zadeh J, Fenton S, Hancock K, Millett J, de Wardener HE: Evidence that the hypothalamus may be a source of a circulating Na,K-ATPase inhibitor. J Endocrinol 1983;98:221–226.

22 Vasdev S, Longerich L, Johnson E, Brent D, Gault MH: Dehydroepiandrosterone sulphate is a digitalis-like factor in plasma of healthy human adults. Res Commun Chem Pathol Pharmacol 1985;49:387–399.

23 Kelly RA, O'Hara DS, Mitch WE: Identification of Na,K-ATPase inhibitors in human plasma as non-esterified fatty acids and lysophospholipids. J Biol Chem 1986;261:1170–1171.

24 MacGregor GA, Fenton S, Alaghband-Zadeh J, Markandu N, Roulston JE, de Wardener HE: Evidence for a raised concentration of a circulating sodium transport inhibitor in essential hypertension. Br Med J 1981;283:1355–1357.

25 Millett JA, Holland SM, Alaghband-Zadeh J, de Wardener HE: Na,K-ATPase-inhibiting and glucose-6-phosphate dehydrogenase-stimulating activity of plasma and hypothalamus of the Okamoto spontaneously hypertensive rat. J Endocrinol 1986; 108:69–73.

26 Holland SM, Millett J, Alaghband-Zadeh J, de Wardener HE, Ferrari P, Bianchi G: Cytochemically assayable Na^+,K^+-ATPase inhibition by Milan hypertensive rat plasma. Hypertension 1987;9:498–503.

27 Holland SM, Millett J, Alaghband-Zadeh J, de Wardener HE, Pamnani M, Haddy F: Cytochemically detectable glucose-6-phosphate dehydrogenase-stimulating/Na,K^+-ATPase-inhibiting activity of plasma and hypothalamus in reduced renal mass hypertension. Am J Hypertens 1991;4:318–320.

28 de Wardener HE, MacGregor GA: Dahl's hypothesis that a saluretic substance may be responsible for a substantial rise in arterial pressure: its possible role in essential hypertension. Kidney Int 1980;18:1–9.

29 Millett JA, Holland SM, Alaghband-Zadeh J, de Wardener HE: Extraction and characterization of a cytochemically assayable Na^+/K^+-ATPase inhibitor/glucose-6-phosphate dehydrogenase stimulator in the hypothalamus and plasma of man and the rat. J Endocrinol 1987;112:299–303.

30 Lichtstein D, Samuelov S: Membrane potential changes induced by the ouabain-like compound extracted from mammalian brain. Proc Natl Acad Sci USA 1982;79: 1453–1456.

31 Akugawa K, Hara N, Tsukada Y: Partial purification and properties of the inhibition of Na,K-ATPase and ouabain-binding in bovine central nervous system. J Neurochem 1984;42:775–780.

32 Morgan K, Foord SM, Spurlock G, Charalambous BM, Dieguez C, Scanlon MF, Mir MA: Release of an active sodium transport inhibitor (ASTI) from rat hypothalamic cells in culture. Endocrinology 1984;115:1642–1644.

33 Haber E, Haupert GT: The search for a hypothalamic Na,K-ATPase inhibitor. Hypertension 1987;9:315–324.

34 Illescas M, Ricote M, Mendez E, Robles RG, Sancho JM: Partial purification of a sodium pump inhibitor from bovine adenohypophysis. Its comparison with the natriuretic factor isolated from hypothalamus. Clin Exp Theory Pract 1988;10(suppl 1):301–307.

35 Takahashi H, Matsusawa M, Ikegaki I, Suga K, Nishimura M, Yoshimura M, Yamada H, Sano Y: Digitalis-like substance is produced in the hypothalamus but not the adrenal gland in rats. J Hypertens 1988;6:(suppl):S345–S347.

36 Goto A, Yamada K, Iohii M, Yoshioka M, Ishiguro T, Eguchi C, Sugimoto T: Existence of a polar digitalis-like factor in mammalian hypothalamus. Biochem Biophys Res Commun 1989;161:953–958.

37 Nakamura K, Gerold M, Thoenen H: Experimental hypertension in the rat: reciprocal changes in norepinephrine turnover in heart and brain stem. Naunyn Schmiedebergs Arch Pharmacol 1971;268:125–139.

38 Ciriello J: Contribution of forebrain mechanisms in the maintenance of deoxycorticosterone acetate-salt hypertension. Clin Exp Hypertens [A] 1988;10(suppl 1):169–178.
39 Black J, Waeber B, Bresnahan MR, Gavras I, Gavras H: Blood pressure response to central and/or peripheral inhibition of phenylethanolamine-N-methyltransferase in normotensive and hypertensive rats. Circ Res 1981;49:518–524.
40 Elghozi J-L, Mazeaud M, Le Quan-Bui KH, Aragon I, Meyer P, Devynck MA: Regional distribution of noradrenaline and dopamine-β-hydroxylase in the brain of spontaneously hypertensive rats. Clin Exp Hypertens [A] 1982;4:461–475.
41 Oparil S, Yang R-H, Jin H, Wyss JM, Chen Y-F: Central mechanisms of hypertension. Am J Hypertens 1989;2:474–482.
42 Meldrum MJ, Westfall TC: Comparison of norepinephrine release in hypertensive rats. I. Hypothalamic and brainstem tissues. Clin Exp Hypertens [A] 1986;8:201–219.
43 Brezenoff H: Brain acetyl choline and hypertension: a compelling association. Trends Autonom Pharmacol 1985;3:203–212.
44 Winternitz SR, Wyss JM, Oparil S: The role of the posterior hypothalamic area in the pathogenesis of hypertension in the spontaneously hypertensive rat. Brain Res 1984;324:51–58.
45 Saavedra JM, Grobecker H, Axelrod J: Changes in central catecholaminergic neurones. Circ Res 1978;42:529–534.
46 Takeda K, Buñag RD: Augmented sympathetic nerve activity and pressor responsiveness in DOCA hypertensive rats. Hypertension 1980;2:97–101.
47 Westfall TC, Meldrum MJ: Alterations in the release of norepinephrine at the vascular neuroeffector junction in hypertension. Annu Rev Pharmacol Toxicol 1985; 25:621–641.
48 Takeda K, Bunag RD: Sympathetic hyperactivity during hypothalamic stimulation in spontaneously hypertensive rats. J Clin Invest 1978;62:642–648.
49 Esler M, Jennings G, Lambert G: Noradrenaline release and the pathophysiology of primary human hypertension. Am J Hypertens 1989;2:140–146.
50 Julius S, Schork N, Schork A: Sympathetic hyperactivity in early stages of hypertension: the Ann Arbor data set. J Cardiovasc Pharmacol 1988;12(suppl 3):S121–S129.
51 Anderson EA, Sinkey CA, Lawton WJ, Mark AL: Elevated sympathetic nerve activity in borderline hypertensive humans. Evidence from direct intraneural recordings. Hypertension 1989;14:177–183.
52 Haeusler G, Finch L, Thoenen H: Central adrenergic neurons and the initiation and development of experimental hypertension. Experientia 1972;28:1200–1203.
53 Kubo T, Hashimoto M: Effects of intraventricular and intraspinal 6-hydroxydopamine on blood pressure of DOCA-saline hypertensive rats. Arch Int Pharmacodyn 1979;238:50–59.
54 Kubo T, Hashimoto M: Effects of intraventricular and intraspinal 6-hydroxydopamine on blood pressure of spontaneously hypertensive rat. Arch Int Pharmacodyn 1978;232:166–176.
55 Van den Buuse M, Versteeq DHG, De Jong W: Brain noradrenaline and the development of hypertension: the effect of treatment with central 6-hydroxydopamine or DSP-4. Clin Exp Pharmacol Physiol 1986;13:469–476.

56 Vargas HM, Brezenoff HE: Suppression of hypertension during chronic reduction of brain acetylcholine in spontaneously hypertensive rats. J Hypertens 1988;6:739–745.
57 Giuliano R, Brezenoff HE: Increased central cholinergic activity in rat models of hypertension. J Cardiovasc Pharmacol 1987;10:113–122.
58 Brezenoff HE, Caputi AP: Intracerebroventricular injection of hemicholinium-3 lowers the blood pressure in conscious spontaneously hypertensive rats but not in normotensive rats. Life Sci 1980;26:1037–1045.
59 de Wardener HE: From a circulating Na,K-ATPase inhibitor to a hypothalamic G6PD stimulator. Ther Res, in press.
60 de Wardener HE: The primary role of the kidney and salt intake in the aetiology of essential hypertension: part I. Clin Sci 1990;79:193–200.
61 de Wardener HE: The primary role of the kidney and salt intake in the aetiology of essential hypertension: part II. Clin Sci 1990;79:289–297.
62 de Wardener HE, Millett J, Holland S, MacGregor GA, Alaghband-Zadeh J: Ouabain-like Na,K-ATPase inhibitor in the plasma of normotensive and hypertensive humans and rats. Hypertension 1987;10(suppl 1):I-51–I-56.
63 Vargas H, Ringdahl B: Intracisternal injection of hemicholinium-3 does not lower blood pressure in the adult spontaneously hypertensive rat. Neurosci 1990;16:553.
64 Brezenoff HE, El-Fakahany E, Yong-Fu Xiao: The effect of 4-DAMP mustard on blood pressure and muscarinic binding in the spontaneously hypertensive rat. Annals Neurosci 1990;1:24–33.
65 Brezenoff HE: Cardiovascular responses to interhypothalamic injection of carbechol and certain cholinesterase inhibitors. Neuropharmacol 1972;11:637–644.
66 Brezenoff HE, Wirecki TS: The pharmacological specificity of muscarinic receptors in the post hypothalamus. Life Sci 1970;9:99–109.
67 Buccafusco JJ, Brezenoff HE: Pharmacologic study of a cholinergic mechanism within the rat posterior hypothalamic nucleus which mediates a hypertensive response. Brain Res 1979;165:295–310.
68 Philippu A: Involvement of hypothalamic catecholamines in the regulation of the blood pressure. Trends in Pharmacol Sci 1980;1:376–378.
69 Takeda K, Bunag RD: Sympathetic hyperactivity during hypothalamic stimulation in spontaneously hypertensive rats. J Clin Invest 1978;62:642–648.
70 Nelson DO, Boulant JA: Altered CNS neuroanatomical organization of spontaneously hypertensive (SHR) rats. Brain Res 1981;226:119–130.

Prof. H.E. de Wardener, MD, Research Laboratories, Charing Cross Hospital, Fulham Palace Road, London W6 8RF (UK)

Berlyne GM (ed): The Kidney Today. Selected Topics in Renal Science.
Contrib Nephrol. Basel, Karger, 1992, vol 100, pp 188–206

Malnutrition in Continuous Ambulatory Peritoneal Dialysis and Use of Intraperitoneal Amino Acids

Nicholas V. Dombros[a], *George E. Digenis*[b], *Dimitrios G. Oreopoulos*[c]

[a] Peritoneal Dialysis Unit, First Department of Internal Medicine, AHEPA General Hospital and University of Thessaloniki, Greece;
[b] Department of Therapeutics, Alexandra Hospital and University of Athens, Greece, and
[c] Toronto Hospital and Department of Medicine, Toronto, Ont., Canada

For many years chronic protein and energy malnutrition and muscle wasting has been recognized in both nondialyzed [1] and dialyzed uremic patients [1, 2]. The progress of dialysis was advanced considerably by the introduction of continuous ambulatory peritoneal dialysis (CAPD), which quickly established itself as a successful alternative to hemodialysis. As of early 1991, 55,000 patients were being maintained on chronic peritoneal dialysis world-wide [3]. The results of CAPD are improving and large numbers of patients are maintained on this therapy for long periods (5–10 years). As a result, the long-term effects of CAPD have started to emerge and the profession is paying close attention to such complications of CAPD such as peritonitis, exit-site infection, loss of ultrafiltration, renal osteodystrophy and sclerosing peritonitis. Parallel to dialysis, the field of clinical nutrition has advanced and, as a result of careful attention, is being paid to the long-term nutrition of uremic patients [4]. Chronic calorie and protein undernutrition provides the background to many aspects of the uremic syndrome including susceptibility to infection, impaired wound healing, anemia, decreased strength, cardiovascular dysfunction, sexual impairment, progressive neuropathy, fatigue, generalized weakness, failure of rehabilitation and poor quality of life [5].

In patients undergoing CAPD, some [7] but not all workers have shown a correlation between malnutrition [6] or hypoalbuminemia and increased rate of peritonitis and prolonged hospital stay. Thus, Fenton et al. [8] showed that though the mortality among the malnourished CAPD patients was increased, they did not demonstrate that nutritional state was an independent risk factor. Furthermore, they found no association between malnutrition and the subsequent development of peritonitis.

Nutritional assessment is essential for the detection of protein-energy malnutrition for the evaluation of dietary requirements, for the monitoring of the effects of the treatment and for the development of alternative nutritional therapies. However, we have no single definite test by which to assess nutritional status [9] as is indicated by the plethora of tests now applied for that purpose. Nutritional assessment relies on medical history, physical examination, evaluation of nutrient intake, anthropometric measurements, biochemical monitoring, assessment of lean body mass and percent body fat carried out by such techniques as measurement of bioelectrical impedance [10], computerized axial tomography [11], magnetic resonance imaging [12], ultrasonography [13], protein catabolic rates, nitrogen balance studies, measurement of total body nitrogen (TBN) by neutron activation analysis, etc. Newer measurements and techniques include the measurement of serum fibronectin, which is said to be a sensitive, reliable and simple index of nitrogen balance [14], of serum insulin-like growth factors (IGF-1) [15] and a combination of urea and creatinine kinetics, that is, Kt/V urea and Kt/V creatinine [16]. Some of these tests are not widely available because they are complex or require expensive equipment.

Recently, Fenton et al. [8] and Young et al. [17] have studied a clinical assessment tool, the subjective global assessment, which is based on a history and physical examination for assessment of the nutritional status in CAPD patients (see below).

Evidence of Malnutrition among CAPD Patients

In 1981, Williams et al. [18] reported that, during their first year, 13 patients on CAPD showed a decline in daily dietary protein and calorie intake from 1.46 to 1.06 g/kg ($p < 0.01$) and from 36.5 to 28.9 kcal/kg ($p < 0.01$), respectively. Their TBN fell significantly from 1,768 to 1,498 g ($p <$

0.01). Muscle mass and total body fat remained unchanged while average body weight increased from 61.5 to 66.7 kg ($p < 0.001$).

In a 1983 study, Heide et al. [19], in a nutritional follow-up of 20 CAPD patients for 18 and 24 months, again showed a decrease in TBN, increase in body weight and a decrease in oral protein intake over time.

In 1985, Schilling et al. [20] reported a long-term nutritional study of 26 patients, who had been on CAPD for 2 years; 13 of them had been on CAPD for 3 years. Protein intake was stable at about 1.0 g/kg/day, serum albumin 3.5 g/dl and total protein around 6.3 g/dl. In this group of patients, TBN, after a decrease in the first year to 85% of the predicted normal for women and to 75% for men, remained stable thereafter. These results indicated that, on a daily protein intake of about 1.0 g/kg, most CAPD patients will lose body nitrogen.

To determine the prevalence of malnutrition during CAPD, Fenton et al. [8] assessed the nutritional status of 118 patients on three occasions over a 6-month period, using 'subjective global assessment'. Although this assessment is subjective, it seems to yield results that are as reliable as some of the above-mentioned objective assessments. They found that 10 (42%) of the 24 patients assessed during their first 3 months on CAPD were malnourished. Of the 94 patients, who had been on CAPD for more than 3 months, 17 (18%) were malnourished. Of those 27 patients ranked as malnourished at their first assessment regardless of the duration of CAPD, 14 (52%) improved, 8 (29.5%) remained unchanged while 5 (18.5%) dropped out of the study. These data indicated that, regardless of the duration of CAPD, patients who become malnourished have a 50% chance to return to normal nutritional state with conventional management. However, approximately 10% of patients remain malnourished throughout.

In another study, Young et al. [17] examined the nutrition of 224 patients from six centers in Europe and North America to assess the prevalence of protein-energy malnutrition. This study was also based on 'subjective nutritional assessment' and used 21 variables derived from history and clinical examination, or anthropometry and biochemistry. These workers found that 18 (8%) patients were severely malnourished, 73 (32.6%) were mildly to moderately malnourished, and the remained 133 (59.4%) showed no evidence of malnutrition.

Salusky et al. [21] assessed the nutrition of 24 clinically stable children who had been on CAPD for 6.6 ± 4.4 months. Their energy intake tended to be more suppressed than their protein intake. Compared to age- and sex-matched normal controls, these CAPD children had a lower height, weight and midarm muscle circumference. Their serum total protein, albumin, transferrin and many amino acids were decreased while serum triglycerides and cholesterol were elevated. These observations suggest that children undergoing CAPD will have shorter stature and poor nutritional status. These abnormalities may reflect low protein and energy intake and dialysate protein losses.

In a study of 35 CAPD patients, Pollock et al. [22] showed a significant nitrogen depletion to 88.2% of normal in men and 87.5% in women on CAPD. TBN correlated significantly with calculated dietary protein intake. Finally, Nelson et al. [23] found that nondiabetic women on CAPD had anthropometric measurements considerably smaller than those of healthy adults.

This review suggests that malnutrition is common among CAPD patients and we believe that more accurate tests and measurements will show that its true incidence is higher.

Causes of Malnutrition in CAPD

The main causes of malnutrition in CAPD patients can be grouped under three headings: (1) inadequate intake; (2) increased losses, and (3) increased utilization as in catabolic illnesses.

Inadequate Nutrient Intake

Calorie malnutrition is unusual among adult CAPD patients because of the amount of glucose (80–361 g/day) absorbed from the dialysis solution [24]. In contrast, the decreased protein intake of uremia may be further aggravated by CAPD, because of associated factors such as uremic toxicity due to inadequate dialysis, constipation, nausea, vomiting, emotional depression, or loss of appetite, particularly for proteins associated with abdominal distension due to the increase in intraabdominal pressure. Indeed in some patients oral intake may increase if they receive their meals after draining and before infusing the new dialysis solution. On the other hand, glucose has an anorectic effect. In a study of nonuremic rabbits [25], food consumption decreased dramatically, immediately after the initiation of CAPD. Such appetite suppression was more pronounced with hypertonic than with isotonic solutions.

However, in CAPD patients, anorexia is not total, because these patients primarily avoid such protein sources as meat, poultry and fish. Thus while the overall spontaneous energy intake seems to be sufficient, the spontaneous protein intake is low [26]. In CAPD patients the average dietary protein intake is about 1.0 g/kg/day and the average total energy intake (including absorption of dialysate glucose) about 30 kcal/kg/day [19, 27]. Although some patients may adapt and achieve nitrogen balance with this rather low intake, many CAPD patients will be in a negative balance.

Protein and Amino Acid Losses in the Dialysate Effluent

Loss of proteins through the dialysate in patients undergoing CAPD is one of the method's principal drawbacks. The loss of a particular protein may approach or exceed the amount of that protein which healthy persons synthesize: in dialyzed patients, protein synthesis is limited by such factors as reduced availability of amino acids and enzyme cofactors, that is, vitamins and minerals, which themselves may be depleted during peritoneal dialysis.

Protein removal via peritoneal dialysis involves both normal and abnormal proteins, in the latter case it is therapeutic [28]. Two such abnormal proteins removed by peritoneal dialysis are IgG paraprotein in primary amyloidosis [29], and IgG paraprotein [30] and Bence-Jones IgA protein [31] in multiple myeloma.

The variety and amount of protein found in peritoneal effluent reminds us of the proteins lost in the urine of patients with nephrotic syndrome. However, apart from the hyperlipidemia, which is common in both groups, CAPD patients rarely have the other complications of the nephrotic patient such as hypercoagulability and thrombosis.

Many metabolic disturbances have not been identified or studied adequately in the CAPD patient but, in the long run, the success of peritoneal dialysis will depend chiefly on proper nutrition [32].

On average, the daily loss of protein through the dialysate is in the order of 6–9 g; however, there is a large interindividual variation from 3 to 20 g/day [33, 34]. During or after peritonitis, this quantity may be increased by 50–100%. Despite these larger losses, serum total proteins are maintained at low normal levels (~6.5 g/dl) even after prolonged periods (more than 5 years) on CAPD [35]. The major fraction of total dialysate protein is albumin (48–65%) but IgG accounts for the 15% [33, 34].

Albumin losses average 5–6 g/day, while the serum albumin levels usually are below normal at about 3.5 g/dl. In CAPD patients, Kaysen and Schoenfeld [36] found a decreased catabolism and increased synthesis of albumin. Thus, these patients seem to maintain albumin homeostasis which produces normal plasma albumin mass, total albumin mass and normal albumin distribution. This may explain why only few patients develop severe hypoalbuminemia even during long-term treatment with CAPD [35].

The amount of protein lost in CAPD depends on the type of dialysis, the dwell time, the size of protein, its serum concentration, the patient's clinical status, the presence of peritonitis, etc. [37]. Low-molecular-weight proteins will be depleted more rapidly, possibly too rapidly, to permit the patient's biosynthetic system to compensate. These losses may add to the stresses already affecting the dialyzed patient [32].

Amino Acid Losses

Daily total amino acid losses average 2.5 g [38] and range from 1.2 to 3.4 g/day. Essential amino acids account for 29% of total amino acid loss into the dialysate [39]. These losses average only 5.3% of the total dialysate nitrogen losses. Although the loss of certain amino acids may aggravate deficiencies of specific amino acids in CAPD patients, the total daily amino acid losses which are not large can be replaced easily by oral intake.

In CAPD patients, plasma amino acid profiles show decreased levels of mainly essential amino acids – valine, leucine, lysine, tryptophan and histidine, but nonessential amino acids – tyrosine, serine and taurine also are decreased. On the other hand, several nonessential amino acids such as citrulline, proline, glutamate, half cystine may be increased [35]. The non-fasting plasma amino acid profile tends to reflect recent intake of protein and other nutrients. On the other hand, abnormalities in fasting levels reflect the metabolic and nutritional derangements of uremia rather than depletion due to dialysate amino acid losses [34].

Skeletal muscle contains the largest pool of intracellular amino acids and, in the long term, muscle amino acid levels may provide a better index of nutritional state than the plasma amino acid level. Lindholm et al. [40] found that, of the essential amino acids, only tyrosine had a significantly low concentration in the muscle intracellular compartment, while lysine was significantly increased. However, Graziani et al. [41] found low muscle tyrosine, leucine, and phenylalanine concentrations.

Of the nonessential amino acids, tyrosine and serine [41] and taurine [40] were found to be decreased. With regards to taurine, its low muscle intracellular concentration in CAPD patients [40] indicates impaired synthesis because the serum contained adequate concentrations of its precursors, the sulfuric amino acids methionine and cysteine. These authors postulate that low muscle taurine concentration might contribute to such symptoms as fatigue, arrhythmias, cardiomyopathy and growth retardation in children. In contrast with the above finding, Metcoff et al. [42] found that, in their patients, CAPD is associated with a normal intracellular amino acid pool and consequently with normal protein synthesis.

Lipid Abnormalities in CAPD

Several studies have demonstrated the hyperlipidemic effect of CAPD, which becomes apparent within the first few months. Hypertriglyceridemia [43] and hypo-HDL-cholesterolemia [43, 44] have been reported since the early days of CAPD. The continuous absorption of dialysate glucose and the resulting elevated insulin levels provoke an increased hepatic synthesis of VLDL and consequently an increase in circulating triglycerides.

Removal of circulating triglycerides is limited, in part at least because of the loss, through the dialysate, of the potent, protein-bound lipoprotein-lipase activator, heparin sulfate [45]. Other workers [46, 47] have reported that hypertriglyceridemia is associated with a low serum-carnitine level. Between 60 and 80% [48, 49] of patients undergoing long-term CAPD have hypertriglyceridemia. About 15–30% of patients on CAPD develop hypercholesterolemia de novo during their first year [48, 50], due to increased levels of both VLDL and LDL cholesterol [51].

In summary, about 8% CAPD patients have severe protein malnutrition while 23–38% have mild malnutrition [17]. In addition, CAPD patients develop lipid abnormalities and some become obese. In CAPD patients the main causes of malnutrition are inadequate protein intake, increased protein and amino acid losses through the peritoneum, and superimposed catabolic illnesses (e.g. peritonitis). Other causes include inadequate dialysis, endocrine abnormalities, psychosocial problems, the use of steroids, etc.

Prevention and Treatment of Malnutrition in CAPD Patients

To maintain good nutrition in CAPD patients one must supply adequate nutrients for daily required intake and compensate for peritoneal losses. In the presence of superimposed catabolic situations, such as peritonitis, other infections and the administration of steroids, one should supplement the nutrient supply. The management of the already malnourished CAPD patient includes dietetic intervention, oral protein supplements, partial or total parenteral nutrition and intraperitoneal infusion of amino acids.

Following elegant metabolic balance studies, Kopple and Blumenkrantz [52, 53] made a proposal regarding the daily dietary requirements for CAPD patients (table 1). Adherence to these guidelines should enable

Table 1. Daily dietary allowances for CAPD patients: tentative proposal[a]

Protein	1.2–1.3 g/kg[b]
Energy (oral and dialysate)	35–42 kcal/kg[b]
Carbohydrate (oral)	35% of ingested calories
Fat	remainder of ingested nonprotein calories
Polyunsaturated:saturated fatty acid ratio	1.5:1.0
Total fiber	20–25 g
Calcium	1,000–1,400 mg
Phosphorus	800–1,200 mg[c]
Magnesium	200–300 mg
Potassium	70–80 mEq
Sodium and water	as tolerated by water balance and serum sodium
Supplemental vitamins[d]	
Ascorbic acid	70–100 mg
Pyridoxine hydrochloride (B_6)	5–10 mg
Folic acid	1 mg
Other water-soluble vitamins	recommended daily allowances for normal persons
Vitamins A, E, and K	none
Vitamin D	varies

[a] Adapted from Kopple and Blumenkrantz [53].
[b] Per kilogram of mean weight of normal persons of the same age, height, and sex.
[c] Phosphate binders are often needed.
[d] Refers to vitamin intake in addition to vitamins present in foods.

most CAPD patients to maintain satisfactory nutritional status and reduce their overall mortality on dialysis.

A long-term longitudinal study has demonstrated that with good nutritional support involving close surveillance of the patients' diet and close teamwork with a trained dietitian, body weight, body mass index, midarm circumference, and midarm muscle circumference increased in the first year and this was maintained for at least 4 years thereafter [54]. However, other investigators [18–20] have not been able to achieve these successes.

It is important to identify those CAPD patients who are at high risk for malnutrition and to monitor them carefully so that appropriate action may be taken. Such patients are the elderly, those with poor dietary protein intake, those with a metabolic, primary, renal disease such as diabetes mellitus and amyloidosis, those with severe or prolonged peritonitis or other intercurrent catabolic illness and finally those undergoing surgery.

As the number of elderly patients on CAPD increases, we must increase our vigilance. The recommended daily dietary allowances should be adjusted for the elderly because their normal protein intake often falls below these recommendations. In one study in elderly nonuremic patients the authors found that the average daily protein intake for females over 60 years of age was 55 g and for males over 65 years was 67 g [55].

Since TBN and other sophisticated measurements, or muscle biopsy are not practical on any large scale, for routine use we must depend on careful anthropometric measurements, dietary histories and careful clinical assessment bearing in mind their shortcomings. In everyday practice, the best guide to dietary compliance may be serial measurements of serum proteins, albumin, transferrin, urea, anthropometric measurements and serial calculations of daily protein intake from the urea generation rate.

Use of Intraperitoneal Amino Acids

All commercial peritoneal dialysis solutions contain glucose as an osmotic agent to obtain ultrafiltration, even though glucose is less than ideal as an osmotic agent. Its continuous absorption from the dialysate (500–800 kcal/day or 30–40% of total energy intake) and the resulting elevated plasma glucose levels produce hyperinsulinemia, hypertriglyceridemia, hypercholesterolemia, hypo-HDL-cholesterolemia and, in some patients, obesity. Loss of ultrafiltration capacity of the peritoneal membrane is due – at least in part – to extensive use of hyperosmotic solutions with a high glucose content. In relation to malnutrition in CAPD, the most serious disadvantage of dialysis with glucose-containing dialysis solutions is anorexia and the loss through the peritoneum of various nutrients such as proteins, amino acids, trace elements and vitamins, leading to hypoalbuminemia, hypoproteinemia, reduced serum transferrin and decreased lean body mass, etc.

These side effects of glucose and the relative frequency of protein malnutrition in CAPD patients inspired a search for alternatives to glucose as an osmotic agent in peritoneal dialysis. An amino acid solution was viewed as an ideal osmotic alternative for CAPD because such a solution would provide nutrients, could treat, or at least prevent, protein depletion and at the same time avoid the glucose-induced side effects.

Gjessin [56] was the first to add amino acids to the dialysate, in an effort to compensate for losses during peritoneal dialysis. However, investigators at Toronto Western Hospital, who were the first to administer amino acids in CAPD, suggested that these could be used as the sole osmotic agent while testing this hypothesis in anephric rabbits on CAPD. Thus, Oreopoulos et al. [43] found that an amino acid solution had an osmotic effect similar to that of glucose of double concentration (i.e. 1% amino acid similar to that of 2% glucose). Furthermore, the urea-nitrogen and creatinine clearances also were similar to those achieved by a glucose dialysate.

Following this first observation in animals, this same group [57] studied 6 nondiabetic adult CAPD patients who were infused intraperitoneally with 2 liters of a dialysis solution containing 2% of a mixture of essential and nonessential amino acids. The amino acid solution was prepared from Travasol – a non-electrolyte-containing amino acid mixture designed for the intravenous provision of nutritional support. At the end of the 6-hour infusion, 80–90% of the amino acids had been absorbed and there were no evident side effects. Subsequently other studies [58–61] in both adults and children done to assess the therapeutic potential and immediate metabolic consequences of such solutions produced similar results and conclusions. These studies showed that amino acid dialysate provoked no short-term metabolic complications and had an osmotic activity slightly higher to glucose.

These satisfactory results of single cycle use demanded verification with observations over a longer period. In 1983, Oren et al. [62] from Toronto Western Hospital studied 6 stable CAPD patients, who had no evidence of malnutrition, using a 1% amino acid solution alternating with a glucose-containing dialysate (2 liters of each twice a day) over 4 weeks. There was a significant increase in BUN from 64 to 102 mg/dl, in TBN from 1,333 to 1,380 g, serum transferrin from 175 to 222 mg/dl and an increase in anion gap from 15.1 to 17.3. During the study, total body potassium, serum albumin, triglycerides, insulin, glucagon, electrolytes, anthropometric measurements and daily ingestion of calories and protein did not change significantly. The 1% amino acid solution's ultrafiltration effect was intermediate between that obtained with 1.5 and 2.5% glucose dialysates. As in the short-term study by Williams et al. [57], the amino acid solution was prepared from Travasol. The fasting plasma levels of histidine, tryptophan and glycine increased significantly while that of alanine decreased. Finally, the levels of several essential amino acids remained below normal. Recently, Bernard et al. [63] administered one 1.1% amino acid solution intraperitoneally daily over a period of 16–35 days to 20 malnourished stable CAPD patients with steadily low protein (0.8 g/kg/day) and energy (25–30 kcal/kg/day) intake. In the first 7 patients, mean nitrogen balance increased from $+8 \pm 20$ to $+30 \pm 10$ mg N/kg/day ($p < 0.01$). Net protein anabolism, estimated using a single-dose, ^{15}N-glycine technique, rose from 0.18 ± 0.06 to 0.27 ± 0.14 g protein/kg/day ($p = 0.1$), serum transferrin and midarm muscle circumference increased significantly suggesting, according to the authors, utilization of amino acid nitrogen for protein synthesis. There was a decrease in mean total CO_2 and a significant increase in BUN from 52.9 ± 14.5 to 86.7 ± 16.5 mg/dl ($p < 0.05$). These results suggest that the nutrition of malnourished CAPD patients with a low protein intake could be improved by the administration of amino acids intraperitoneally. Contrary to the previous findings in February 1991, Fenton et al. [64] described a study of 8 CAPD patients who had low protein catabolic rate (below 0.8 g/kg/day) and received a supplement of 2 liters of 1.25 or 2.5% amino acid solution once daily. All patients developed severe gastrointestinal symptoms requiring either withdrawal or reduction of the amino acid concentration during the first week ($n = 4$) or between days 16 and 45 ($n = 4$) of the study. Blood urea doubled while the protein catabolic rate rose from 0.68 to 1.07 g/kg/day in 1 month. Only 3 of the 8 patients reached target PCR. These authors attributed the gastrointestinal symptoms to the rapid increase in protein intake and the development of uremia.

Three papers [65–67] have reported experience with administration of amino acid solutions to CAPD patients for an intermediate period (8–12 weeks). In an interesting study, Arfeen et al. [65] measured circulating intermediary metabolites, hormones and plasma amino acids in 7 nondiabetic CAPD patients, before and after 8 weeks during which 2 of the 4 dextrose exchanges were replaced by a 1% amino acid solution. Mean 24-hour concentrations of plasma total and essential amino acids were significantly higher after the amino acid dialysate. Mean 24-hour concentrations of the branched-chain amino acids leucine, isoleucine and valine also were significantly higher. Also serum albumin increased. After the amino acid supplement, the 24-hour profiles and mean 24-hour concentrations of blood glucose, serum insulin and triglycerides, plasma nonesterified fatty acids, plasma 3-hydroxybutyrate and plasma alanine were unchanged. Plasma bicarbonate decreased (from 21 to 16 m*M*/l). In agreement with other studies, these authors concluded that the use of a 1% amino acid solution over an 8-week period in CAPD patients improves the plasma amino acid profile but produces a metabolic acidosis. The other endocrine and metabolic abnormalities of uremia remain unchanged. No anthropometric measurements were performed during this study.

In a 12-week study, Pederson et al. [66] alternated amino acid solutions with glucose solutions (two of each per day) in 6 stable CAPD patients with adequate protein intake (1.2–1.5 g/kg/day) and no evidence of malnutrition. During the study, plasma valine, isoleucine, leucine, and serine concentrations increased while methionine decreased. These authors suggested that the increase in plasma three-branched-chain amino acids might stimulate protein synthesis but they could not demonstrate that this happened because these patients had no anthropometric assessments. Unchanged were serum concentrations of cholesterol, HDL, LDL, glucose and glucosylated hemoglobin, albumin, transferrin and phosphate. Serum urea increased slightly, while serum triglycerides increased significantly. Thus the amino acid solution, which was well tolerated, induced some beneficial effects but only in plasma amino acid concentrations.

In another 12-week study, Young et al. [67] substituted the daily morning exchange with a commercial 1% amino acid solution (Baxter '151') in 8 CAPD patients with plasma albumin less than 3.5 g/dl. The fasting plasma amino acid concentrations remained constant. Plasma transferrin increased significantly. Cholesterol and apolipoprotein 8 decreased. Albumin, apolipoprotein A, IgG, IgA, or prealbumin did not change significantly. Statistically significant increases in urea and decreases in bicarbonate had no discernible clinical correlations. The protein and daily energy intakes were adequate and did not change nor did body weight, fat or arm muscle circumference. In general, the amino acid solution was well tolerated and produced a modest nutritional benefit and some reduction in hyperlipidemia. Thus in these 8- to 12-week studies, amino acid solutions were well tolerated but had no measurable or significant effect on the nutrition of CAPD patients.

Based on previous experience [25, 43, 57, 62], the Toronto Western Hospital group proceeded to long-term studies. In 1985, Schilling et al. [68], studied 3 CAPD patients infused with either one 2% or two 1% amino acid solutions intraperitoneally daily for 5–6 months. These Travasol-based amino acid solutions were administered usually in the morning. By the end of the study, 1 patient, who before the study was malnourished and had a low protein intake (0.4 g/kg/day), low serum albumin (2.5 g/dl) and low TBN (nitrogen index of 0.56, normal 0.80–1.20), showed a positive nitrogen balance and an increase in his serum albumin (3.0 g/dl) and TBN (nitrogen index of 0.64). The other 2 patients, who before the study had normal nitrogen indices and reasonably satisfactory daily protein intake (0.9 and 0.95 g/kg), demonstrated suppression of appetite while the TBN remained unchanged. It is interesting that, in 1 of these 2 patients, the calculated nitrogen balance became negative at the end of the 6 months. In these 3 patients, cholesterol plasma levels remained unchanged while in 2 patients triglycerides were not altered, but in 1 patient they rose almost threefold.

In another long-term study from the same group, Dombros et al. [69] studied the effect of an intraperitoneal Travasol-based amino acid solution given to 5 CAPD patients for 6 months. All subjects had a low oral protein intake (<0.8 g/kg/day) and/or a low serum albumin (<3.5 g/dl). They infused 2 liters of 1% amino acid solution overnight (dwell time 8 h) in an effort to avoid its anorexic effect demonstrated after daytime use of amino acids. As expected, by the end of the first month BUN had increased from 22.04 to 28.06 m*M*/l and it remained elevated throughout the study (26.6 m*M*/l at 6 months). The first 2 of these 5 patients received a 2% amino acid solution at the beginning of the study; within 1 month both developed an increase in BUN from 20.4 to 40.8 m*M*/l and symptoms of severe uremia such as anorexia, vomiting, malaise, pruritus, restlessness and abdominal cramps. All of these symptoms subsided when the amino acid concentration was reduced from 2 to 1%. One other patient who started with a 2% amino acid solution developed similar symptoms and refused to continue, while another patient dropped out of the study because of abdominal cramps, restlessness, and pruritus experienced during the first infusion of the 1% solution. On follow-up other parameters such as body weight, skinfold thickness (of biceps, triceps, subscapular, suprailiac and midarm circumference), oral protein intake, hematocrit, protein, albumin, transferrin, cholesterol triglycerides and calculated protein balance remained unchanged. Fasting plasma amino acid levels also were unchanged. The average total (oral + amino acid) protein intake remained unchanged – 1.17 ± 0.34 g/kg/day (the first month) and 1.19 ± 0.31 g/kg/day (the sixth month). Oral caloric intake alone – 19.62 ± 3.87 kcal/kg/day (at the beginning) and 22.4 ± 7.00 (at the end of the study) remained lower than that recommended – $>$30 kcal/kg/day (table 1). The average TBN declined from 1.746 to 1.554 kg but this decline did not reach statistical significance ($p > 0.05$). In 3 of these 5 patients, TBN was measured again for a third time 6 months after the end of the study, while on their ordinary diet and not receiving amino acid solutions; it remained unchanged (from 1.460 to 1.480) in 1 patient and showed a slight further decrease in 2 others from 1.930 and 1.450 at the end of the study to 1.760 and 1.290 kg, respectively, 6 months later.

These patients may not have been able to anabolize the absorbed amino acids, as indicated by the unchanged serum albumin, transferrin and TBN because of their low caloric intake. These findings agree with those of Young et al. [67] who in a 12-week study observed only modest nutritional benefits.

However, these two studies contradict a study from Italy by Bruno et al. [70] that used a 1% amino acid solution (Baxter '151') – the same one used by Young et al. [67] in 6 well-nourished patients for 6 months. This prospective cross-over study infused 2 liters of the amino acid solution just before the noon meal. Calculated nitrogen balance, which had been negative (–1.3 g/day), became positive (+3.1 g/day) at the end of the 6-month period. Plasma cholesterol, triglycerides, HDL cholesterol decreased significantly. Patients who were already overweight gained more weight, both in fat and lean mass. Plasma urea levels increased and pH and bicarbonate decreased significantly ($p < 0.05$). Plasma protein, albumin and transferrin did not change. The concentrations of all essential amino acids, except isoleucine, increased towards normal range. Total plasma essential and nonessential amino acids also increased. All these parameters returned towards basal values when the amino acids were discontinued. Differences between the results obtained in this and in the studies of Young et al. [67] and Dombros et al. [69] may be explained in part by the following factors: (1) The Italian patients were overweight and their weight increased further while receiving infusions of amino acids; mean weight was 73.7 kg; in the other two studies the mean weight was 65.9 and 68.5 kg, respectively. (b) The amino acid solution was administered at different times – morning in Young's, before lunch in Bruno's and overnight in Dombros' study. It is possible that before the intraperitoneally administered and absorbed amino acids can be used in protein synthesis, the patient needs the caloric load provided by lunch. (c) There were substantial differences in the composition of the amino acid solutions; that of Dombros et al. [69] is based on Travasol, a solution formulated for total parenteral nutrition in patients with normal renal function [71]. The new solution, Baxter '151', used by Bruno et al. [70] and Young et al. [67], has a higher percentage of essential amino acids than the Travasol-based one (60 vs. 39%) and has a greater proportion of branched-chain amino acids (48 vs. 40%). Also it is higher in arginine (9.7 vs. 4.9%) but is lower in alanine plus glycine (13.2 vs. 41.6%). Although this has not been reported, the solution Baxter '151' contains tryptophan, an essential amino acid, in a concentration of 2.5% (wt/wt) of the total amino acids. However, the solution used in the Italian study was the same as that used by Young et al. [67] who did not observe the same beneficial effects.

In conclusion, evidence to date indicates that amino acid (1%) containing solutions are well tolerated, provide adequate ultrafiltration and dialysis and can lead to a positive nitrogen balance when infused for short periods (3–4 weeks). Their beneficial effect over long periods (3–6 months) remains to be established; further study is needed to identify the conditions of administration and the kinds of patients who will benefit from the chronic administration of these solutions.

References

1 Kopple JD: Dietary requirements; in Massry SG, Sellers AL (eds): Clinical Aspects of Uremia and Dialysis. Springfield, Thomas, 1976, p 453.

2 Marckmann P: Nutritional status of patients on hemodialysis and peritoneal dialysis. Clin Nephrol 1988;29:75–78.

3 Nolph KD: Peritoneal dialysis in 1991; in La Greca G, et al (eds): Peritoneal Dialysis. Milano, Wichtig Editore, 1991, pp 3–7.

4 Blagg CR: Importance of nutrition in dialysis patients. Am J Kidney Dis 1991;17: 458–461.

5 Blumenkrantz MJ, Salusky IB, Schmidt RW: Managing the nutritional concerns of the patient undergoing peritoneal dialysis; in Nolph KD (ed): Peritoneal Dialysis, ed 2. Boston, Nijhoff, 1985, pp 345–401.

6 Lee HY, Kim YK, Kang SW, et al: Influence of nutritional status on CAPD peritonitis. Yonsei Med J 1990;31:65–70.

7 Young GA, Young IB, Young SM, et al: Nutrition and delayed hypersensitivity during continuous ambulatory peritoneal dialysis in relation to peritonitis. Nephron 1986;43:177–186.

8 Fenton SSA, Johnston N, Delmore T, et al: Nutritional assessment of continuous ambulatory peritoneal dialysis patients. Trans Am Artif Intern Organs 1987;33:650–653.

9 Baker JP, Detsky AS, Wesson DE, et al: Nutritional assessment. A comparison of clinical judgement and objective measurements. N Engl J Med 1982;306:969–972.

10 Presta E, Wang J, Harrison GG, et al: Measurements of total body electrical conductivity: A new method for estimation of body composition. Am J Clin Nutr 1983; 37:735–739.

11 Heymsfield SB, Olafson RP, Kutner MH, et al: A radiographic method of quantifying protein-caloric undernutrition. Am J Clin Nutr 1979;32:693–702.

12 Lewis DS, Rollwitz WL, Bertrant HA, et al: Use of NMR for measurement of total body water and estimation of body fat. J Appl Physiol 1986;60:836–840.

13 Heymsfield SB, Rolandelli R, Casper K, et al: Application of electromagnetic and sound waves in nutritional assessment. J Parenter Enteral Nutr 1987;11:S64.

14 Hou FF, Zhang X, Wang AL, et al: Fibronectin in patients with chronic renal failure undergoing dialysis. Nephron 1990;55:45–48.

15 Jacob V, Le Carpentier JE, Salzano S, et al: IGF-1, a marker of undernutrition in hemodialysis patients. Am J Clin Nutr 1990;52:39–44.

16 Keshaviah PR, Nolph KD: A kinetic approach to nutritional assessment. Periton Dial Int 1991;11(suppl 1):abstr 132.

17 Young GA, Kopple JD, Lindholm B, et al: Nutritional assessment of CAPD patients: an international study. Am J Kidney Dis, in press.

18 Williams P, Kay R, Harrison J, et al: Nutritional and anthropometric assessment of patients on CAPD over one year: Contrasting changes in total body nitrogen and potassium. Periton Dial Bull 1981;1:82–87.

19 Heide B, Pierratos A, Khanna R, et al: Nutritional status of patients undergoing continuous ambulatory peritoneal dialysis. Periton Dial Bull 1983;3:138–141.

20 Schilling H, Wu G, Petit J, et al: Nutritional status of patients on long-term CAPD. Periton Dial Bull 1985;5:12–18.

21 Salusky IB, Fine RN, Nelson D, et al: Nutritional status of childern undergoing continuous ambulatory peritoneal dialysis. Am J Clin Nutr 1983;38:599–611.

22 Pollock CA, Allen BJ, Warden RA, et al: Total body nitrogen by neutron activation in maintenance dialysis. Am J Kidney Dis 1990;16:38–45.

23 Nelson EE, Hong CD, Pesce AL, et al: Anthropometric norms in dialysis population. Am J Kidney Dis 1990;16:32–37.

24 Grodstein GP, Blumenkrantz MJ, Kopple JD, et al: Glucose absorption during ambulatory peritoneal dialysis. Kidney Int 1981;19:564–567.

25 Oreopoulos DG, Marliss EB, Anderson GH, et al: Nutritional aspects of CAPD and the potential use of amino acid containing dialysis solutions. Periton Dial Bull 1983; 3:S10–S12.

26 Bonzel KE, Bonatz K, Senghaos C, et al: Nutritional aspects in children on CAPD;

in Maher JF, et al (eds): Frontiers in Peritoneal Dialysis. New York, Field, Rich & Assoc Inc, 1986, pp 421–428.
27 Lindholm B, Bergstrom J: Nutritional management of patients undergoing peritoneal dialysis; in Nolph KD (ed): Peritoneal Dialysis, ed 3. Dordrecht, Kluwer Academic Publishers, 1988, pp 230–260.
28 Giacchino F, Quarello F, Pellerey M, Piccoli G: Continuous ambulatory peritoneal dialysis improves immunodeficiency in uremic patients. Nephron 1983;35:209–210.
29 Stone WJ, Latos DL, Lankford PG, Baker AS: Chronic peritoneal dialysis amyloidosis, renal failure and factor X deficiency. South Med J 1978;71:764–767.
30 Yium J, Martinez-Maldonado M, Eknoyan G, Suki WN: Peritoneal dialysis in the treatment of renal failure in multiple myeloma. South Med J 1971;64:1403–1405.
31 Russell JA, Fitzharris BM, Corrhingham R, et al: Plasma exchange vs. peritoneal dialysis for removing Bence-Jones protein. Br Med J 1978;ii:1397.
32 Dulaney Hatch FE: Peritoneal dialysis and loss of proteins: A review. Kidney Int 1984;26:253–262.
33 Blumenkrantz MJ, Gahl GM, Kopple JD, et al: Protein losses during peritoneal dialysis. Kidney Int 1981;19:593–602.
34 Lindholm B, Bergstrom J: Protein and amino acid metabolism in patients undergoing continuous ambulatory peritoneal dialysis. Clin Nephrol 1988;30:S59–S63.
35 Dombros N, Digenis G, Abraham G, et al: Long-term (> 5 years') experience with CAPD. 8th National Conference on CAPD, Kansas City, Feb 10–12, 1988, abstr 53.
36 Kaysen GA, Schoenfeld PY: Albumin homeostasis in patients undergoing continuous ambulatory peritoneal dialysis. Kidney Int 1984;25:107–114.
37 Young GA, Brownjohn AM, Parsons FM: Protein losses in patients receiving continuous ambulatory peritoneal dialysis. Nephron 1987;45:196–201.
38 Dombros N, Oren A, Marliss EB, et al: Plasma amino acid profiles and amino acid losses in patients undergoing CAPD. Periton Dial Bull 1982;2:27–32.
39 Kopple JD, Blumenkrantz MJ, Jones MR, et al: Plasma amino acid levels and amino acid losses during continuous ambulatory peritoneal dialysis. Am J Clin Nutr 1982;36:395–402.
40 Lindholm B, Alverstrand A, Bergstrom J: Muscle free amino acids in patients treated with CAPD; in Maher JF, et al (eds): Frontiers in Peritoneal Dialysis. New York, Field, Rich & Assoc Inc, 1986, pp 400–464.
41 Graziani G, Cantaluppi A, Casati S, et al: Branched chain and aromatic free amino acids in plasma and skeletal muscle of uremic patients undergoing hemodialysis and CAPD. Int J Artif Organs 1984;7:85–88.
42 Metcoff J, Pederson J, Fable J III, et al: Protein synthesis cellular amino acids and energy levels in CAPD patients. Kidney Int 1987;32(suppl 2):S136–S144.
43 Oreopoulos DG, Crassweller P, Katirtzoglou A, et al: Amino acids as an osmotic agent (instead of glucose) in continuous ambulatory peritoneal dialysis; in Legrain M (ed): Continuous Ambulatory Peritoneal Dialysis. Amsterdam, Excerpta Medica, 1979, pp 335–340.
44 Oreopoulos DG, Clayton S, Dombros N, et al: Nineteen months' experience with continuous ambulatory peritoneal dialysis. Proc Eur Dial Transplant Assoc 1979;11:178–183.

45 Felts JM, Staprans I, Zacherle B: Mechanisms of hyperlipidemia in chronic renal failure and in nephrotic syndrome. Proc Ann Contractors Conf. AKCUP NIAMDD, NIH 1978;11:71.
46 Buonchristiani U, DiPaolo N, Carobi C, et al: Carnitine depletion with CAPD; in Gahl GM, et al (eds): Advances in Peritoneal Dialysis. Amsterdam, Excerpta Medica, 1981, pp 441–445.
47 Bazzato G, Lucatello S, Landini S, et al: Intraperitoneal *L*-carnitine administration: Successful therapeutic approach of hypertriglyceridemia in patients on double-bag system CAPD (abstract). 2nd Int Symp on Periton Dial, Berlin, June 1981, pp 16–19.
48 Ramos JM, Heaton A, McGurk JG, et al: Sequential changes in serum lipids and their subfractions in patients receiving continuous ambulatory peritoneal dialysis. Nephron 1983;35:20–23.
49 Nolph KD, Ryan KL, Prowant B, Twardowski Z: A cross-sectional assessment or serum vitamin D and triglyceride concentrations in a CAPD population. Periton Dial Bull 1984;4:232–237.
50 Lindholm B, Norbeck HE: Serum lipids and lipoproteins during continuous ambulatory peritoneal dialysis. Acta Med Scand 1986;220:143–151.
51 Gokal R, Ramos JM, McGurk JG, et al: Hyperlipidaemia in patients on continuous ambulatory peritoneal dialysis; in Gahl GM, et al (eds): Advances in Peritoneal Dialysis. Amsterdam, Excerpta Medica, 1981, pp 430–433.
52 Blumenkrantz MJ, Kopple JD, Moran JK, Coburn JW: Metabolic balance studies and dietary protein requirements in patients undergoing CAPD. Kidney Int 1982; 21:849–861.
53 Kopple JD, Blumenkrantz MJ: Nutritional requirements for patients undergoing continuous ambulatory peritoneal dialysis. Kidney Int 1983;24(suppl 16):S295–S302.
54 Bennett SE, Walls J: Nutritional requirements of CAPD; in Moyal (ed): Dietetics in the 1990s. London, Libby Eurotext, 1989, pp 193–196.
55 Davies L: A study of the nutrition and well-being of elderly people at home. Three score and then? London, Heinemann Medical, 1981, pp 97–106.
56 Gjessin J: Addition of amino acid to peritoneal dialysis fluid. Lancet 1968;ii:812.
57 Williams P, Marliss EB, Anderson GH, et al: Amino acid absorption following intraperitoneal administration in CAPD patients. Periton Dial Bull 1982;2:124–130.
58 Goodship TJ, Lloyd S, McKenzie PW, et al: Short-term studies on the use of amino acids as an osmotic agent in continuous ambulatory peritoneal dialysis. Clin Sci 1987;73:471–478.
59 Lindholm B, Werynski A, Bergstrom J: Peritoneal dialysis with amino acid solutions: Fluid and solute transport kinetics. Artif Organs 1988;12:2–10.
60 Hanning RM, Balfe JW, Zlotkin SH: Effect of amino acid containing dialysis solutions on plasma amino acid profiles in children with chronic renal failure. J Pediatr Gastroenterol Nutr 1987;6:942–947.
61 Hanning RM, Balfe JW, Zlotkin SH: Effectiveness and nutritional consequences of amino acid-based vs. glucose-based dialysis solutions in infants and children receiving CAPD. Am J Clin Nutr 1987;46:22–30.

62 Oren A, Wu G, Anderson GH, et al: Effective use of amino acid dialysate over four weeks in CAPD patients. Periton Dial Bull 1983;3:66–73.
63 Bernard D, Kopple JD, Brunori G, et al: Nutritional benefits of intraperitoneal amino acids in CAPD patients (abstract). 6th Int Congr on Nutrition and Metabolism in Renal Diseases, Harrogate, Aug 1991.
64 Fenton S, Cummings A, Richardson R, et al: Low protein catabolic rate: Can it be corrected with intraperitoneal amino acid supplements? Periton Dial Int 1991;(suppl 1):abstr 79.
65 Arfeen S, Goodship TH, Kirkwood A, Ward MK: The nutritional/metabolic and hormonal effects of 8 weeks of continuous ambulatory peritoneal dialysis with a 1% amino acid solution. Clin Nephrol 1990;33:192–199.
66 Pedersen FB, Dragsholt C, Laier E, et al: Alternate use of amino acid and glucose solutions in CAPD. Periton Dial Bull 1985;5:215–218.
67 Young GA, Dibble JB, Hobson SM, et al: The use of an amino acid-based CAPD fluid over 12 weeks. Nephrol Dial Transplant 1989;4:285–292.
68 Schilling H, Wu G, Petit J, et al: Effects of prolonged CAPD amino acid containing solutions in three patients. Proc 5th Annu CAPD Conf, 1985, pp 49–55.
69 Dombros N, Prutis K, Tong M, et al: Six-month overnight intraperitoneal amino acid infusion in continuous ambulatory peritoneal dialysis patients. No effect on nutritional status. Periton Dial Int 1990;10:79–84.
70 Bruno M, Bagnis C, Marangella M, et al: CAPD with an amino acid dialysis solution: A long-term cross-over study. Kidney Int 1989;35:1189–1194.
71 Anderson GH, Patel DG, Jeejeebhoy KN: Design and evaluation by nitrogen balance and blood aminograms of an amino acid mixture for total parenteral nutrition of adults with gastrointestinal disease. J Clin Invest 1987;53:904–912.

D.G. Oreopoulos, MD, PhD, Peritoneal Dialysis Unit, Division of Nephrology, Toronto Western Hospital, 399 Bathurst Street, Toronto, Ont. M5T 2S8 (Canada)

Berlyne GM (ed): The Kidney Today. Selected Topics in Renal Science.
Contrib Nephrol. Basel, Karger, 1992, vol 100, pp 207–235

Lessons of Peritoneal Ultrastructure: From an Inert Dialyzing Sheet to a Living Membrane

Lazaro Gotloib, Avshalom Shostak[1]

Department of Nephrology and Kornach Laboratory for Experimental Nephrology, Central Emek Hospital, Afula, Israel

'Le second précepte était de diviser chacune des difficultés que j'examinerais, en autant de parcelles qu'il se pourrait, et qu'il serait requis pour les mieux résoudre ...'

Descartes, Discours de la méthode (1637)

In spite of early work by Cunningham [1] and Putnam [2], the peritoneum has classically been considered an inert, semipermeable dialyzing membrane [3, 4]. Passive molecular diffusion through porous cellulose membranes is restricted only by pore and molecular radii [5], the processes as a whole tending towards equilibrium. Living membranes, however, show selective permeability [6]. Death of cells forming a biological barrier will cause loss of this selectivity, making the membrane just an inert obstacle which will allow passive diffusion.

The concept of peritoneal inertness was questioned by several investigators [7–11]. Facing this controversy, we decided to explore the ultrastructure of the peritoneum, hoping that a wider knowledge of its anatomy would eventually lead to a deeper understanding of its physiology. This review will discuss three morphological aspects which, in our opinion, can stimulate further functional research: (1) the net negative charge of meso-

[1] The authors wish to thank the City Council of Upper Nazareth, Israel, for its continuing support.

thelial and microvascular endothelial layers; (2) the presence of fenestrated capillaries in submesothelial interstitium, and (3) absorption of albumin from the peritoneal cavity, as well as transmesothelial and translymphatic endothelial transport of albumin-gold complexes.

Discussion

Net Negative Charge

Glycocalyx

Studies done on perfused frog mesenteric capillaries demonstrated that fixed negative charges within microvascular pathways restricted the transport of medium-sized negative proteins, relative to positively charged solutes of the same size [12]. As for other microvascular beds, the luminal aspect of peritoneal capillary endothelium is covered by glycocalyx (fig. 1, 2 – left inset) [13]. This structure appears as a regular, well-organized polymeric carpet, whose major components are sialoconjugates, proteoglycans and acidic polymers. They provide a fibrous network and a nonthrombogenic surface [14], and impart a net negative charge to the luminal plasmalemma [15]. The fact that endothelial cells, as well as polymorphonuclears [16], erythrocytes [17] (fig. 2 – inset) and platelets [18] also have negative surface charge, creates a situation in which they repel

---

Fig. 1. The endothelial cell (e) of this rat heart blood capillary shows RR-decorated glycocalyx (open arrow). Black star = microvascular lumen; b = subendothelial basement membrane. × 64,000. *Inset:* Rat peritubular fenestrated capillary. Open arrow points to endothelial cells (e) glycocalyx. Black star = microvascular lumen; b = subendothelial basement membrane; i = interstitial space. RR. × 24,500.

Fig. 2. Renal fenestrated peritubular capillary. The rat was perfused 24 h after induction of systemic sepsis. Basement membrane (large arrow) is devoid of PEI-labelled anionic charges. A few tracer particles (small arrow) decorate the luminal aspect of the endothelial cell (e). t = Tubular cell; r = erythrocyte. × 29,970. *Left inset:* Erythrocyte (r) from healthy rat, showing plasmalemmal glycocalyx (arrow). l = Microvascular lumen; e = endothelial cell of heart blood capillary. RR and 0.2% tannic acid. × 37,960. *Right inset:* Renal peritubular capillary from normal rat. Anionic sites (curved arrows) are periodically distributed along both aspects of the basement membrane. Short arrow = endothelial cell; r = erythrocyte; t = tubular cell. PEI 2%. × 24,500.

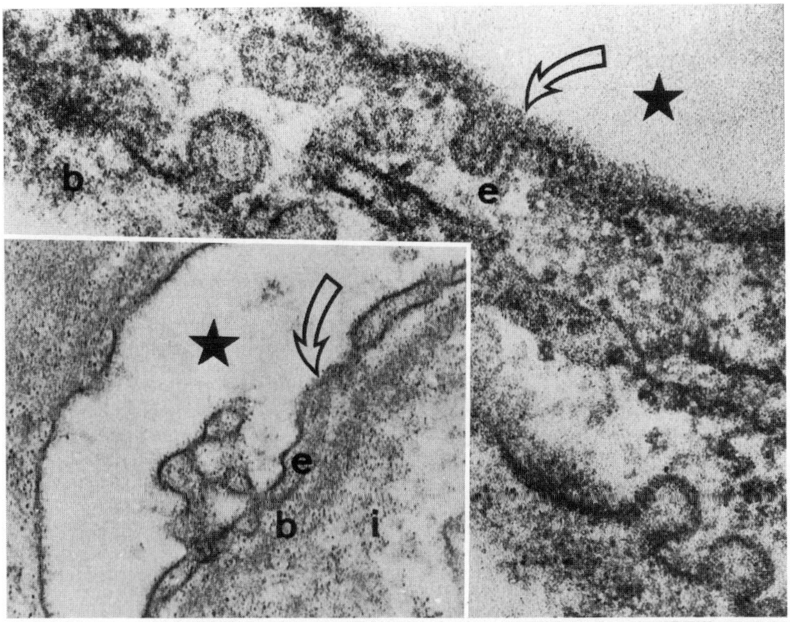

1

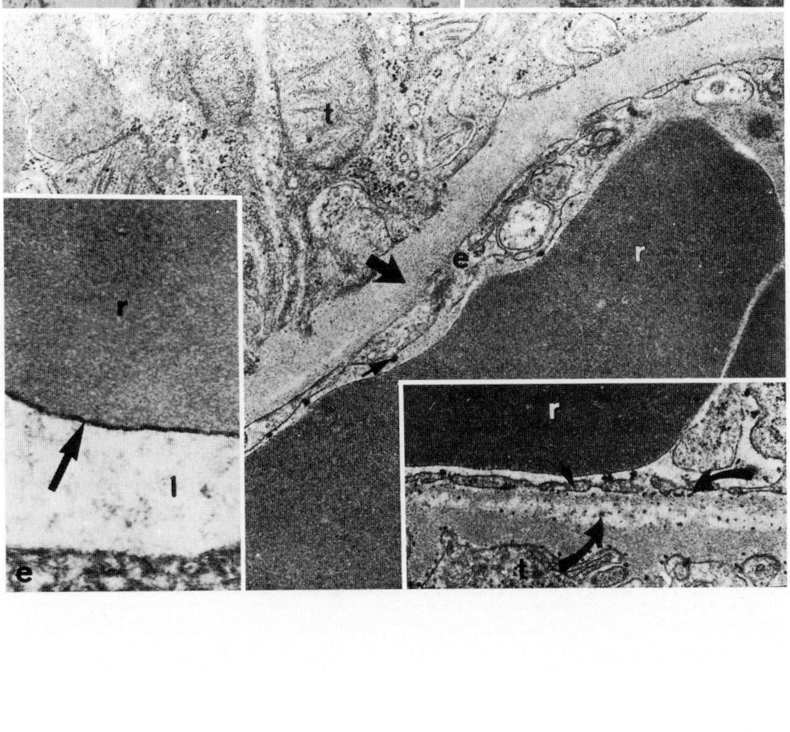

2

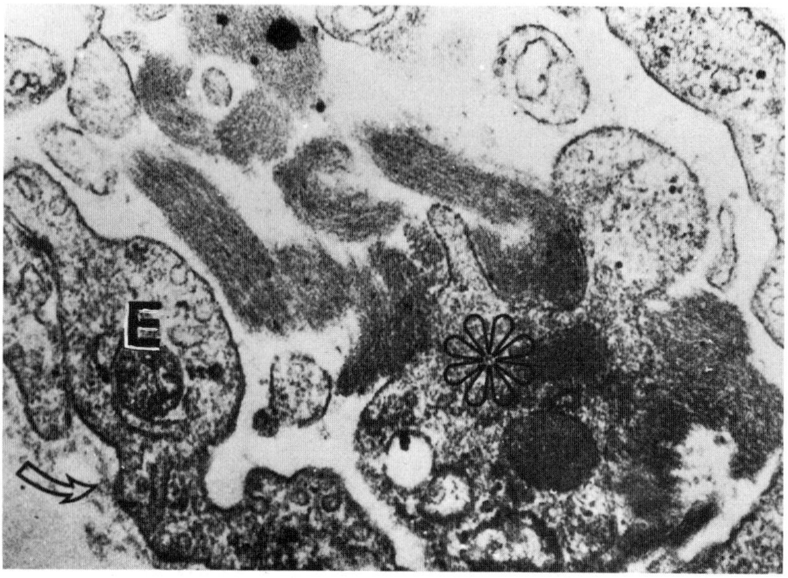

Fig. 3. Myocardium from septic rat (24 h). Microvascular lumen is occupied by a microthrombus (asterisk). Subendothelial basement membrane shows occasional anionic sites (open arrow). E = Endothelial cell. PEI 2%. × 18,100.

each other, insuring blood fluidity at the blood-endothelial interface [19]. Furthermore, the glycocalyx behaves as a size, shape and charge barrier regulating the passage of small and large molecules, especially anionic plasma proteins, across the microvascular wall [20].

Serum albumin, which has net negative charge, can also bind anions as a result of the presence of cationic groups [21]. Consequently, the glycocalyx is visualized as a meshwork which, by trapping circulating proteins, will eventually tighten its mesh, rendering the underlying endothelium less accessible to water and water-soluble molecules [21], as well as to anionic macromolecules [22].

The glycocalyx meshwork has also been observed on the luminal aspect of mesothelial cells of visceral, diaphragmatic and parietal peritoneum [23]; as well as covering the mesothelial cell plasmalemma (fig. 4 – large inset) and that of microvilli (fig. 4 – small inset). Free floating human mesothelial cells, harvested from patients on CAPD, also show ruthenium-red decorated glycocalyx (fig. 5) covering their entire surface. Conversely,

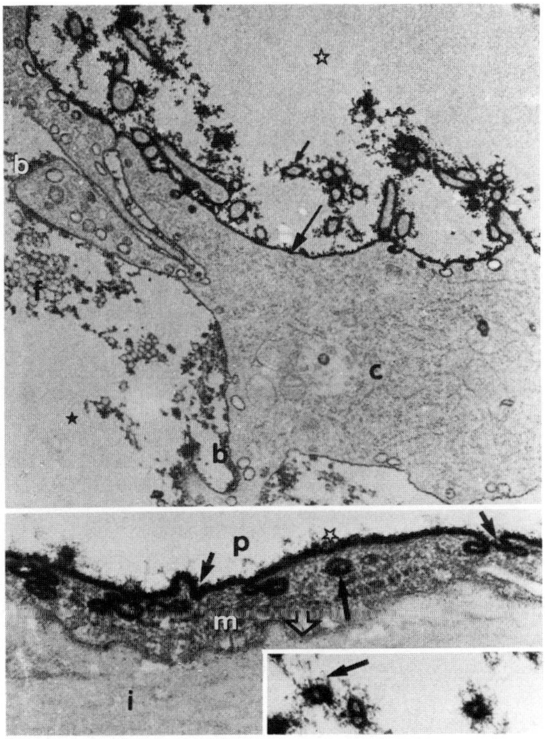

Fig. 4. Mesentery from septic rat, 24 h after induction of *E. coli* peritonitis. Glycocalyx of microvilli (short arrows) and plasmalemma (long arrow) of mesothelial cell (c). Both submesothelial basement membrane (b) and RR-labelled anionic sites are almost completely absent. Note severity of interstitial edema (black star). Open star = abdominal cavity; f = collagen fibers. RR. × 16,230. *Large inset:* Mesentery of normal rat. Glycocalyx (open star) is decorated by RR, and so are the limiting membranes of coated pits (short arrows) and coated vesicles (long arrows). Submesothelial anionic sites (open arrow) appear along the basement membrane. p = Peritoneal space; m = mesothelial cell; i = interstitial space. RR. × 27,400. *Small inset:* Mesentery of normal rat. Cross-sectioned microvilli show RR-decorated glycocalyx (arrow). RR. × 31,350.

the coating of implanted cells is asymmetric, appearing only on the luminal aspect of the cell's surface (fig. 4 – large inset).

The functional implications of mesothelial cells' glycocalyx, with regard to their permeability properties as a layer, are similar to those described for microvascular endothelium.

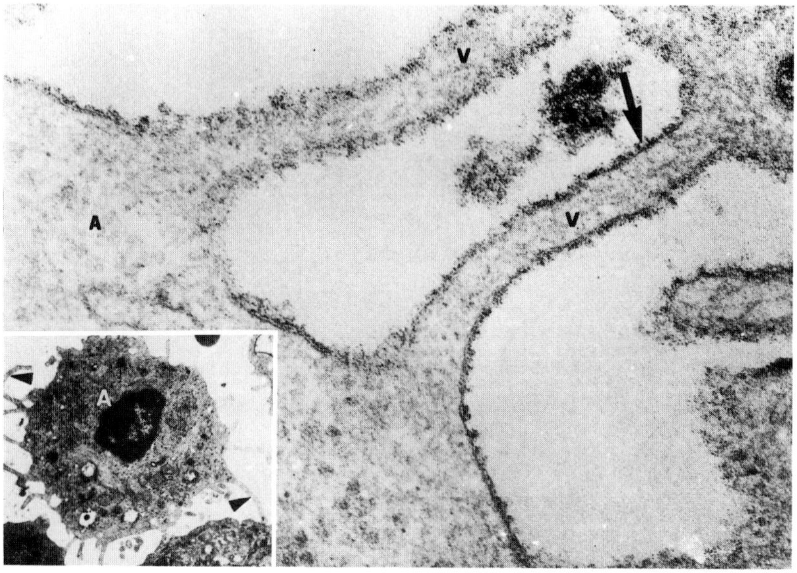

Fig. 5. Human free-floating mesothelial cell (A) showing microvilli (V), whose luminal plasmalemma has RR-labelled glycocalyx (arrow). × 63,500. *Inset:* Human free-floating mesothelial cell (A) with microvilli (arrows). × 3,540.

Experimental septic peritonitis has been shown to induce a substantial decrease in ruthenium-red decoration of luminal glycocalyx in mesothelial (fig. 6–12) and endothelial cells; in the peritoneum, and in other microvascular beds (fig. 6 – inset, 10, 11 13) [24, 25]. In capillaries, this phenomenon will facilitate formation of microthrombi (fig. 3) as well as the development of capillary endothelial cell edema observed during acute inflammatory reactions [26].

Basement Membrane

The biochemical composition of basement membranes is only partially known. Several proteoglycans, which have been identified in different capillary beds, furnish basement membranes with most of their strong negative charge. Labelling these substances with electron dense cationic tracers demonstrated their presence as anionic sites in glomerular (fig. 7) [27, 28] and peritubular capillaries (fig. 2 – right inset, 7 – inset) [29, 30], in pancreatic fenestrated capillaries (fig. 8 – lower inset) [31, 32], in myocar-

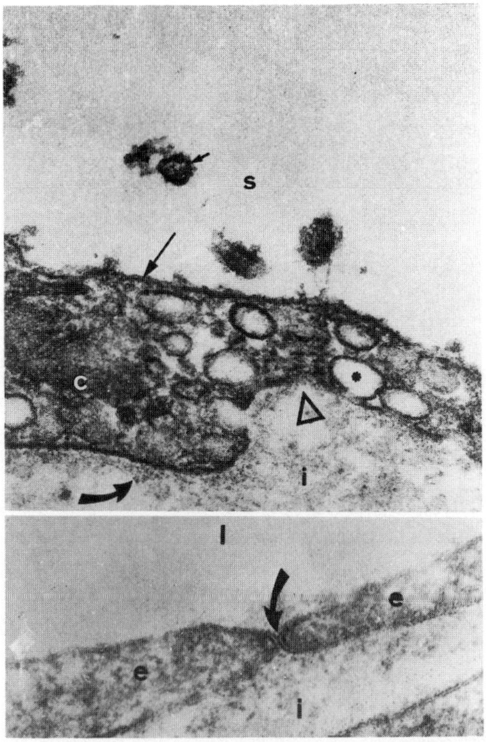

Fig. 6. Rat mesentery, 5 days after induction of *E. coli* peritonitis. Plasmalemma of both microvilli (short arrow) and mesothelial cell (long arrow) illustrates the classical trilaminar structure and lack of glycocalyx. The same is seen on the limiting membrane of pinocytotic vesicles (asterisk). Submesothelial basement membrane is absent (open arrowhead) or disorganized, and with few anionic sites (curved arrow). Interstitium (i) is still edematous. s = Peritoneal space; c = mesothelial cell. RR and 0.5% tannic acid. × 47,120. *Inset:* Rat mesenteric postcapillary venule, 24 h after induction of *E. coli* peritonitis. A tight junction (arrow) separates microvascular lumen (l) from edematous interstitial space (i). Notice absence of subendothelial basement membrane and anionic sites, as well as lack of glycocalyx on luminal aspect of endothelial cells (e). RR and 0.5% tannic acid. × 40,150.

dial microvessels of rat (fig. 9 – upper inset), in capillaries of mouse mesenteric, diaphragmatic and parietal peritoneum (fig. 10) [13], as well as along the submesothelial basement membrane (fig. 4) [23].

It has been shown that the main role of anionic sites in glomerular capillaries is to restric the transmural passage of negative macromolecules

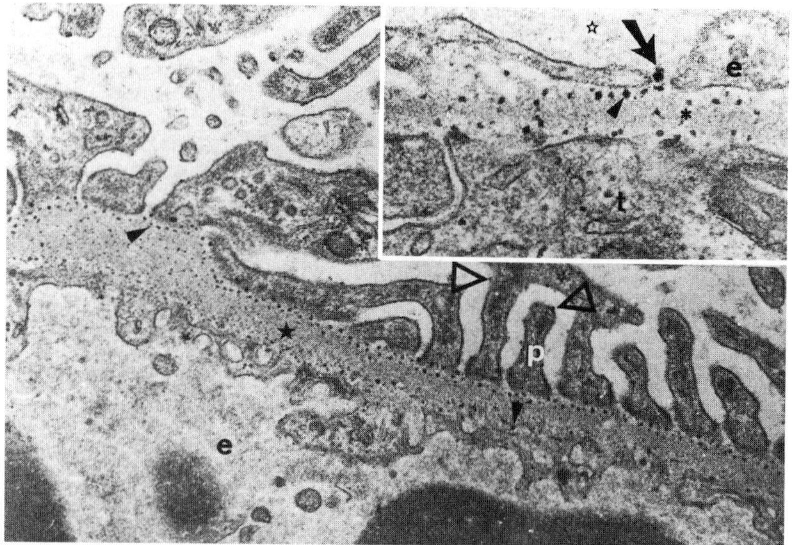

7

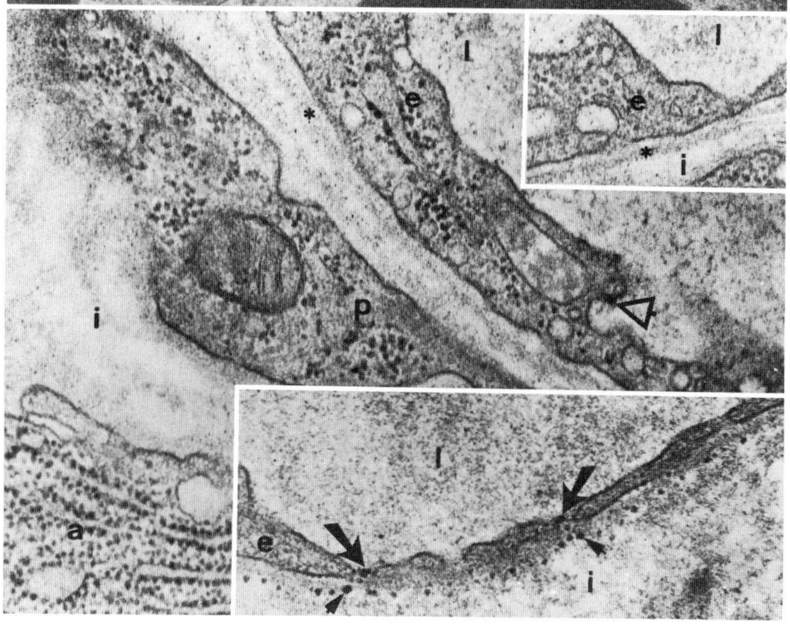

8

Fig. 7. Kidney glomeruli from normal rat. Basement membrane (star) shows anionic sites periodically distributed along both lamina rara interna and externa (black arrowheads). Tracer particles also appear on foot processes (open arrowheads). e = Endothelial cell; p = foot processes. PEI 2%. ×24,500. *Inset:* Partial view of kidney peritubular fenestrated capillary, from healthy rat. Basement membrane (asterisk) shows anionic sites

[27, 33]. Enzymatic removal of heparan-sulfate anionic sites from the glomerular capillary basement membrane significantly increased the leakage of anionic plasma albumin [34]. This concept has been extended to other microvascular beds. Perfusion of rabbit's ear showed that negative dextran sulfate is less readily transported from the capillary lumen to the interstitial space than neutral dextran of comparable size [35]. Another study, done on capillaries of rat skeletal muscle, measured the transfer of two different radiolabelled isozymes of LDH, of similar molecular weight (140 kD) but different electrostatic charge. Results showed that permeability for the positively charged LDH was three times higher than that for the negative isozyme [36]. Studies carried out on perfused frog mesenteric capillaries established that the presence of fixed negative charges within transmicrovascular pathways restricts the transport of negative intermediate-sized proteins, relative to positive solutes of the same size [12].

We have recently observed that neutralization of anionic sites by intraperitoneal protamine induced a significant increase in protein passage to the peritoneal cavity during isosmotic peritoneal dialysis in rabbits [37]. Neutralization of protamine with heparin prevented this enhanced peritoneal permeability to anionic proteins, indicating that availability of anionic sites is the crucial limiting factor. Conversely, protamine did not significantly affect the transfer of small neutral molecules.

It appears, therefore, that the main role of microvascular and mesothelial anionic fixed charges (i.e., restriction of the transmural passage of anionic protein) is especially evident when they are absent. This is also the

---

(arrowheads), periodically distributed along both luminal and abluminal aspects. Tracer particles also label the luminal aspect of the fenestral diaphragm (arrow). e = Endothelial cell; open star = microvascular lumen; t = tubular cell. PEI 2%. × 49,950.

Fig. 8. Rat pancreatic postcapillary venule, 24 h after induction of *E. coli* sepsis. Basement membrane (asterisk) of endothelial cell (e) is devoid of anionic sites. Tracer particles decorate inner membrane of a coated pit (open arrow). Interstitial space (i) is edematous. l = Microvascular lumen; p = perithelial cell; a = acinar cell. PEI 2%. × 24,500. *Upper inset:* Pancreatic fenestrated capillary, from PEI-perfused rat, 24 h after induction of sepsis. There are no anionic sites along subendothelial basement membrane (asterisk). l = Capillary lumen; i = interstitial space. PEI 2%. × 24,500. *Lower inset:* Pancreatic fenestrated capillary from PEI-perfused normal rat. Anionic sites are regularly distributed along subendothelial basement membrane (short arrows). Luminal aspects of fenestral diaphragms is also decorated by tracer (long arrows). l = Capillary lumen; i = interstitial space. PEI 2%. × 24,500.

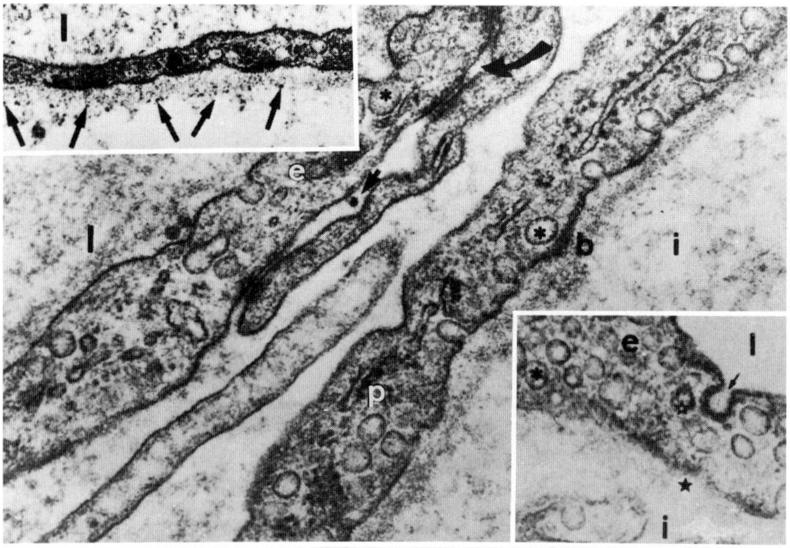

9

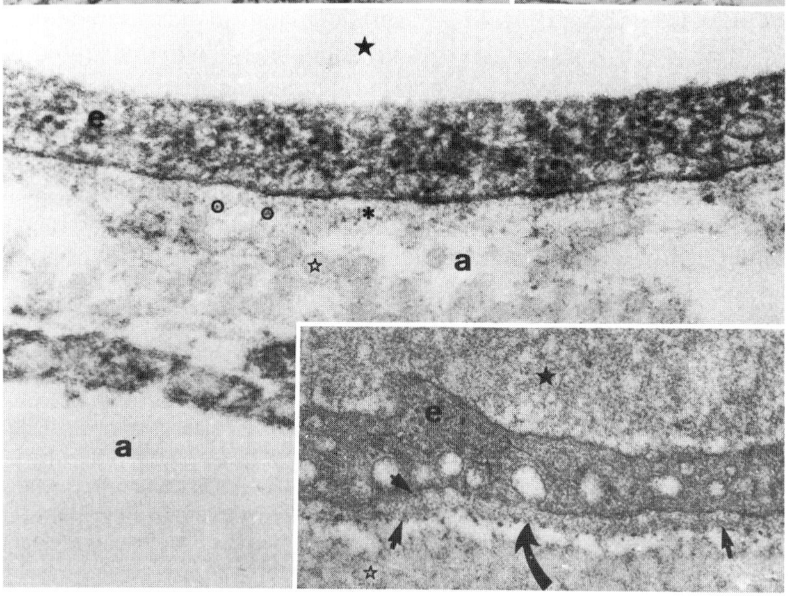

10

Fig. 9. Myocardial postcapillary venule from PEI-perfused rat, 24 h after induction of *E. coli* sepsis. Basement membrane and anionic sites under the endothelial cell (e) are missing. Tracer particles (short arrow) appear traversing the interendothelial cell junction (long arrow). Basement membrane (b) under perithelial cell (p) is also devoid of anionic sites. Coated vesicles appear in both endothelial and perithelial cells. Interstitium (i) is edematous. l = Microvascular lumen. PEI 2%. × 29,970. *Upper inset:* Heart blood capil-

case in acute inflammatory reactions, during which the development of protein-rich interstitial edema is a consequence of abnormally increased microvascular permeability to anionic plasma proteins [38]. Majno and Palade [39] demonstrated that inflammatory mediators substantially increase microvascular permeability to macromolecules. Under the influence of histamine, postcapillary venules showed large junctional gaps, whose formation was ascribed to endothelial cell contraction. This prediction of a functional large pore system, located at interendothelial cell junctions and regulating microvascular permeability to macromolecules, is still debated [40–45].

It has been suggested that interendothelial gaps are not empty holes. Instead, they contain material which, acting as a barrier to the movement of fluid and macromolecules [40, 41, 46], appears to be an extension of the negatively charged fiber-matrix of the glycocalyx [21]. Subsequent experiments in frog mesenteric microvessels showed that increased capillary permeability to albumin could be reduced in spite of the persistence of large interendothelial gaps [47].

Generalized acute inflammatory reaction secondary to endotoxemia can induce a 200% increase on peritoneal uptake of albumin [48]. Several mediators of inflammation like PAF [49–51], neutrophilic and eosinophilic cationic proteins [52] and TNF [53], significantly increase the permeability to anionic plasma proteins in both, microvascular beds and endothelial cell monolayers [54]. This results from neutralization of anionic fixed charges by the aforementioned cationic mediators of inflam-

---

lary from RR-perfused normal rat. Anionic sites (arrows) are evident along subendothelial basement membrane. l = Microvascular lumen. RR. × 24,500. *Lower inset:* Myocardial capillary from RR-perfused rat, 24 h after induction of sepsis. Subendothelial basement membrane and anionic sites are absent (black star). Coated pit (arrow), coated vesicle (open star) and pinocytotic vesicle (asterisk) appear in the endothelial cell (e). l = Capillary lumen; i = interstitium. RR. × 29,970.

Fig. 10. Diaphragmatic capillary from RR-perfused rat, 24 h after induction of *E. coli* sepsis. A few anionic sites (open circles) can be seen along a disorganized basement membrane (asterisk). Endothelial cell (e) lacks glycocalyx. Interstitium (a) is grossly edematous. Black star = capillary lumen; open star = collagen fibers. RR. × 38,140. *Inset:* Diaphragmatic capillary from RR-perfused healthy rat. Basement membrane (curved arrow) shows anionic sites (short arrows), periodically distributed along both aspects of the basement membrane. Black star = capillary lumen; e = endothelial cell; open star = collagen fibers. RR. × 38,140.

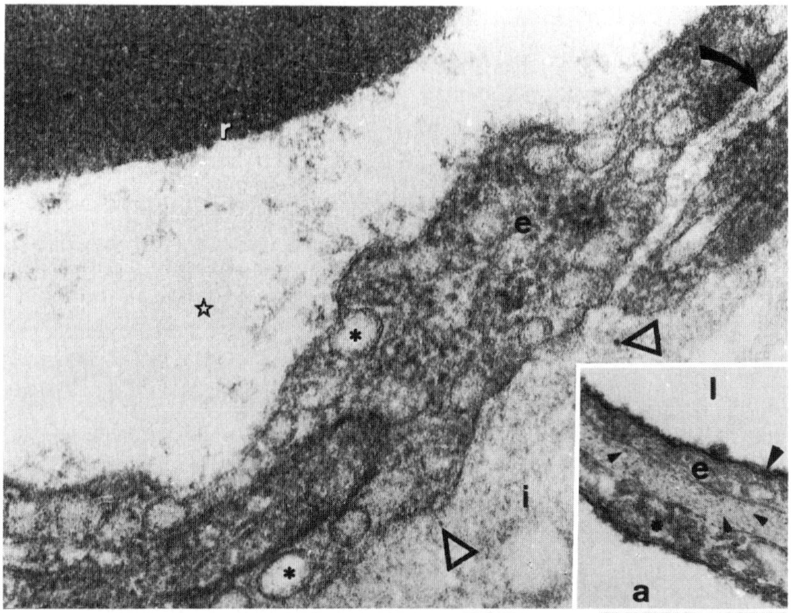

11

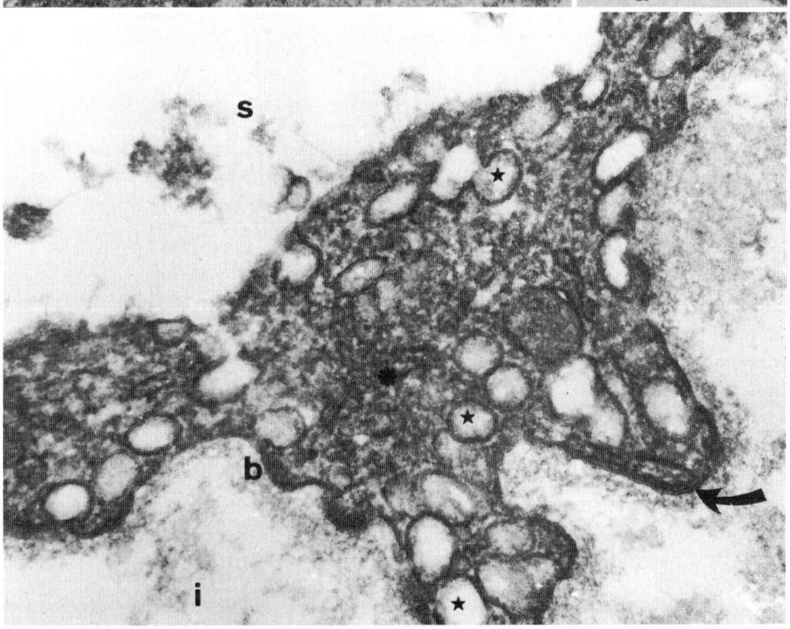

12

13

Fig. 11. Lung capillary from RR-perfused septic rat. Luminal aspect of the endothelial cell (e) has no glycocalyx. Threads of basement membrane are sporadically seen (curved arrow), and there remain a few anionic sites (open arrowheads). Interstitium (i) is edematous. r = Erythrocyte; open star = capillary lumen; asterisks = pinocytotic vesicles. RR and 0.5% tannic acid. × 49,950. *Inset:* Lung capillary from RR-perfused healthy rat. Microvascular endothelial cell (e) shows RR-labelled luminal glycocalyx (large arrowhead). Anionic sites (small arrowheads) are present along both aspects of the basement membrane. l = Capillary lumen; asterisk = alveolar cell; a = alveolar space. RR and 0.5% tannic acid. × 24,500.

Fig. 12. Mesentery from RR-perfused rat, 5 days after induction of *E. coli* peritonitis. Luminal border of mesothelial cell (asterisk) is devoid of glycocalyx. Limiting membrane of pinocytotic vesicles (stars) and cell plasmalemma (arrow) show trilaminar structure. Basement membrane (b) lacks anionic sites. Interstitial tissue (i) is still edematous. s = Peritoneal space. RR and 0.5% tannic acid. × 38,140.

Fig. 13. Postcapillary mesenteric venule from RR-perfused rat, 5 days after induction of *E. coli* peritonitis. Endothelial cell (e) has no glycocalyx. The trilaminar structure of plasmalemma can be seen in endothelial and perithelial (p) cells, as well as in the limiting membrane of pinocytotic vesicles (open stars). Black star and asterisk point to the missing basement membrane. Anionic sites are also absent. l = Microvascular lumen; i = interstitial space. RR and 0.5% tannic acid. × 49,950.

mation. In this context, studies done in our laboratory showed that acute generalized inflammatory reaction, secondary to septic peritonitis, significantly reduced the density distribution of anionic sites normally present along the subendothelial basement membrane of several rat capillary beds: diaphragmatic skeletal muscle (fig. 10) [24], myocardium (fig. 3, 9, 9 – lower right inset) [26], fenestrated capillaries of pancreas (fig. 8, 8 – upper inset), lung continuous capillaries (fig. 11), and even at the level of kidney peritubular capillaries (fig. 2) [Gotloib and Shostak, unpubl. observations]. Peritonitis also induced a significant reduction of the negative charge in submesothelial mesenteric (fig. 6 – inset, 13) and diaphragmatic capillaries (fig. 10 – inset) [24], as well as along the basement membrane laying under mesenteric mesothelium (fig. 4, 6) [25]. Concurrently with the reduction in negative charge, all examined tissues showed substantial amounts of interstitial edema (fig. 4, 6, 8–12) and, at times, swelling of capillary endothelial cells, as well as signs of mitochondrial damage.

It can be deduced from these observations that neutralization of peritoneal anionic sites by means of protamine sulfate creates a situation similar to that observed during peritonitis in peritoneal dialysis patients: large protein losses to the dialysate, without significant change in clearances of small solutes [55]. This reasoning is further supported by the observation that intravenous injection of protamine sulfate to nephrectomized rats induced a 60% loss of plasma albumin from the intravascular compartment to the interstitial space, mimicking again the permeability changes occurring during generalized acute inflammatory reaction [56]. Thus, the normal selectivity of the capillary wall is lost, and the microvascular membrane behaves like nonselective artificial membranes currently used for plasmapheresis. Neither the Starling theory [57] nor the static two-pore-sizes model of capillary permeability [58, 59] can explain the transvascular passage of anionic plasma proteins during acute inflammation. So, as stated by Winn et al. [60], it is clear that conventional conventional mathematical formulas developed to describe microvascular permeability do not always explain experimental data. Such observations raise the possibility that factors other than the traditional ones might be included in equations that relate transperitoneal movement of solutes, especially proteins, to the primary driving forces. Consequently, we consider that electrical charge of the mesothelial layer, microvascular wall and plasma proteins, should be incorporated into the equations describing transfer of solvent and charged solutes across the peritoneum. This concept applies particularly to peritoneal permeability to proteins during peritonitis.

Fenestrated Capillaries

The presence of fenestrated capillaries in the submesothelial tissue (fig. 14) adds another question to the already intricate puzzle of peritoneal permeability. The observed incidence of this special type of microvessel was 2% in human parietal peritoneum, 29% in rabbit diaphragmatic submesothelium [61], and 27% in mouse mesentery [62]. The diameter of fenestrae observed in these studies ranged between 60 and 85 nm, well within the range (50–120 nm) measured in other capillary beds [63]. Most fenestral openings were closed by diaphragms (fig. 14). It has been speculated that, at least in fenestrated capillary beds, fenestrae could represent the large pore system [59, 64]. However, it has been demonstrated that renal glomerular, peritubular and intestinal fenestrated capillaries markedly restrict the passage of macromolecules [29, 65, 66]. This limitation in permeability was shown to depend not only on molecular size, but mainly on the net charge of the permeant molecule.

Additional studies showed the presence of anionic charges on the endothelial cell glycocalyx, at the luminal aspect of fenestral diaphragms, and along the subendothelial basement membranes of several fenestrated capillary beds: kidney glomeruli (fig. 7) [33], renal peritubular (fig. 2 – right inset, 7 – inset) [29], intestinal [30, 32] and pancreatic microvessels (fig. 8 – lower inset) [31, 32]. Hence, negative fixed charges have been observed along the subendothelial basement membrane as well as at the luminal aspect of endothelial cells of mouse mesenteric fenestrated capillaries (fig. 15) [67]. The luminal aspect of fenestral diaphragms was also decorated by cationized ferritin (fig. 15 – inset) showing a 3:1 ratio of labelled:unlabelled diaphragms. This data is consistent with previous studies which suggested that fenestrae are relatively impermeable to anionic macromolecules [66, 68] and, consequently, that protein permeability of fenestrated capillaries is no higher than that shown by continuous capillaries [69, 70]. This charge discrimination is effective for molecules as small as 2 nm (approximate molecular weight: 2 kD).

Conversely, permeability of fenestrated capillaries to neutral solutes like vitamin B_{12} and inulin is 15–28 times higher than that of continuous muscular microvessels, whereas hydraulic conductance and small solute permeability of fenestrated endothelia appear to be substantially higher than those of most continuous endothelia [70]. At this point it is pertinent to recall that, in peritoneal dialysis, the dialysance ratio of inulin to urea is 20 times higher than the observed with cuprophane membranes [71].

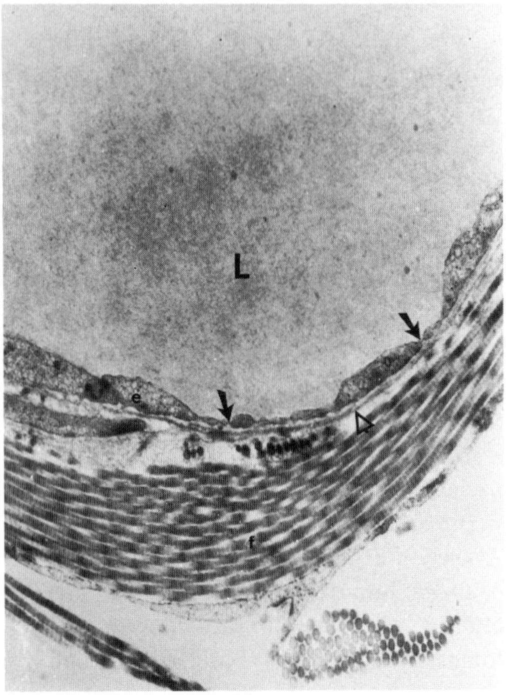

14

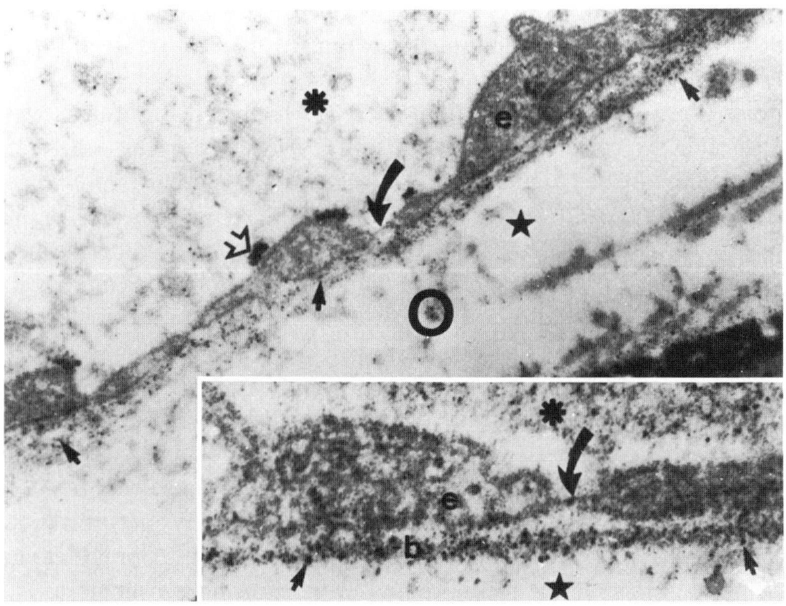

15

Therefore, these permeability characteristics of fenestrated capillaries could be related to the higher peritoneal permeability to middle molecular size solutes compared with that achieved during hemodialysis through cellulose membranes [72, 73]. This ability of the peritoneum was a basic element in the conception of CAPD [74].

Absorption of Albumin from the Peritoneal Cavity: Transmesothelial and Translymphatic Endothelial Transport of Albumin-Gold Complexes

The diaphragmatic capillary net is organized as a plexus along the submesothelial surface. A prominent feature of diaphragmatic lymphatics is the presence of flattened, elongated cisternae, whose endothelial layer is thin and occasionally shows tight junctions [75, 76]. Adjacent cells usually overlap, leaving an open interface which, at times, can be as wide as 12 μm. The glycocalyx of cisternal endothelium becomes heavily decorated with cationized ferritin, which also appears along intercellular clefts [76].

Stomata (fig. 18) can be found only between mesothelial cells overlaying lacunae. Together, mesothelial and lymphatic endothelial cells line the channel which communicates the abdominal cavity and the lacunar lumen [75, 76]. Approximately 2–6% of the whole junctional system of lymphatic capillaries consists of open junctions (fig. 16) with gaps of up to 100 nm, which could account for a substantial part of lymphatic permeability [77]. A subendothelial basement membrane may be present, but shows numer-

---

Fig. 14. Diaphragmatic fenestrated capillary from healthy rabbit. Black arrows point to fenestral diaphragms. Endothelial cell (e) lies on continuous basement membrane (open arrow head). L = Capillary lumen; f = collagen fibers. × 31,280.

Fig. 15. Mesenteric fenestrated capillary from a mouse injected with cationized ferritin. Endothelial cell (e) shows clumps of ferritin in restricted areas of the luminal aspect (open arrow). Curved black arrow points to fenestral diaphragm, devoid of tracer. Subendothelial basement membrane shows periodically ordered tracer particles, along its three layers (short black arrows). Interstitial collagen fibers are also decorated by the tracer (black circle). Asterisk = Capillary lumen; black star = interstitial space. × 25,260. *Inset:* Partial view of another fenestrated capillary, at higher magnification. Curved arrow indicates cationized ferritin decorating the luminal aspect of fenestral diaphragm. Again, basement membrane (b) shows anionic sites along its three layers (short arrows). Asterisk = Capillary lumen; black star = interstitial space. × 63,500.

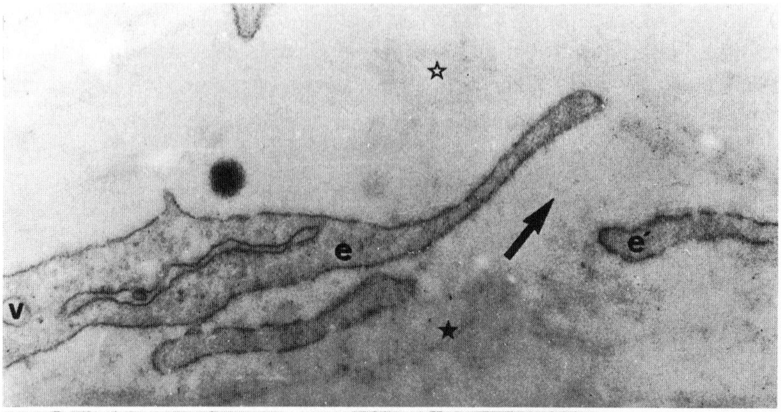

16

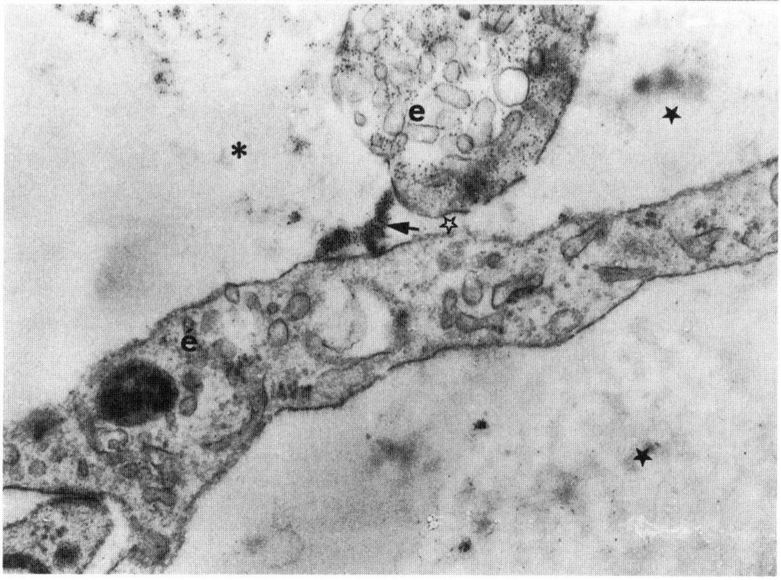

17

Fig. 16. Lymphatic capillary of mice mesentery. Two overlapping endothelial cells (e and é) demarcate an open intercellular junction (arrow), through which the interstitial space (black star) and the capillary lumen (open star) communicate. Notice lack of subendothelial basement membrane. v = Pinocytotic vesicle. × 24,500.

Fig. 17. Mesenteric lymphatic capillary from a mouse injected intraperitoneally with cationized ferritin. Tracer particles (arrow) can be seen entering the microvascular lumen (asterisk), through the open interendothelial junction (open star). Again, note the absence of basement membrane and anionic sites. e and é = Lymphatic endothelial cells; black stars = interstitial space. × 29,970.

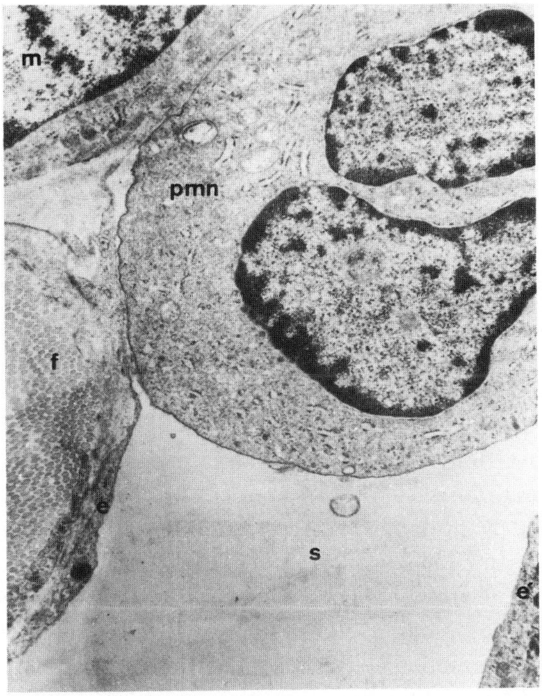

Fig. 18. Diaphragmatic peritoneum from healthy, nonperfused rabbit. A leukocyte finds its way through a peritoneal stoma (s), limited by lymphatic endothelial and mesothelial (m) cells. f = Collagen fibers; pmn = polymorphonuclear cell. × 11,710.

ous interruptions [77]. Anionic fixed charges are rarely found along the abluminal aspect of lymphatic capillary endothelial cells (fig. 17) [78]. Diaphragmatic lymphatic lacunae and capillaries from the peritoneal lymphatic network drain into a system of small caliber lymph vessels, furnished with semilunar valves.

Since Starling [79], it has been accepted that besides the removal of excess interstitial fluid, the lymphatic system has the special function of absorbing protein from the interstitium as well as from serous cavities. The mechanisms involved in lymph formation are in part the result of hydraulic [80] and/or oncotic forces [77], acting across the initial lymphatics. Assuming that the interstitial pressure is negative [81], any rise will also increase the initial lymphatic flow. This theory is also supported by experimental studies which show close correlation between the rate of fluid absorption from the peritoneal cavity and intraabdominal pressures [82–85] well within the range of values observed during clinical peritoneal dialysis [86, 87]. The scarcity of stomatal and lacunar subendothelial basement membrane, as well as the absence of subendothelial anionic sites, is expected to facilitate the passage of negative albumin from the interstitial space into the lacunar lumen. The same mechanism of hydrostatic pressure has been invoked to explain the transmesothelial absorption of protein through parietal and visceral peritoneum [85].

However, other investigators [88–90] have proposed the vesicular theory of lymph formation, which holds that pinocytotic vesicles provide the major route for transendothelial transport of albumin. It has also been proposed that this mechanism could, in turn, create the osmotic gradient eventually needed for further fluid flow into lymphatic capillaries [77]. The vesicular theory found additional support from in vitro studies showing the temperature dependence of protein transport across lymphatic endothelium at 4 °C. In this experimental situation, known to substantially restrain pinocytosis [91], the rate of protein transport by vesicles was significantly reduced [92]. Pinocytosis has also been shown to be involved in the transfer of macromolecules across mesothelial cells of mouse omentum [93]. This could explain the fact that intraperitoneal concentration of albumin, as well as molecular size of plasma proteins, have not been found to be determinants in the kinetics of transperitoneal macromolecular absorption [94].

Albumin-gold complexes (Alb-Au) have been recently used for investigating receptor-mediated endocytosis of albumin in blood capillary endothelia [95, 96]. In an effort to verify the existence of this mechanism in peritoneum, we examined by means of transmission electron microscopy, tissue obtained from mice after intraperitoneal injection of Alb-Au complexes [97]. We observed that particles of the tracer were taken in by mesothelial and lymphatic endothelial cells. This phenomenon was seen in mesenteric as well as in parietal and diaphragmatic peritoneum [97]. The study showed different stages in the process of transmesothelial absorption of albumin, all of them occurring 10 min after intraperitoneal injection of Alb-Au. The marker was present in coated pits and coated vesicles, in plasmalemmal vesicles, as well as in multivesicular bodies of mesothelial cells (fig. 9, 20 – inset). At the same time (10 min after IP injection), Alb-Au complexes were observed in the submesothelial interstitial space (fig. 19–21 – insets), in the proximity of lymphatic capillaries, in plasmalemmal vesicles and multivesicular bodies of lymphatic endothelial cells; as well as in the

19

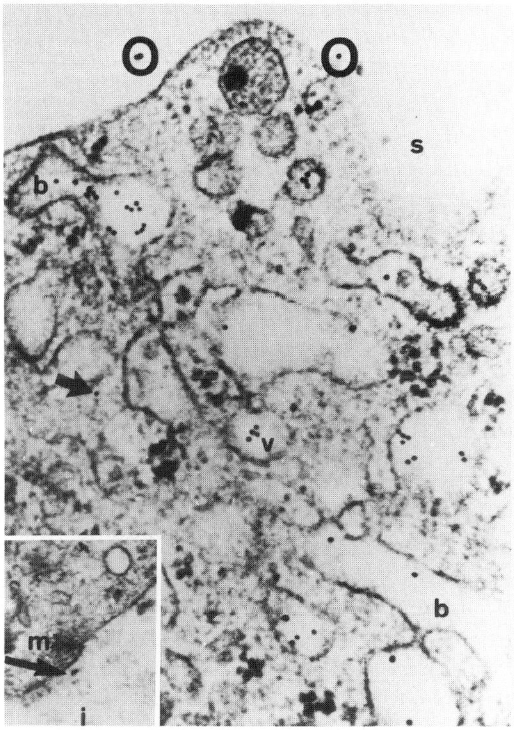

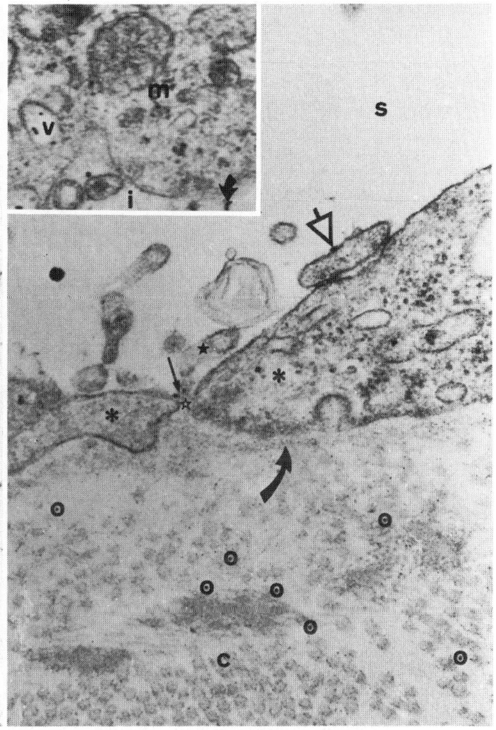

20

Fig. 19. Mouse mesentery, 10 min after intraperitoneal injection of albumin-gold complexes. Tracer particles can be seen near the mesothelial cell luminal aspect (open circles), in multivesicular bodies (b), cell cytoplasm (arrow) and pinocytotic vesicles (v). × 45,160. *Inset:* Another area of the same sample, showing gold tracer in the submesothelial space (arrow). m = Mesothelial cell; i = interstitium. × 22,170.

Fig. 20. Mouse mesentery, 10 min after intraperitoneal injection of albumin-gold complexes. Tracer particles are adsorbed on the luminal surface of microvilli (open arrow), on the glycocalyx network (straight arrow), facing an intermesothelial cell junction, and between the collagen fibers located in the submesothelial interstitium (open circles). s = Peritoneal space; black star = microvilli; asterisks = mesothelial cells; curved arrow = basement membrane; c = interstitial collagen fibers. × 22,170. *Inset:* Another area of the same sample, showing gold complexes on the internal aspect of a pinocytotic vesicle (v), and in the submesothelial interstitium (arrow). m = Mesothelial cell; i = interstitial space. × 27,100.

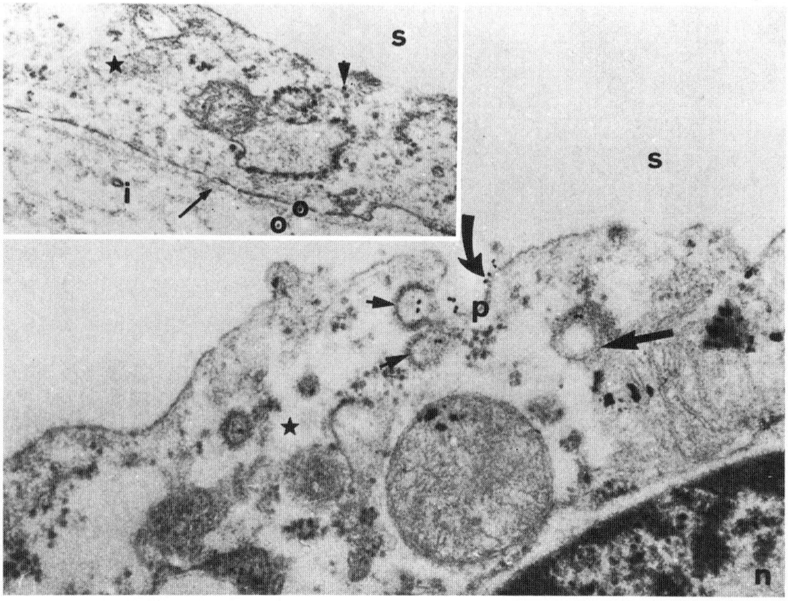

Fig. 21. Mouse parietal peritoneum from anterior abdominal wall, 10 min after intraperitoneal injection of albumin-gold complexes. Tracer particles are adsorbed on luminal plasmalemma (curved arrow) of mesothelial cell (black star), on internal aspect of a coated vesicle (long arrow), as well as in coated pits (short arrows). s = Peritoneal space; n = mesothelial cell nucleus; p = coated pit. ×24,500. *Inset:* Another area of the same sample, showing gold complexes in mesothelial cell cytoplasm (short arrow), and in interstitium on both aspects of the basement membrane (open circles). s = Peritoneal space; star = mesothelial cell; long arrow = basement membrane; i = interstitial space. ×29,970

lumen of lymphatic capillaries (fig. 2). Alb-Au complexes were not observed within mesothelial junctions. We concluded that at least part of the peritoneal absorption of the ligand took place transcellularly, by a mechanism of receptor-mediated endocytosis [98–100]. This implies that both mesothelial and lymphatic endothelial cells of diaphragmatic, visceral and parietal peritoneum are actively involved in the process of albumin absorption. It should be recalled that the diaphragmatic contribution to the total peritoneal surface area is just around 4% [101]. This, in addition to the transcytosis of albumin, suggests that the widely accepted 90% absorption of macromolecules through diaphragmatic lymphatics [83] may well have been an overestimation. Unfortunately, we do not yet have new quantitative data.

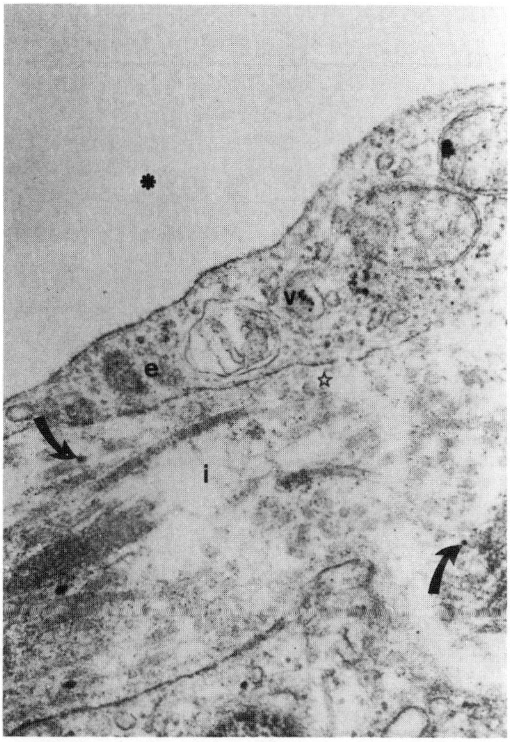

Fig. 22. Lymphatic capillary from mouse diaphragmatic peritoneum, 10 min after intraperitoneal injection of albumin-gold complexes. Tracer particles can be seen in a pinocytotic vesicle (v), and in submesothelial interstitium (curved arrows). Subtle threads of basement membrane (open star) are seen occasionally. Asterisk = Microvascular lumen; e = lymphatic endothelial cell; i = interstitial space. ×22,170.

Conclusions

This partial review of new data, mainly emerging from ultrastructural studies, has tried to point out both the complexity of, and our uncertainty about most mechanisms governing peritoneal permeability. Interpretation of whole organ studies, in our case peritoneal dialysis, is based on the objectionable assumption that the entire system can be analyzed as a single capillary model. This problem, first exposed more than 10 years ago [102], includes some of the major still unanswered questions which outstanding scientists of this century have been dealing with: capillary permeability, formation of lymph, and microvascular endothelial cell participation in the phenomenon of acute inflammation. Finding the adequate answers is one of the most stimulating challenges presented by nature to the nephrologist.

'... but vouchsafe to me the strength,
the leisure and the ambition ever to extend my knowledge.
For art is great, but the mind of man is ever expanding ...'
Maimonides, Daily prayer of a physician (12th century)

References

1 Cunningham RS: Studies on absorption from serous cavities. III. The effect of dextrose upon the peritoneal mesothelium. Am J Physiol 1920;53:488–495.
2 Putnam TJ: The living peritoneum as a dialyzing membrane. Am J Physiol 1922/1923;63:548–563.
3 Schechter AJ, Cary MK, Carpentieri AL, Darrow DC: Changes in the composition of fluids injected into the peritoneal cavity. Am J Dis Child 1933;46:1015–1026.
4 Aune S, Scand J: Transperitoneal exchange. 1. Peritoneal permeability studied by transperitoneal plasma clearance of urea, PAH, inulin and serum albumin in rabbits. Gastroenterology 1970;5:85–97.
5 Renkin EM: Filtration, diffusion and molecular sieving through porous cellulose membranes. J Gen Physiol 1955;88:225–243.
6 Stetten D: Thermodynamic, kinetic and biological stability. Am J Med 1952;13: 251–254.
7 Odor L: Observations of the rat mesothelium with the electron and phase microscopes. Am J Anat 1954;95:433–465.
8 Shear L, Harvey JD, Barry KG: Peritoneal sodium trnsport enhancement by pharmacologic and physical agents. J Lab Clin Med 1966;67:181–188.
9 Cascarano J, Rubin AD, Chick WI, Zweifach BW: Metabolically induced permeability changes across mesothelium and endothelium. Am J Physiol 1964;206:373–382.
10 Rasio EA: Metabolic control of permeability in isolated mesentery. Am J Physiol 1974;226:962–968.
11 Gotloib L, Oreopoulos DG: Transfer across the peritoneum: passive or active? Nephron 1981;29:201–202.
12 Adamson RH, Huxley VH, Curry FE: Single capillary permeability to proteins having similar size but different charge. Am J Physiol 1988;254:H304–H312.
13 Gotloib L, Bar Sella P, Jaichenko J, Shostak A: Ruthenium-red stained polyanionic fixed charges in peritoneal microvessels. Nephron 1987;47:22–28.
14 Turner MR, Gluch G, Michel CC: The effects of catonized ferritin and native ferritin upon the filtration coefficient of single frog capillaries. Microvasc Res 1983;25: 205–222.

15 Luft JH: Fine structure of capillary and endocapillary layer as revealed by ruthenium-red. Fed Proc 1966;25:1173–1183.
16 Bangham, AD: The adhesiveness of leukocytes with special reference to zeta potential. Ann NY Acad Sci 1964;116:945–949.
17 Danon D: Biophysical Aspects of Red Cell Aging. 11th Congr Int Soc Hematology. New York, Grune & Stratton, 1966, pp 394–405.
18 White JG: Ultrastructural modifications in platelet membranes and cytoskeleton following activation. Blood Cells 1983;9:237–261.
19 Simionescu M: Cellular organization of the alveolo-capillary unit: structural-functional correlation; in Said (ed): The Pulmonary Circulation and Acute Lung Injury. New York, Futura Publishing, pp 13–36.
20 Schneeberger EE, Hamelin M: Interactions of serum proteins with lung endothelial glycocalyx: its effect on endothelial permeability. Am J Physiol 1984;247:H206–H217.
21 Curry FE, Michel CC: A fiber-matrix model of capillary permeability. Microvasc Res 1980;20:96–109.
22 Simionescu M, Simionescu N, Santoro F, Palade GE: Differentiated microdomains of the luminal plasmalemma of murine muscle capillaries: Segmental variations in young and old animals. J Cell Biol 1985;100:1396–1407.
23 Gotloib L, Shostak A, Jaichenko J: Ruthenium-red stained anionic charges of rat and mice mesothelial cells and basal lamina: the peritoneum is a negatively charged dialyzing membrane. Nephron 1988;1:65–70.
24 Gotloib L, Shostak A, Jaichenko J, Galdi P: Decreased density distribution of mesenteric and diaphragmatic microvascular anionic charges during murine abdominal sepsis. Resuscitation 1988;16:179–192.
25 Gotloib L, Shostak A, Jaichenko J: Loss of mesothelial electronegative fixed charges during murine septic peritonitis. Nephron 1989;51:77–83.
26 Gotloib L, Shostak A, Galdi P, Jaichenko J, Fudin R: Loss of microvascular charges accompanied by interstitial edema in septic rats' heart. Circ Shock, in press.
27 Chang RLS, Deen WM, Robertson ChR, Brenner BM: Permselectivity of the capillary wall. IV. Restricted transport of polyanions. Kidney Int 1975;8:212–218.
28 Kanwar YS, Linker E, Farquhar MG: Increased permeability of the glomerular basement membrane to ferritin after removal of glycosaminoglycans (heparin sulfate) by enzyme digestion. J Cell Biol 1980;86:688–693.
29 Deen WM, Ueki IF, Brenner BM: Permselectivity of renal peritubular capillaries to neutral dextran and endogenous albumin. Am J Physiol 1976;231:283–291.
30 Charonis AS, Wissig SL: Anionic sites in basement membranes. Differences in their electrostatic properties in continuous and fenestrated capillaries. Microvasc Res 1983;25:265–285.
31 Simionescu M, Simionescu N, Palade GE: Preferential distribution of anionic sites on the basement membrane and the abluminal aspect of the endothelium in fenestrated capillaries. J Cell Biol 1982;95:425–436.
32 Bankston PW, Milici AJ: A survey of the binding of polycationic ferritin in several fenestrated capillary beds: Indication of heterogeneity in the luminal glycocalyx of fenestral diaphragms. Microvasc Res 1983;26:36–48.
33 Kanwar YS, Farquhar MG: Anionic sites in the glomerular basement membrane. J Cell Biol 1979;81:137–153.

34 Rosenzweig LJ, Kanwar YS: Removal of sulfated (heparan sulfate) or non-sulfated (hyaluronic acid) glycosaminoglycans results in increased permeability of the glomerular basement membrane to ^{125}I bovine serum albumin. Lab Invest 1982;47: 177–184.

35 Areekul S: Reflection coefficients of neutral and sulfate substituted dextran molecules in the isolated perfused rabbit ear. Uppsala J Med Sci 1969;74:129–138.

36 Haraldsson B: Physiological studies of macromolecular transport across capillary walls. Acta Physiol Scand 1986;128(suppl 553):1–40.

37 Galdi P, Shostak A, Jaichenko J, Fudin R, Gotloib L: Protamine sulfate induces enhanced permeability to proteins. Nephron 1991;57:45–51.

38 Grega GJ, Adamski SW, Dobbins SE: Physiological and pharmacological evidence for the regulation of permeability. Fed Proc 1986;45:96–100.

39 Majno G, Palade GE: Studies on inflammation. 1. The effect of histamine and serotonin on vascular permeability: an electron microscopic study. J Biophys Biochem Cytol 1961;11:571–605.

40 Renkin EM, Carter RD, Joyner WL: Mechanism of the sustained action of histamine and bradykinin on transport of large molecules across capillary walls in the dog paw. Microvasc Res 1974;7:49–60.

41 Diana JN, Long SC, Yao H: Effect of histamine on equivalent pore radius in capillaries of isolated dog hindlimbs. Microvasc Res 1972;4:413–437.

42 Renkin EM: Some consequences of capillary permeability to macromolecules: Starling's hypothesis reconsidered. Am J Physiol 1986;250:H706–H710.

43 Grega GJ, Adamski SW, Svensjo E: Is there evidence for venular large junctional gap formation in inflammation? Microcirc Endothelium Lymphatics 1985;2:211–233.

44 Bottaro D, Shepro D, Hechtman HB: Intimal and microvessel endothelial cell barrier function in vitro. Microvasc Res 1986;32:389–398.

45 Welles SL, Shepro D, Hechtman HB: Vasoactive amines modulate actin cables, stress fibers and surface area in culture bovine endothelium. J Cell Physiol 1985; 123:337–342.

46 Clough GF, Michel CC, Phillips ME: Inflammatory changes in permeability and ultrastructure of single vessels in the frog mesenteric microcirculation. J Physiol 1988;395:99–114.

47 Michel C, Blumberg S, Clough G: Hydroxyethyl-rutosides (HR) reduce permeability of frog mesenteric microvessels. Phlebology 1990;5(suppl 1):3–7.

48 Van Lambalgen AA, Rasker MTE, Van Den Bos GC, Thijs LG: Effects of endotoxemia on systemic plasma loss and hematocrit in rats. Microvasc Res 1988;36:291–304.

49 Benveniste J: Platelet-activating factor, a new mediator of anaphylaxis and immune complex deposition from rabbit and human basophils. Nature 1974;249:581–582.

50 Camussi G, Tetta C, Coda R, Segoloni GP, Vercellone A: Platelet activating factor-induced loss of glomerular-anionic charges. Kidney Int 1984;25:73–81.

51 Bessin P: PAF-acether and leukotriene participation in acute circulatory shock. Pharmacol Res Commun 1986;18(suppl):139–149.

52 Janoff A, Zweifach BW: Production of inflammatory changes in the microcirculation by cationic proteins extracted from lysosomes. J Exp Med 1964;120:747–764.

53 Beutler B, Cerami A: Cachectin: More than a tumor necrosis factor. N Engl J Med 1987;316:379–385.

54 Peterson MW, Stone P, Shasby DM: Cationic neutrophil proteins increase transendothelial albumin movement. J Appl Physiol 1987;62:1521–1530.

55 Rubin J, Nolph KD, Arfania D, et al: Follow-up of hourly peritoneal clearances with and without nitroprusside augmentation in patients undergoing CAPD. Kidney Int 1979;16:619–633.

56 Vehaskari VM, Chang CTC, Stevens JK, Robson AM: The effects of polycations on vascular permeability in the rat. A proposed role for charge sites. J Clin Invest 1984; 73:1053–1061.

57 Starling EM, Tubby AM: On absorption from and the secretion into the serous cavities. J Physiol (Lond) 1894;16:140–155.

58 Pappenheimer JR: Passage of molecules through capillary walls. Physiol Rev 1953; 33:387–423.

59 Grotte G: Passage of dextran molecules across the blood-lymph barrier. Acta Chir Scand 1956;(suppl 211):1–84.

60 Winn R, Nadir B, Gleisner J, Stothert J, Hildebrandt J: Deficiencies in pore-membrane models of microvascular fluid and solute transudation. J Appl Physiol 1981; 51:1574–1580.

61 Gotloib L, Shostak A, Bar Sella P, Eiali V: Fenestrated capillaries in human parietal and rabbit diaphragmatic peritoneum. Nephron 1985;41:200–202.

62 Gotloib L, Shostak A, Jaichenko J. Fenestrated capillaries in mice submesothelial mesenteric microvasculature. Int J Artif Organs 1989;12:20–24.

63 Wolf JR: Ultrastructure of the terminal bed as related to function; in Kaley, Altura (eds): Microcirculation. Baltimore, University Park Press, 1977, p 96.

64 Majno G: Ultrastructure of the vascular membrane; in Hamilton, Dow (eds): Handbook of Physiology, vol III, sect 2: Circulation. Washington, Am Physiol Soc, 1965, pp 2293–2375.

65 Brenner BM, Hostetter TH, Humes DH: Glomerular permselectivity: barrier function based on discrimination of molecular size and charge. Am J Physiol 1978;234: F455–F460.

66 Granger DN, Taylor AE: Permselectivity of intestinal capillaries. Physiologist 1980; 23:47–52.

67 Gotloib L, Shostak A, Jaichenko J, Galdi P, Pudin R: Anionic fixed charges in the fenestrated capillaries of the mouse mesentery. Nephron 1990;55:419–422.

68 Bearer EL, Orci L: Endothelial fenestral diaphragms: a quick-freeze, deep-etch study. J Cell Biol 1985;100:418–428.

69 Renkin EM: Multiple ways of capillary permeability. Circ Res 1977;41:735–743.

70 Levick JR, Smaje LH: An analysis of the permeability of a fenestra. Microvasc Res 1987;33:233–256.

71 Henderson LW: The problem of peritoneal membrane area and permeability. Kidney Int 1973;3:409–410.

72 Babb AL, Johansen PJ, Stand MJ, Tenckhoff H, Scribner BH: Bidirectional permeability of the human peritoneum to middle molecules. Proc Eur Dial Transplant Assoc 1973;10:247–261.

73 Babb AL, Popovich RP, Christopher TG, Scribner BH: The genesis of the square meter-hour hypothesis. Trans Am Soc Artif Internal Organs 1971;17:81–91.
74 Popovich RP, Moncrief JW, Decherd JF, Bomar JB, Pyle WK: Preliminary verification of the low dialysis clearance hypothesis via a novel equilibrium peritoneal dialysis technique. Absr Am Soc Artif Internal Organs 1976;5:64.
75 Tsilibarry EC, Wissig SL: Absorption from the peritoneal cavity. SEM study of the mesothelium covering the peritoneal surface of the muscular portion of the diaphragm. Am J Anat 1977;149:127–133.
76 Leak LV: Distribution of cell surface charges on mesothelium and lymphatic endothelium. Microvasc Res 1986;31:18–30.
77 Casley-Smith JR: Lymph and lymphatics; in Kaley, Altura (eds): Microcirculation. University Park Press, Baltimore, 1981, vol 4, pp 423–502.
78 Jones WR, O'Morchoe CCC, Jarosz HM, O'Morchoe GJ: Distribution of charged sites on lymphatic endothelium. Lymphology 1986;19:5–14.
79 Starling EH: On the absorption of fluid from the connective tissue spaces. J Physiol (Lond) 1896;16:140–155.
80 Allen L: Volume and pressure changes in terminal lymphatics. Am J Physiol 1938; 123:3–4.
81 Guyton AC: A concept of negative interstitial pressure based on pressures in implanted perforated capsules. Circ Res 1963;12:399–414.
82 Allen L, Vogt E: A mechanism of lymphatic absorption from serous cavities. Am J Physiol 1937;119:776–782.
83 Courtice FC, Steinbeck AW: The lymphatic drainage of plasma from the peritoneal cavity of the cat. Aust J Exp Biol Med Sci 1950;28:161–169.
84 Zink J, Greenway CV: Control of ascites absorption in anesthetized cats: effects of intraperitoneal pressure, protein and furosemide diuresis. Gastroenterology 1977; 73:1119–1124.
85 Flessner JM, Parker RL, Sieber SM: Peritoneal lymphatic uptake of fibrinogen and erythrocytes in the rat. Am J Physiol 1983;244:H89–H96.
86 Gotloib L, Mines M, Garmizo AL, Varka I: Hemodynamic effects of intraabdominal pressure in peritoneal dialysis. Periton Dial Bull 1981;1:41–43.
87 Gotloib L, Garmizo AL, Varka I, Mines M: Reduction of vital capacity due to increased intra-abdominal pressure during peritoneal dialysis. Periton Dial Bull 1981;1:63–64.
88 Albertine KH, O'Morchoe ChCC: Renal lymphatic ultrastructure and translymphatic transport. Microvasc Res 1980;19:338–351.
89 Dobbins WO, Rollins EL: Intestinal mucosal lymphatic permeability: an electron microscopic study of endothelial vesicles and cell junctions. J Ultrastruct Res 1970; 33:29–59.
90 Jones WR, O'Morchoe ChCC, O'Morchoe PJ: The organization of endocytotic vesicles in lymphatic endothelium. Microvasc Res 1983;25:286–299.
91 Casley-Smith JR: Endocytosis: the different energy requirements for the uptake of particles by small and large vesicles into peritoneal macrophages. J Microsc 1969;90: 15–30.
92 O'Morchoe ChCC, Jones WR III, Jarosz HM, O'Morchoe PJ, Fox LM: Temperature dependence of protein transport across lymphatic endothelium in vitro. J Cell Biol 1984;98:629–640.

93 Fedorko ME, Hirsch JG: Studies on the transport of macromolecules and small particles across mesothelial cells of the mouse omentum. Exp Cell Res 1971;69: 113–127.
94 Regoeczi E, Zaimi O, Chindemi PA, Charlwood PA. Absorption of plasma proteins from peritoneal cavity of normal rats. Am J Physiol 1989;256:E447–E452.
95 Ghitescu L, Fixman A, Simionescu M, Simionescu N: Specific binding sites for albumin restricted to plasmalemmal vesicles of continuous capillary endothelium: receptor mediated transcytosis. J Cell Biol 1986;102:1304–1311.
96 Milici AJ, Watrous NE, Stukenbrok H, Palade GE: Transcytosis of albumin in capillary endothelium. J Cell Biol 1987;105:2603–2612.
97 Shostak A, Gotloib L, Jaichenko J, Galdi P, Fudin R: Transcytosis of intraperitoneally injected albumin in mice peritoneum (abstract). J Am Soc Nephrol 1990;1: 392.
98 Goldstein JL, Anderson RG, Brown MS: Coated pits, coated vesicles and receptor-mediated endocytosis. Nature 1979;279:679–685.
99 Pastan I, Willingham MC: The pathway of endocytosis; in Pastan, Willingham (eds): Endocytosis. New York, Plenum Press, 1985, pp 1–44.
100 Salzman NH, Maxfield FR: Intracellular fusion of sequentially formed endocytotic compartments, J Cell Biol 1988;106:1083–1091.
101 Esperanca MJ, Collins DL: Peritoneal dialysis efficiency in relation to body weight. J Pediatr Surg 1966;1:162–169.
102 Nolph KD: The Peritoneal Dialysis System. Contrib Nephrol. Basel, Karger, 1979, vol 17, pp 44–49.

Lazaro Gotloib, MD, Department of Nephrology and Kornach Laboratory for Experimental Nephrology, Central Emek Hospital, Afula 18101 (Israel)

Berlyne GM (ed): The Kidney Today. Selected Topics in Renal Science.
Contrib Nephrol. Basel, Karger, 1992, vol 100, pp 236–253

Vascular Endothelin and Renal Disease Progression

Norberto Perico[a, b], *Giuseppe Remuzzi*[a, b]

[a] Mario Negri Institute for Pharmacological Research and
[b] Division of Nephrology, Ospedali Riuniti di Bergamo, Bergamo, Italy

Background

The discovery that endothelins, a family of peptides of endothelial origin [1, 2], induce renal vasoconstriction [3–8] and stimulate mesangial cell proliferation [9], raised the possibility that such peptides participate in renal disease progression [10].

Endothelin Genes, Synthetic Pathway and Regulation

Three distinct genes encoding three endothelin (ET) isoforms, endothelin-1 (ET-1), endothelin-2 (ET-2) and endothelin-3 (ET-3), have been identified by Southern blot analysis of genoma of various animals [11]. ET_s are highly homologous 21-amino acid peptides with two intrachain disulfide rings [2, 10, 12] (fig. 1). Human and porcine ET-1 have identical sequences. Human and mouse ET-2 differ from one another by one amino acid and differ from ET-1 by two and three amino acids, respectively. Human and rat ET-3 have identical sequences, but differ from ET-1 by six amino acids. All these isopeptides have a high degree of sequence homology with sarafotoxins, peptide toxins isolated from snake venom [12, 13], suggesting that genes encoding a snake venom toxin have evolved into genes encoding an important mammalian regulatory peptide.

The original report by Yanagisawa et al. [2] has proposed a hypothetical biosynthetic pathway of ET in cultured porcine aortic-endothelial cells which occurs through a proteolytic processing of the specific prohor-

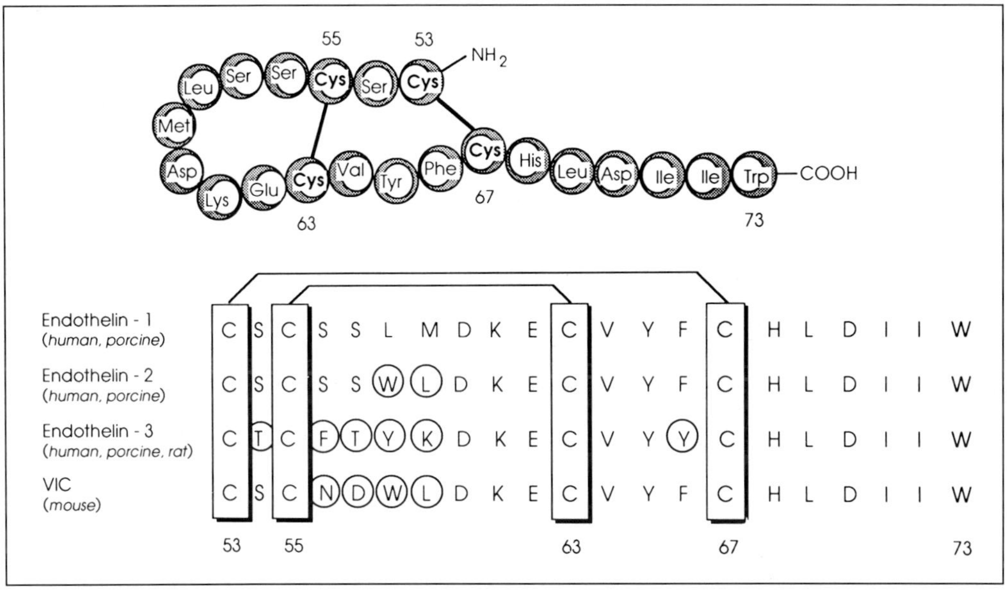

Fig. 1. Structure and sequence homology of ET isopeptides. *Top:* Proposed conformation of mature, 21-amino acid ET isopeptides. All native ETs have two intramolecular disulfide bonds. *Bottom:* Amino acid sequence homology of the three isoforms of ET and the vasoactive intestinal contractor (VIC) from different species. Bars represent disulfide bridges. The circles indicate amino acid sequence changes compared with the ET-1 sequence.

mone, preproendothelin (fig. 2). According to this pathway the 203-residue prepro-ET is proteolytically cleaved by dibasic pair-specific endopeptidases and carboxypeptidases to form a 39-amino acid intermediate pro-ET (big endothelin). Big-ET is then converted to the mature ET via a putative protease, named ET-converting enzyme, that cleaves the precursor fragment through an unusual proteolytic processing between Trp^{21} and Val^{22}. The final product is the 21-amino acid peptide ET containing amino acids Cys^{53} to Trp^{73}.

The vasoconstrictor activity of Big-ET-1 is about 100-fold lower in vitro [14], but comparable in vivo [14, 15] to that of mature peptide. This suggests that conversion of Big-ET to ET is essential for bioactivity of mature peptide and the putative converting enzyme, most likely a membrane bound neutral metalloproteinase [16, 17], resides within the vessels.

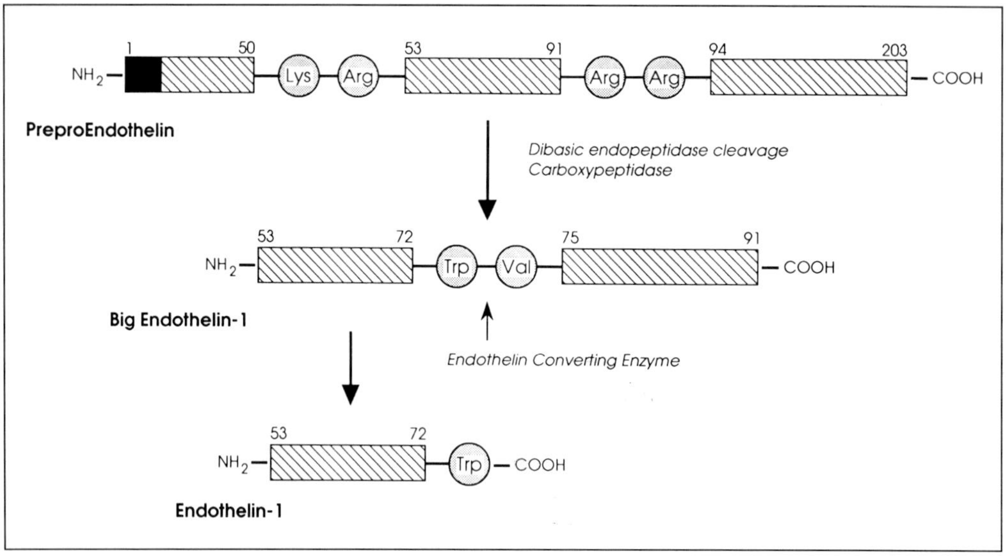

Fig. 2. The proposed pathway of ET biosynthesis. Dark box represents putative signal sequence. The porcine/human prepro-ET, containing 203 amino acids, is believed to be cleaved to the 39-amino acid residue intermediate (big ET-1) by dibasic endopeptidases and carboxypeptidases. Big ET-1 is then converted to the mature ET-1 by a specific endopeptidase referred to as ET-converting enzyme, which cleaves the precursor peptide at the Trp^{73}-Val^{74} bond.

The cellular origin of prepro-ET have been extensively investigated. Yanagisawa et al. [2] first reported that prepro-ET mRNA was constitutively expressed in cultured aortic endothelial cells, but then failed to detect ET gene expression in microvascular endothelium of brain, heart, lung or kidney. Later on, however, measurable levels of prepro-ET mRNA were found in many tissues from various animal species [18]. In this respect, focussing on the kidney, prepro-ET mRNA expression has been shown in cultured bovine glomerular capillary endothelial cells associated with the release of the corresponding peptide in the cell supernatant, suggesting that renal microvasculature is capable of producing ET [19]. Of particular interest, a transcript of identical molecular weight has been found in RNA obtained from resting rat [20] and human [21] mesangial cells in culture, and from several cultured tubular epithelial cell lines [22], indicating that renal cells other than glomerular endothelial cells constitutively express prepro-ET gene. These observations have been recently confirmed by Ujiie et al. [23] who investigated the distribution of ET-1 mRNA and ET-1 production from microdissected nephron segments in the rat. They have been able to detect polymerase chain reaction product of prepro-ET-1 mRNA in glomeruli and inner medulla collecting duct, but not in other nephron segments.

Prepro-ET-1 mRNA expression and the rate of peptide release in cultured cells is enhanced by a variety of agents and mechanical stimuli including thrombin, transforming growth factor-β (TGF-β) [24], angiotensin II, epinephrine [2], arginine vasopressin, bradykinin, interleukin-1 [11, 25–28], the calcium ionophore A23187, ionomycin, phorbol esters, and hemodynamic shear stress [29].

Receptor Subtypes

ET acts via cell-surface receptors, whose binding is rapid, specific and saturable [30]. However, the dissociation rate of ET from its receptor is low and may not be entirely reversible [31–33] providing a possible explanation for the characteristically long-lasting vasoconstrictor activity of the mature peptide [34]. Specific binding sites have been identified in several organs, including lung, heart, kidney, intestine, adrenal gland and brain [35–39]. Despite the fact that the overall density is somewhat less in the kidney than in lung and heart [35, 38, 39], autoradiographic studies using ^{125}I-labeled ET-1 reveal high affinity binding to glomeruli, vasa recta bundles, and the inner medulla, but with lesser extent to proximal tubules [35, 37, 40]. The discovery of various ET isoforms and the diversity of biological functions would predict the existence of more than one ET receptor. In this respect, saturation and competitive displacement binding studies have indicated that receptor subtypes exist with tissue-specific distribution [34, 41–43]. Lack of linearity of ET-1 binding to mesangial cell membrane suggested the possibility of multiple classes of ET receptors [44] and recently at least two ET receptor subtypes have been cloned and sequenced [45, 46]. The receptor subtype cloned by Arai et al. [46] from bovine tissues, indicated as ET_A receptor, is highly specific for ET-1 and probably represents the vascular smooth muscle ET receptor. By contrast, the second receptor subtype (ET_B receptor) isolated from rat lung [45] does not discriminate between the three ET isoforms and is probably expressed on endothelial cells. More recently, evidence has been provided by cross-linking of ^{125}I-ET-1 to intact rat mesangial cells [47] of two receptors of 60 and 73 kDa with high affinity for the three ET isotypes or for ET-1 and

ET-2 respectively. The regulation of ET receptor number, activity and turnover and the physiological significance of receptor subtypes remains ill defined, however. Such information will help designing specific antagonists that may allow to define selective responses to the various isoforms and finding pharmacologic strategies for ET-mediated diseases.

Extrarenal and Renal Action

ETs contract vascular smooth muscle cells [2] and have strong inotropic [48] and chronotropic [49] effects on the myocardium. The smooth muscle cell contraction is largely the result of increasing the inositol phosphate turnover in conjunction with a rise in intracellular free calcium that regulate actin-myosin interactions and generate tension [30]. Intravenous bolus injection of ET in animals elicits a biphasic systemic pressor response, characterized by an initial rapid and transient systemic vasodilation (ET-1 > ET1 > ET2) [50] followed by sustained rise in arterial blood pressure [51, 52]. While the sustained pressor response is due to a direct vasoconstrictor effect of the peptide, the initial transient drop in blood pressure has been attributed to the release of vasodilator substances, including prostacyclin and endothelium-derived relaxing factor (EDRF), from vascular endothelium [53]. Of note, a regional selectivity of ET-induced vasoactive effects has been shown with vasodilation being more prominent in the muscle, whereas vasoconstriction mostly localized in mesenteric and renal vessels [54].

Besides these vasoactive properties, ET induces mitogenesis of several types of cells in culture including vascular smooth muscle [55, 56] and mesangial cells [4, 9] by up-regulating the expression of proto-oncogenes, c-fos [9, 55], c-myc [55], and VL-30 [57]. All these inducible genes convert short-term, transmembrane signals into long-term responses requiring transcriptional regulation of target genes, and are activated by other pro-mitogenic agents including epidermal growth factor and phorbol ester [58]. Multiple pathways mediate nuclear signaling by ET, the most plausible being both the increase in intracellular [Ca^{2+}] and protein kinase C activity [30]. In many tissues cells expressing prepro-ET gene are proximal to target cells bearing ET binding sites. These include smooth muscle cells, fibroblasts or pericytes [2, 30, 36, 59]. Thus, ETs are most likely local hormones with a paracrine activity on nearly all cells.

While the renal vascular bed is highly sensitive to the vasoconstrictor activity of exogenously administered ET, no data are available on whether endogenously synthesized ET contributes to maintain renal function in normals. Systemic infusion of ET-1 to intact animals transiently reduces arterial blood pressure followed by a sustained hypertension with an intense dose-dependent renal vasoconstriction [2, 5, 6, 60–67]. This results in a reduction of renal plasma flow (RPF) with variable effects on glomerular filtration rate (GFR) depending on the dose infused [5]. However, if ET-1 is infused directly into the renal artery, RPF and GFR decline in parallel [5, 63, 67]. At glomerular level, infusion of mild pressor dose leads to an increase in efferent and afferent arteriolar resistance, the former being proportionally greater so that mean glomerular capillary hydraulic pressure increases [5]. Increasing the dose of ET up to 300 pmol resulted in an intense constriction of both afferent and efferent arterioles with a consequent marked fall in glomerular plasma flow and SNGFR [4].

Attempts to prevent the renal vasoconstrictor response to exogenous ET with pharmacological manipulations, suggest that the renal vascular contraction is at least in part due to activation of voltage-sensitive Ca^{2+} channels [63]. Whether arachidonic acid metabolites via cyclo-oxygenase enzyme pathway participate in mediating such a phenomenon remains still controversial [68].

The 'Remnant Glomerulus': Adaptive Changes in Unaffected Capillaries Within a Glomerulus with Segmental Sclerosis

Increasing evidence is available for a possible role of ET in acute renal failure, which includes the effect of an anti-ET antibody or a specific ET receptor antagonist in ameliorating acute renal vasoconstriction induced by transient bilateral renal artery occlusion [69, 70] or by ciclosporin infusion [71, 72] in rat, and the enhanced plasma levels of ET upon exposure of animals to renal ischemia [73], ciclosporin [74] or endotoxin [75, 76]. By contrast, whether ET participates in the sequence of events underlying progressive deterioration of renal function in chronic nephropathies is still a matter of investigation. Studies aimed at clarifying the progressive nature of renal injury in experimental disease of the kidney have indicated that glomerular capillary hypertension is the key determinant for the progressive deterioration of renal function quite independently from the nature of the initial insult [77, 78]. Alternative mechanisms have also been proposed which include activation of coagulation and intraglomerular thrombosis [79], intrarenal calcium deposits [80] and glomerular hypertrophy [81]. Recent attention focussed on the possibility that altered glomerular permeability to proteins plays a role in glomerulosclerosis [82–84]. Actually the intense protein reabsorption activity of proximal tubular cells that occurs when glomerular permselective properties are lost [85] led to an abnormal amount of protein [83] and lipids [86] into the renal interstitium. This causes an inflammatory reaction [83, 86] which in most models precedes and may actually promote interstitial fibrosis and segmental glomerulosclerosis [83, 84, 87–89]. Segmental areas of glomerulosclerosis result from the occlusion of some glomerular capillaries where extracellular matrix accumulates while remaining open capillaries – 'remnant glomerulus' – undergo major hemodynamic changes due to blood flow redistribution. We propose that ET synthesis is altered in such circumstances and this would contribute to progressive ischemia of other capillaries and to the extension of the sclerotic process to the entire glomerular tuft.

Does the Intrarenal Synthesis of ET Contribute to the Development of Glomerular Obsolescence?

Hemodynamic Pathway

Mechanical forces generated by the redistribution of blood flowing under pressure may profoundly influence the biology of endothelial cell lining the glomerular capillaries. Actually the endothelium transduces hemodynamic force signals to smooth muscle cells through the release of vasoactive substances that ultimately modulate the local vascular tone [90, 91]. However, how endothelial cells sense fluid forces and couple the initial mechanical deformation of the plasma membrane to the release of vasoactive substance is not yet known. In vitro findings that controlled levels of fluid shear stress on the endothelium regulates the transmembrane flow of potassium ions [92] through the activation of endothelial potassium channel [93] raises the possibility that membrane hyperpolarization promotes the release of vasoactive substances. Whether this occurs directly or by enhancing intracellular free Ca^{2+} is a matter of current investigations [90, 94]. Of interest, the rise in intracellular Ca^{2+} has been recognized as one of the stimuli inducing the release of prostacyclin and EDRF (EDRF/nitric oxide) [93]. Besides vasorelaxants, mechanical forces are capable of promoting the release from vascular endothelium of soluble factors with vasoconstrictor activity [95]. This is considered a local response for keeping blood flow relatively constant over a wide range of arterial perfusion pressures. Thus, low fluid shear stress of 5 dyn/cm^2 on cultured porcine aortic endothelial cells stimulates prepro-ET gene expression and the release of the mature peptide into cultured medium [29]. In different experimental settings, Sharefkin et al. [96] have recently showed that application of high laminar shear stress (25 dyn/cm^2) for 24 h to primary cultures of human umbilical vein endothelial cells suppressed both the precursor gene mRNA transcript level and the rate of ET-1 release into culture medium. This discrepancy may reflect both species differences and/or the degree of the applied mechanical forces. At glomerular level, one can postulate that ET synthesis is altered in the remnant capillaries because of the redistribution of blood flow and the associated changes in shear forces on microvascular endothelium. ET-1 generated by glomerular endothelial cell would act in paracrine fashion on mesangial cells which have ET receptors [47]. Actually in vitro experiments have documented that ET stimulates mitogenesis of cultured rat mesangial cells coupled with the increased expression of proto-oncogenes c-fos and c-myc [9] that are presumably involved in regulating DNA replication.

Proliferating mesangial cells up-regulate extracellular matrix protein gene expression which may favor the development of glomerular scarring [97, 98]. Further support to this possibility is the observation that in rat mesangial cells ET increased expression of mRNA for collagen type I, III and IV, and laminin [99]. Whether the accumulation of extracellular matrix is a consequence of enhanced translation of the proteins or of an inhibition of degradation is open to speculation. Besides being mitogenic, ET activates mesangial cell phospholipase A_2 and promotes arachidonate release from membrane phospholipids [100]. This leads to excessive formation of TxA_2 [101] which is also a stimulus for up-regulating extracellular matrix protein gene expression [102]. Thus, TxA_2-mediated extracellular matrix protein accumulation may represent an additional pathway by which ET may promote glomerulosclerosis.

Inflammatory Pathway

Besides glomerular endothelial cells, resting human mesangial cells in culture express prepro-ET mRNA [21]. The constitutive expression of the gene was associated with the release of the corresponding protein into the culture supernatant, suggesting that mesangial cells possess the machinery to translate RNA information into the mature ET [21]. The demonstration of abundant and specific receptors for ET on mesangial cells [44] associated with their ability to release the active peptide would suggest that proliferation of mesangial cells induced by ET may also occur by an autocrine

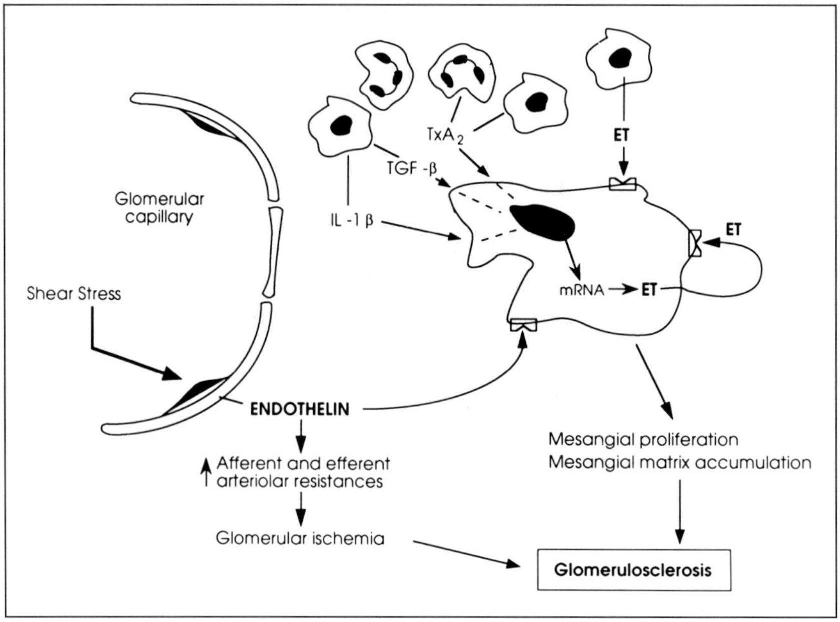

Fig. 3. Proposed mechanisms by which the increased intrarenal synthesis of ET might contribute to the development of glomerular obsolescence. Major hemodynamic changes due to blood flow redistribution in the 'remnant glomerulus' might promote the release of ET from glomerular endothelial cells through mechanical forces. Released ET enhanced afferent and efferent arteriolar resistances causing persistent glomerular hypoperfusion which, in turn, could result in a further downstream glomerular damage that is irreversible. Besides the hemodynamic pathway, ET generated by glomerular endothelial cells would act in paracrine fashion on mesangial cells through specific receptors. Moreover, local mediators from inflammatory cells recruited into the interstitium during the process of renal disease progression stimulate mesangial cells to increase ET synthesis and to proliferate in response to ET. Proliferating mesangial cells up-regulate extracellular matrix protein production, which may favor the development of glomerular scarring.

mechanism. More importantly, expression of prepro-ET mRNA was stimulated up to 3- to 8-fold over steady-state levels when mesangial cells were incubated for 6 h with TGF-β, the chemically stable TxA_2 mimetic U-46619, and thrombin, but not with interleukin-1β (IL-1β) [21]. Since in the mesangial cells IL-1β has been reported to induce granulocyte-macrophage colony-stimulating factor and IL-6-specific mRNA transcripts [103], the existence of different and selective pathways of mesangial cell activation were hypothesized. The stimulatory effect of these proinflammatory agents on prepro-ET gene expression was coupled which the increased synthesis and release of ET in the cell supernatant [21]. Similar findings were recently obtained in rat mesangial cells in culture [20].

Increased gene expression and production of ET in mesangial cells exposed to TGF-β, TxA_2 analogue, and thrombin may have theoretical implications for the progression of glomerular injury [104]. It is tempting to speculate that inflammatory cells recruited into the interstitium during the process of renal disease progression release substances which stimulate mesangial cells to increase ET synthesis and proliferate in response to ET. Furthermore, resting or stimulated cultured human macrophages produce and release ET as documented by various studies using different methodologies [105]. Given the fact that in the progressive renal diseases macrophages are a major cellular component of the interstitial infiltrate [106] found in close proximity to sclerotic glomeruli, one can postulate that macrophage-derived ET also participates in the injury.

Altogether, these findings support the working hypothesis that ET generated by various cell sources upon chemical and mechanical stimuli may play an important role in glomerular remodeling of progressive renal diseases (fig. 3).

In vivo Evidence for an Increased Renal ET Synthesis Associated with Renal Disease Progression

The issue of whether enhanced ET synthesis takes place in vivo in glomerular diseases has been recently addressed in rats after renal mass ablation, a model of chronic renal disease characterized by compensatory glomerular hemodynamic alterations and progressive glomerulosclerosis [78, 106]. In these animals, while no changes were found in the plasma levels, the urinary excretion rate of ET was significantly increased 45 days after surgery compared to sham-operated rats (fig. 4) [10]. Whether the enhanced urinary ET excretion may reflect extrarenal or renal synthesis of the peptide has not been conclusively established so far. Since the estimated plasma half-time ($t^{1/2}$) after a single intravenous injection of ET is <1 min [107], the changes in plasma ET levels are probably of limited pathophysiological significance and do not allow a conclusion of the extrarenal origin of the urinary ET. That the kidney contributes to the enhanced urinary excretion of ET is supported by the fact that renal corti-

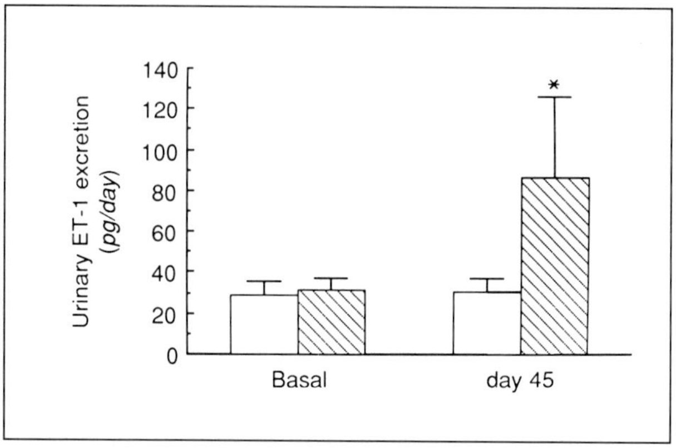

Fig. 4. Urinary excretion of ET-1 in rats before (basal) and 45 days after sham operation (□) or renal mass ablation (▨). Values are mean ± SD. * $p < 0.01$ vs. sham at the same time point and vs. basal [from ref. 10].

cal tissue and isolated glomeruli from animals with renal mass reduction generate more ET than control kidney [10]. Moreover, after a 50-min intravenous infusion of radiolabeled ET to normal rats and animals with renal mass ablation, less than 0.3 and 0.03%, respectively, of total infused radioactivity was recovered in the urine [10], suggesting that urinary excretion of ET may not reflect events occurring in the systemic circulation but rather its renal synthesis. The pathobiological significance of the enhanced renal production of ET in renal disease progression has been recently reinforced by the demonstration of a parallel increase over an 80-day follow-up in glomerular ET gene expression and in the percentage of sclerotic glomeruli in rats injected with puromycin aminonucleoside and uninephrectomy [108].

Few human data are also available. Thus, urinary excretion of ET-1 is increased in patients with chronic progressive nephropathies, including membranous proliferative glomerulonephritis, IgA nephropathy, minimal change disease, focal glomerulosclerosis and lupus nephritis [109]. Of interest to the purpose of the present review, among such diseases the highest values were found in a single patient affected by focal glomerulosclerosis. Besides that, evidence in humans of a cause-effect relationship between abnormal ET generation and development of sclerotic lesion is still missing. However, immunohistochemical studies performed in patients with symptomatic atherosclerosis showed an increased local concentration of ET in atherosclerotic aorta [110]. Whether this also applies to glomerulosclerosis is presently under investigation.

Despite the fact that increasing evidence suggests that ET synthesis is stimulated in renal disease states, only the development of effective pharmacologic interventions, namely specific ET receptor blockade, will ultimately clarify the role of ET in the process of progressive deteriorating renal function that characterizes most experimental and human nephropathies.

References

1 Masaki T, Kimura S, Yanagisawa M, Goto K: Molecular and cellular mechanism of endothelin regulation. Implication for vascular function. Circulation 1991;80:219–233.

2 Yanagisawa M, Kurihara H, Kimura S, et al: A novel potent vasoconstrictor peptide produced by vascular endothelial cells. Nature (Lond) 1988;332:411–415.

3 King AJ, Brenner BM: Endothelium-derived vasoactive factors and renal vasculature. Am J Physiol 1991;260:R653–R662.

4 Badr KF, Murray JJ, Breyer MD, Takahashi K, Inagami T, Harris RC: Mesangial cell, glomerular and renal vascular responses to endothelin in the rat kidney. J Clin Invest 1989;83:336–342.

5 King AJ, Brenner BM, Anderson S: Endothelin: a potent renal and systemic vasoconstrictor peptide. Am J Physiol 1989;256:F1051–F1058.

6 Loutzenhiser R, Epstein M, Hayashi K, Horton C: Direct visualization of effects of endothelin on the renal microvasculature. Am J Physiol 1990;258:F61–F68.

7 Yuan BH, McMurtry IF, Conger JD: Effect of endothelin on isolated perfused rat afferent and efferent arterioles (abstract). Clin Res 1989;37:586A.

8 Rubanyi GM, Parker Botelho LH: Endothelins. FASEB J 1991;5:2713–2720.

9 Simonson MS, Wann S, Mené P, et al: Endothelin stimulates phospholipase C, Na/H exchange, c-fos expression, and mitogenesis in rat mesangial cells. J Clin Invest 1989;83:708–712.

10 Benigni A, Perico N, Gaspari F, et al: Increased renal endothelin production in rats with reduced renal mass. Am J Physiol 1991;260:F331–F339.

11 Inoue A, Yanagisawa M, Kimura S, et al: The human endothelin family: three structurally and pharmacologically distinct isopeptides predicted by three separate genes. Proc Natl Acad Sci USA 1989;86:2863–2867.

12 Kloog Y, Ambar I, Sokolovsky M, Wollberg Z: Sarafotoxin, a novel vasoconstrictor peptide: phosphoinositide hydrolysis in rat heart and brain. Science 1988;242:268–270.

13 Kloog Y, Sokolovsky M: Similarities in mode and sites of action of sarafotoxins and endothelins. Trends Pharmacol Sci 1989;10:212–214.

14 Kimura S, Kasuya Y, Sawamura T, et al: Conversion of big endothelin-1 to 21-residue endothelin-1 is essential for expression of full vasoconstrictor activity: struc-

ture-activity relationships of big endothelin-1. J Cardiovasc Pharmacol 1989; 13(suppl 5):S5–S12.

15 Kashiwabara T, Inagaki Y, Ohta H, et al: Putative precursors of endothelin have less vasoconstrictor activity in vitro but a potent pressor effect in vivo. FEBS Lett 1989; 247:73–76.

16 Okada K, Miyazaki Y, Takada J, Matsuyama K, Yamaki T, Yano M: Conversion of big endothelin-1 by membrane-bound metalloendopeptidase in cultured bovine endothelial cells. Biochem Biophys Res Commun 1990;171:1192–1198.

17 Ikegawa R, Matsumura Y, Tsukahara Y, Takaoka M, Morimoto S: Phosphoramidon, a metalloproteinase inhibitor, suppresses the secretion of endothelin-1 from cultured endothelial cells by inhibiting a big endothelin-1 converting enzyme. Biochem Biophys Res Commun 1990;171:669–675.

18 Parker Botelho LH, Cade C, Phillips PE, Rubanyi GM: Tissue specificity of endothelin synthesis and binding; in Rubanyi GM (ed): Endothelin. New York, Oxford University Press, 1991.

19 Marsden PA, Dorfman DM, Collins T, Brenner BM, Orkin SH, Ballermann BJ: Regulated expression of endothelin-1 in glomerular capillary endothelial cells. Am J Physiol 1991;261:F117–F125.

20 Sakamoto H, Sasaki S, Hirata Y, et al: Production of endothelin-1 by rat cultured mesangial cells. Biochem Biophys Res Commun 1990;169:462.

21 Zoja C, Orisio S, Perico N, et al: Constitutive expression of endothelin gene in cultured human mesangial cells and its modulation by transforming growth factor-β, thrombin, and thromboxane A_2 analogue. Lab Invest 1991;64:16–20.

22 Uchida S, Naruse M, Ogata E, Yanagisawa M, Masaki T, Kurokawa K: Gene expression and synthesis of endothelin in renal tubular epithelial cells in culture (abstract). Kidney Int 1990;37:380.

23 Ujiie K, Tereda Y, Nonoguchi H, Shinohara M, Tomita K, Marumo F: Endothelin-1 mRNA and endothelin-1 production along the rat nephron segments (abstract). J Am Soc Nephrol 1991;2:420.

24 Kurihara H, Yoshizumi M, Takaku M, et al: Transforming growth factor B stimulates the expression of endothelin mRNA by vascular endothelial cells. Biochem Biophys Res Commun 1989;159:1435–1440.

25 Yanagisawa M, Inoue A, Takuwa Y, Mitsui Y, Kobayashi M, Masaki T: The human preproendothelin-1 gene: Possible regulation by endothelial phosphoinositide turnover signaling. J Cardiovasc Pharmacol 1989;13(suppl 5):S13–S17.

26 Emori T, Hirata Y, Ohta K, Shichiri M, Marumo F: Secretory mechanism of immunoreactive endothelin in cultured bovine endothelial cells. Biochem Biophys Res Commun 1989;160:93–100.

27 Yoshizumi M, Kurihara H, Morita T, et al: Interleukin-1 increases the production of endothelin-1 by cultured endothelial cells. Biochem Biophys Res Commun 1990; 166:324–329.

28 Goto K, Kasuya Y, Matsuki N, et al: Endothelin activates the dihydropyridine-sensitive, voltage-dependent Ca^{2+} channel in vascular smooth muscle. Proc Natl Acad Sci USA 1989;86:3915–3918.

29 Yoshizumi M, Kurihara H, Sugiyama T, et al: Hemodynamic shear stress stimulates endothelin production by cultured endothelial cells. Biochem Biophys Res Commun 1989;161:859–864.

30 Simonson MS, Dunn MJ: Cellular signaling by peptides of the endothelin gene family. FASEB J 1990;4:2989–3000.

31 Hirata Y, Yoshimi H, Takaichi S, Yanagisawa M, Masaki T: Binding and receptor down-regulation of a novel vasoconstrictor endothelin in cultured rat vascular smooth muscle cells. FEBS Lett 1988;239:13–17.

32 Clozel M, Fischli W, Guilly C: Specific binding of endothelin on human vascular smooth muscle cells in culture. J Clin Invest 1989;83:1758–1761.

33 Baldi E, Dunn MJ: Saturable high affinity endothelin binding sites in cultured rat mesangial cells (abstract). Kidney Int 1990;37:363A.

34 Yanagisawa M, Masaki T: Endothelin, a novel endothelium-derived peptide. Biochem Pharmacol 1989;38:1877–1883.

35 MacCumber MW, Ross CA, Glaser BM, Snyder SH: Endothelin: visualization of mRNA by in situ hybridization provides evidence for local action. Proc Natl Acad Sci USA 1989,86:7285–7289.

36 MacCumber MW, Ross CA, Snyder SH: Endothelin in brain: receptors, mitogenesis and biosynthesis in glial cells. Proc Natl Acad Sci USA 1990;87:2359–2363.

37 Kohzuki M, Johnston CI, Chai SY, Casley DJ, Mendelsohn FAO: Localization of endothelin receptors in rat kidney. Eur J Pharmacol 1989;160:193–194.

38 Davenport AP, Nunez DJ, Hall JA, Kaumann AK, Brown MJ: Autoradiographical localization of binding sites for porcine [125]endothelin-1 in humans, pigs, and rats: functional relevance in humans. J Cardiovasc Pharmacol 1989;13(suppl 5):S166–S170.

39 Hoyer D, Waeber C, Palacios JM: [125]Endothelin-1 binding sites: autoradiographic studies in the brain and periphery of various species including humans. J Cardiovasc Pharmacol 1989;13(suppl 5):S162–S165.

40 Koseki C, Imai M, Hirata Y, Yanagisawa M, Masaki T: Autoradiographic distribution in rat tissues of binding sites for endothelin: a neuropeptide? Am J Physiol 1989;256:R858–R866.

41 Watanabe H, Miyazaki H, Kondoh M, et al: Two distinct types of endothelin receptors are present on chick cardiac membranes. Biochem Biophys Res Commun 1989; 161:1252–1259.

42 Martin ER, Marsden PA, Brenner BM, Ballermann BJ: Identification and characterization of endothelin binding sites in rat renal papillary and glomerular membranes. Biochem Biophys Res Commun 1989;162:130–137.

43 Kloog Y, Bousso-Mittler D, Bdolah A, Sokolovsky M: Three apparent receptor subtypes for the endothelin/sarafotoxin family. FEBS Lett 1989;253:199–202.

44 Badr KF, Munger KA, Sugiura M, Snajdar RM, Schwartzberg M, Inagami T: High and low affinity binding sites for endothelin on cultured rat glomerular mesangial cells. Biochem Biophys Res Commun 1989;161:776–781.

45 Sakurai T, Yanagisawa M, Takuwa Y, et al: Cloning of a cDNA encoding a nonisopeptide-selective subtype of the endothelin receptor. Nature (Lond) 1990;348:732–735.

46 Arai H, Hori S, Aramori I, Ohkubo H, Nakanishi S: Cloning and expression of a cDNA encoding an endothelin receptor. Nature (Lond) 1990;348:730–732.

47 Martin ER, Brenner BM, Ballermann BJ: Heterogeneity of cell surface endothelin receptors. J Biol Chem, in press.

48 Moravec CS, Reynolds EE, Stewart RW, Bond M: Endothelin is a positive inotropic

agent in human and rat heart in vitro. Biochem Biophys Res Commun 1989;159: 14–18.

49 Ishikawa T, Yanagisawa M, Kimura S, Goto K, Masaki T: Positive chronotropic effects of endothelin, a novel endothelium-derived vasoconstrictor peptide. Pflügers Arch 1988;413:108–110.

50 Warner TD, Mitchell JA, De Nucci G, Vane JR: Endothelin-1 and endothelin-3 release EDRF from isolated perfused arterial vessels of the rat and rabbit. J Cardiovasc Pharmacol 1989;13(suppl 5):S85–S88.

51 Hom GJ, Schutt WA Jr, Rubanyi GM: Potential mechanisms of endothelin-induced transient decreases in arterial pressure in spontaneously hypertensive rats; in Rubanyi GM, Vanhoutte PM (eds): Endothelium-Derived Contracting Factors. Basel, Karger, 1990, pp 98–103.

52 Winquist RJ, Bunting PB, Lumma PK, Garsky VM, Scott AL, Vlasuk GP: Depressor response to endothelin in normotensive and hypertensive rats; in Rubanyi GM, Vanhoutte PM (eds): Endothelium-Derived Contracting Factors. Basel, Karger, 1990, pp 104–109.

53 De Nucci G, Thomas R, D'Orleans-Juste P, et al: Pressor effects of circulating endothelin are limited by its removal in the pulmonary circulation and by the release of prostacyclin and endothelium-derived relaxing factor. Proc Natl Acad Sci USA 1988;85:9797–9800.

54 Gardiner SM, Compton AM, Bennett T: Regional hemodynamic effects of endothelin-1 in conscious, unrestrained, Wistar rats. J Cardiovasc Pharmacol 1989; 13(suppl 5):S202–S204.

55 Komuro I, Kurihara, H, Sugiyama T, Takaku F, Yazaki Y: Endothelin stimulates c-fos and c-myc expression and proliferation of vascular smooth muscle cells. FEBS Lett 1988;238:249–252.

56 Dubin D, Pratt RE, Cooke JP, Dzau VJ: Endothelin, a potent vasoconstrictor, is a vascular smooth muscle mitogen. J Vasc Med Biol 1989;1:150–154.

57 Muldoon L, Rodland KD, Forsythe ML, Magun BE: Stimulation of phosphatidylinositol hydrolysis, diacylglycerol release, and gene expression in response to endothelin, a potent new agonist for fibroblasts and smooth muscle cells. J Biol Chem 1989,264:8529–8536.

58 Rodland KD, Muldoon LL, Dinh T-H, Magun BE: Independent transcriptional regulation of a single VL30 element by epidermal growth factor and activators of protein kinase C. Mol Cell Biol 1988;8:2247–2250.

59 Yoshizawa T, Shinmi O, Giaid A, et al: Endothelin: a novel peptide in the posterior pituitary system. Science 1990;247:462–464.

60 Cairns HS, Rogerson ME, Fairbanks LD, Neild GH, Westwick J: Endothelin induces an increase in renal vascular resistance and a fall in glomerular filtration rate in the rabbit isolated perfused kidney. Br J Pharmacol 1989;98:155–160.

61 Firth JD, Ratcliffe PJ, Raine AEG, Ledingham JGG: Endothelin: an important factor in acute renal failure. Lancet 1988;ii:1179–1181.

62 Hirata Y, Matsuoka H, Kimura K, et al: Renal vasoconstriction by the endothelial cell-derived peptide endothelin in spontaneously hypertensive rats. Circ Res 1989; 65:1370–1379.

63 Katoh T, Chang H, Uchida S, Okuda T, Kurakawa K: Direct effects of endothelin in the rat kidney. Am J Physiol 1990;258:F397–F402.

64 Miller WL, Redfield MM, Burnett JC: Integrated cardiac, renal and endocrine actions of endothelin. J Clin Invest 1989;83:317–320.
65 Pernow J, Bouttier JF, Franco-Cereceda A, Lacroix JS, Natran R, Lundberg JM: Potent selective vasoconstrictor effects of endothelin in the pig kidney in vivo. Acta Physiol Scand 1988;134:373–374.
66 Yokokawa K, Kohno M, Murakawa K, et al: Acute effects of endothelin on renal hemodynamics and blood pressure in anesthetized rats. Am J Hypertens 1989;2: 715–717.
67 Perico N, Plata Cornejo R, Benigni A, Malanchini B, Ladny JR, Remuzzi G: Endothelin induces diuresis and natriuresis in the rat by acting on proximal tubular cells through a mechanism mediated by lipoxygenase products. J Am Soc Nephrol 1991; 2:57–69.
68 Munger KA, Takahashi K, Ebert J, Badr KF: Role for cyclooxygenase products in mediating rat renal responses to endothelin (abstract). Kidney Int 1989;37:375.
69 Kon V, Yoshioka T, Fogo A, Ichikawa I: Glomerular actions of endothelin in vivo. J Clin Invest 1989;83:1762–1767.
70 Chittinandana A, Chan L, Suleymanlar G, Shapiro JI, Schrier RW: Effects of endothelin antagonist on ischemic acute renal failure (abstract). J Am Soc Nephrol 1991; 2:644.
71 Perico N, Dadan J, Remuzzi G: Endothelin mediates the renal vasoconstriction induced by cyclosporine in the rat. J Am Soc Nephrol 1990;1:76–83.
72 Kon V, Sugiura M, Inagami T, Harvie BR, Ichikawa I, Hoover RL: Role of endothelin in cyclosporine-induced glomerular dysfunction. Kidney Int 1990;37:1487–1491.
73 Tomita K, Ujiie K, Nakanishi T: Plasma endothelin levels in patients with acute renal failure. N Engl J Med 1989;321:1127.
74 Fogo A, Hakim RC, Sugiura M, Inagami T, Kon V: Severe endothelial injury in a renal transplant patient receiving cyclosporine. Transplantation 1990;49:1190–1192.
75 Sugiura M, Inagami T, Kon V: Endotoxin stimulates endothelin release in vivo and in vitro as determined by radioimmunoassay. Biochem Biophys Res Commun 1989; 161:1220–1227.
76 Pernow J, Hemsen A, Lundberg JM: Increased plasma levels of endothelin-like immunoreactivity during endotoxin administration in the pig. Acta Physiol Scand 1989;137:317–318.
77 Brenner BM, Meyer TW, Hostetter TH: Dietary protein intake and the progressive nature of kidney disease: The role of hemodynamically mediated glomerular injury in the pathogenesis of progressive glomerular sclerosis in aging, renal ablation, and intrinsic renal disease. N Engl J Med 1982;307:652–659.
78 Hostetter TH, Olson JL, Rennke HG, Venkatachalam MA, Brenner BM: Hyperfiltration in remnant nephrons: a potentially adverse response to renal ablation. Am J Physiol 1981;241:F65–F93.
79 Klahr S, Heifets M, Purkerson ML: The influence of anticoagulation on the progression of experimental renal disease; in Mitch WE, Brenner BM, Stein JH (eds): The Progressive Nature of Renal Disease. New York, Churchill Livingstone, 1986, p 45.

80 Ibels LS, Alfrey AC, Huffer WE: Calcification in end-stage kidneys. Am J Med 1981; 71:33–37.

81 Yoshida Y, Fogo A, Ichikawa I: Glomerular hemodynamic changes vs. hypertrophy in experimental glomerular sclerosis. Kidney Int 1989;35:654–660.

82 Remuzzi G, Bertani T: Is glomerulosclerosis a consequence of altered glomerular permeability to macromolecules? Kidney Int 1990;38:384–394.

83 Bertani T, Cutillo F, Zoja C, Broggini M, Remuzzi G: Tubulo-interstitial lesions mediate renal damage in adriamycin glomerulopathy. Kidney Int 1986;30:488–496.

84 Bertani T, Zoja C, Abbate M, Rossini M, Remuzzi G: Age-related nephropathy and proteinuria in rats with intact kidneys exposed to diets with different protein content. Lab Invest 1989;60:196–204.

85 Maack T: Changes in the activity of acid hydrolases during renal reabsorption of lysozyme. J Cell Biol 1967;35:268–273.

86 Kees-Folts D, Schreiner GF: A lipid chemotactic factor associated with proteinuria and interstitial nephritis induced by protein overload (abstract). J Am Soc Nephrol 1991;2:548.

87 Anderson S, Diamond JR, Karnovsky MJ, Brenner BM: Mechanisms underlying transition from acute glomerular injury to late glomerular sclerosis in a rat model of nephrotic syndrome. J Clin Invest 1988;82:1757–1768.

88 Remuzzi A, Puntorieri S, Battaglia C, Bertani T, Remuzzi G: Angiotensin-converting enzyme inhibition ameliorates glomerular filtration of macromolecules and water and lessens glomerular injury in the rat. J Clin Invest 1990;85:541–549.

89 Anderson S, Rennke HG, Brenner BM: Therapeutic advantage of converting enzyme inhibitors in arresting progressive renal disease associated with systemic hypertension in the rat. J Clin Invest 1986;77:1993–2000.

90 Lansman JB: Going with the flow. Nature (Lond) 1988;331:481–482.

91 Frangos JA, Eskin SG, McIntire LV, Ives CL: Flow effects on prostacyclin production by cultured human endothelial cells. Science 1984;227:1477–1479.

92 Oleson SP, Clapham DE, Davies PF: Haemodynamic shear stress activates a K^+ current in vascular endothelial cells. Nature (Lond) 1988;331:168–170.

93 Cooke JP, Rossitch E Jr, Andon AN, Loscalzo J, Dzau VJ: Flow activates an endothelial potassium channel to release an endogenous nitrovasodilator. J Clin Invest 1991;88:1663–1671.

94 Lansman JB, Hallam TJ, Rink TJ: Single stretch-activated ion channels in vascular endothelial cells as mechanotransducers. Nature (Lond) 1987;325:811–813.

95 Hickey KA, Rubanyi G, Paul RJ, Highsmith RF: Characterization of a coronary vasoconstrictor produced by cultured endothelial cells. Am J Physiol 1985;248: C550–C556.

96 Sharefkin JB, Diamond S, Eskin SG, McIntire LV, Dieffenbach CW: Fluid flow decreases preproendothelin mRNA and suppresses endothelin-1 peptide release in cultured human endothelial cells. J Vasc Surg 1991;14:1–9.

97 Ishimura E, Sterzel RB, Budde K, Kashgarian M: Formation of extracellular matrix by cultured rat mesangial cells. Am J Pathol 1989;134:843–855.

98 Sterzel RB, Lovett DH, Foellmer HG, Perfetto M, Biemesderfer D, Kashgarian M: Mesangial cell hillocks. Nodular foci of exaggerated growth of cells and matrix in prolonged culture. Am J Pathol 1986;125:130–140.

99 Ishimura E, Shouji S, Nishizawa Y, Morii H, Kashgarian M: Regulation of mRNA expression for extracellular matrix by cultured rat mesangial cells (abstract). J Am Soc Nephrol 1991;2:546.
100 Simonson MS, Dunn MJ: Endothelin-1 stimulates contraction of rat glomerular mesangial cells and potentiates β-adrenergic-mediated cAMP accumulation. J Clin Invest 1990;85:790–797.
101 Zoja C, Benigni A, Renzi D, Piccinelli A, Perico N, Remuzzi G: Endothelin and eicosanoid synthesis in cultured mesangial cells. Kidney Int 1990;37:927–933.
102 Bruggeman LA, Horigan EA, Horikoshi S, Ray PE, Klotman PE: Thromboxane stimulates synthesis of extracellular matrix proteins in vitro. Am J Physiol 1991; 261:F488–F494.
103 Zoja C, Bettoni S, Tini ML, Abboud HE, Remuzzi G, Rambaldi A: In vitro regulation of colony stimulating factor genes by interleukin-1β and tumor necrosis factor in human mesangial cells (abstract). Kidney Int 1990;37:205.
104 Simonson MS, Dunn MJ: Endothelin peptides: a possible role in glomerular inflammation. Lab Invest 1991;64:1–4.
105 Ehrenreich H, Anderson RW, Fox CH, et al: Endothelins, peptides with potent vasoactive properties, are produced by human macrophages. J Exp Med 1990;172: 1741–1748.
106 Olson JL, Hostetter TH, Rennke HG, Brenner BM, Venkatachalam MA: Altered glomerular permselectivity and progressive sclerosis following extreme ablation of renal mass. Kidney Int 1982;22:112–126
107 Anggard E, Galfon S, Rae G, et al: The fate of radioiodinated endothelin-1 and endothelin-3 in the rat. J Cardiovasc Pharmacol 1989;13(suppl 5):S46–S49.
108 Osada S, Nakamura T, Ebihara I, Suzuki S: Endothelin gene expression in glomeruli and medulla from focal glomerulosclerosis (abstract). J Am Soc Nephrol 1991;2: 412.
109 Ohta K, Hirata Y, Shichiri M, et al: Urinary excretion of endothelin-1 in normal subjects and patients with renal disease. Kidney Int 1991;39:307–311.
110 Lerman A, Edwards BS, Hallett JW, Heublein DM, Sandberg SM, Burnett JC Jr: Circulating and tissue endothelin immunoreactivity in advanced atherosclerosis. N Engl J Med 1991;325:997–1001.

Giuseppe Remuzzi, MD, Mario Negri Institute for Pharmacological Research, Via Gavazzeni 11, I–24100 Bergamo (Italy)

Berlyne GM (ed): The Kidney Today. Selected Topics in Renal Science.
Contrib Nephrol. Basel, Karger, 1992, vol 100, pp 254–268

Acquired Cysts and Neoplasms of the Kidneys in Renal Allograft Recipients

Isao Ishikawa

Division of Nephrology, Department of Internal Medicine and Division of Clinical Research, Medical Research Institute, Kanazawa Medical University, Uchinada, Japan

The development of acquired cystic disease is observed in 40–90% of dialysis patients [1] while renal cell carcinoma occurs in 1.5% [2]. Since renal transplantation in such patients is usually performed without native nephrectomy, it would be interesting to follow the course of acquired cystic disease subsequent to this surgery.

Acquired Renal Cystic Disease in Transplant Recipients

What Is the Effect of Successful Renal Transplantation on Acquired Cystic Disease Developed during Hemodialysis Treatment?

In 1983, Ishikawa et al. [3] performed a longitudinal study in 7 renal transplant patients who had been on hemodialysis for more than 5 years due to chronic glomerulonephritis. Acquired cysts regressed markedly in 2 of these in the 8–10 months following surgery, while in another, acquired cysts first regressed then reappeared after graft failure. Three of the other patients who were observed only during the posttransplant period were found to have small scarred kidneys with few or no cysts. These results suggest that acquired renal cystic disease regresses rapidly in a fair number of cases after successful renal transplantation; however, cysts may persist in some patients.

In work predating that of Ishikawa et al. [3], Bommer et al. [4] found that 3 out of 4 patients examined by CT scan had acquired cysts after transplantation, while Krempien and Ritz [5] discovered 2 transplant recipients with acquired cysts at autopsy. Both sets of authors [4, 5] concluded that acquired cysts persist even after transplantation. Subsequently, Levine et al. [6] described 6 transplant recipients who displayed significantly shrunken kidneys and no evidence of acquired cysts and attributed this to short duration of hemodialysis and/or the effect of successful renal transplantation. In addition, Vaziri et al. [7] found acquired cysts in 11 out of 22 patients examined either at necropsy (19 patients) or bilateral nephrectomy (3 patients). The group with acquired cysts was characterized by a longer history of dialysis and shorter elapsed time since transplantation. Histologic study showed the cysts to have single-layered cuboidal epithelia or, in a few cases, multilayered epithelia; in 1 patient, cystic kidney was complicated by multifocal renal cell carcinoma. Such findings led these authors to suggest that renal transplantation may retard the evolution of acquired cysts. Minar et al. [8] came to a similar conclusion, i.e. that cystic transformation in dialysis patients may be a reversible process, on the basis of ultrasound examinations that resulted in the detection of acquired cysts in 21 out of 43 dialysis patients (49%), but in only 4 out of 23 transplant recipients (17%). Moreover, Thompson et al. [9] have reported that there is a good correlation between cyst grade and the duration of hemodialysis in patients receiving this treatment, but none between cyst grade and duration of hemodialysis prior to surgery in transplant recipients. This, again, supports the view that acquired cysts regress after renal transplantation.

The main original renal disease in the above-mentioned articles [3–9] was chronic glomerulonephritis and, to a lesser extent, end-stage renal disease of diverse etiology other than autosomal-dominant polycystic kidney disease (ADPKD). The involution of acquired cysts after successful transplantation does not seem to depend on the nature of the original disease. Researchers have noted a rapid regression of acquired cysts within 1–3 months of successful transplantation [2, 11], with the so-called 'contracted' kidneys showing only 1/4–1/5 of the extent of cystic change developed during hemodialysis (fig. 1). On the other hand, Thaysen et al. [10] have reported very slow renal volume reduction in the native kidneys (1.07%/month) after renal allograft in ADPKD patients. Dysplastic kidneys associated with renal cyst development undergo slow volume reduction, as in ADPKD, after successful renal transplantation (fig. 2). However, in the

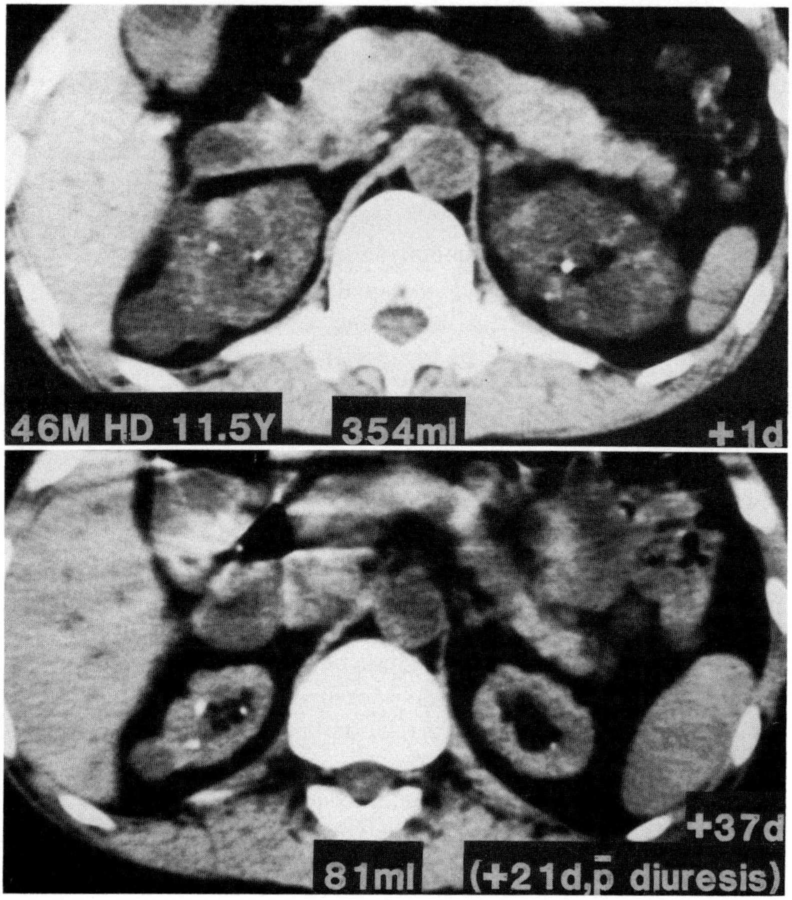

a

b

Fig. 1. a, b Extensive cystic transformation developed during 11.5 years of hemodialysis regresses 37 days after transplantation resulting in a decrease in kidney size to 1/5 that of the pretransplant kidney.

latter two diseases, it is not clear what amount of renal volume is due to uremic acquired cysts complicated with original cystic disease.

Acquired Cysts Regress after Successful Transplantation; However, the Regression Is Not Complete (table 1)

Among studies involving 20 or more renal allograft recipients, Vaziri et al. [7] discovered acquired cysts in 50% of cases (11/22 patients) inves-

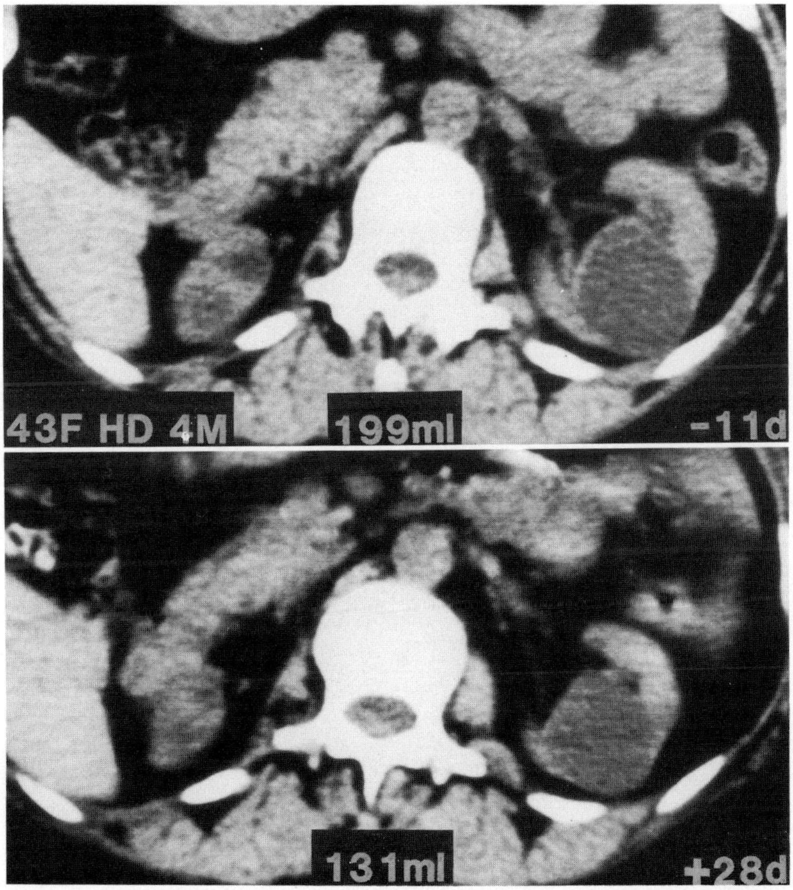

a b

Fig. 2. a, b Cysts complicated with kidney dysplasia regress after successful renal transplantation more slowly than acquired cysts.

tigated by histology, while Minar et al. [8] found cysts in 17% (4/23 patients), Thompson et al. [9] in 41% (18/44 patients), Manns et al. [12] in 37% (14/38 patients) and Ishikawa et al. [13] in 36% (22/61 patients) of patients examined radiologically. In total, 37% of these transplant patients (69/188 patients) revealed acquired cysts after surgery. Although this incidence of acquired cysts in allograft recipients is slightly lower than that seen in dialysis patients (40–90%) [1], there is a marked difference in the

Table 1. Prevalence of acquired renal cysts of the native kidney after renal transplantation

Author/reference	Diagnostic technique	Cases	Mean duration of hemodialysis years	Mean elapsed time since transplantation years	Graft function	Cases with renal cysts		Renal cell carcinomas
						n	%	
Bommer et al., 1980 [4]	CT	4	1.6	2.6	normal	3	75	0
Krempien and Ritz, 1980 [5]	autopsy	2	3.7	prolonged period since transplantation	ND[1]	2	100	0
Ishikawa et al., 1983 [3]	CT	7	6.2	2.9	good in 6 poor in 1	5	71	0
Vaziri et al., 1984 [7]	autopsy and nephrectomy	22	3.7 (+)[2] 0.8 (–)	2.4 (+) 4.3 (–)	? good	11	50	1
Minar et al., 1984 [8]	US	23	1.9	4.3	good	4	17	0
Levine et al., 1984 [6]	CT	6	ND	6.0	normal in 4 mildly impaired in 1 rejection in 1	0	0	0
Thompson et al., 1986 [9]	US	44	2.5	4.2	functioning	18	41	0
Manns et al., 1990 [12]	US	38	3.5 (+)	1.5 (+)	functioning	14	37	0
Ishikawa et al., 1991 [13]	CT	61	2.7	5.3	good	22	36	1[3]
		207				79	38	2

[1] ND: not described.
[2] (+): cases with renal cysts; (–): cases without renal cysts.
[3] Developed before transplantation.

number of cysts in a kidney between the two groups. Several longitudinal studies have demonstrated a marked regression of acquired cysts following transplantation, with only a few cysts persisting in the atrophic, scarred kidney after the operation [2, 3, 11].

Changes in the Native Kidneys in Transplant Recipients with Poor Graft Function. A slowly progressive chronic rejection episode, in which serum creatinine is higher than 3 mg/dl for more than 3 years, is characterized by the presence of acquired cysts in the native kidneys [14] which are very similar to those seen in dialysis patients or in patients with end-stage renal disease. Of 7 patients displaying such slowly progressive rejection, 5 (71.4%) developed acquired cysts in their native kidneys [14], whereas only 22 out of 61 patients (36.1%) with good graft function revealed acquired cysts [13].

Changes in the Native Kidneys in Transplant Recipients with Good Graft Function. In 14 patients with good graft function who had been on hemodialysis for greater than 3 years and followed for more than 3 years after renal allograft, bilateral renal volume was found to be 82.1 ml prior to surgery and 42.9 ml at 3 months, 35.6 ml at 1 year and 33.8 ml at 3 years after the transplantation [15]. Thus, a substantial reduction of renal size was observed as little as 3 months postoperatively.

Our own recent longitudinal study of 61 transplant recipients with good graft function (serum creatinine less than 3 mg/dl), in which a baseline survey performed between 2 months and 1 year after the operation was compared with a second study conducted an average of 5 years later, revealed no visible cysts in the native kidneys of 52.5% (32/61 patients) of the patients, no change in the number of cysts in 14.8% (9/61 patients), a reduction of the number of cysts in 14.8% (9/61 patients) and an increase in the number of cysts in 18% (11/61 patients) [13]; these data are not inconsistent with those of Vaziri et al. [7], because the duration of follow-up after transplantation was very long in our study. The last group, with newly developed cysts, was characterized by a short history of hemodialysis (13.0 ± 17.8 months) and very long elapsed time since transplantation (84.6 ± 17.8 months) (fig. 3) [13]. These cysts may be simple cysts; however, the incidence of simple cysts in the same age group in the general populations is only 0–7% [13]. Therefore, it is probable that there is a separate cystogenic mechanism in patients with good

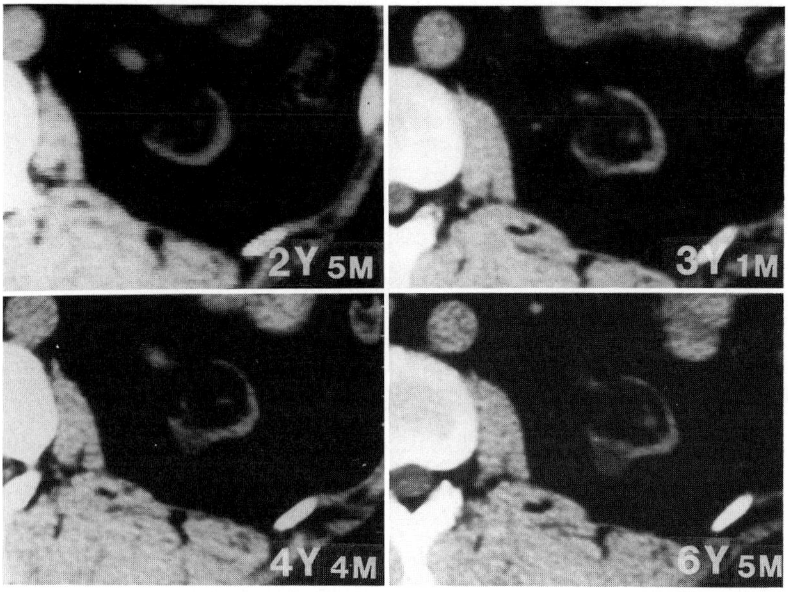

Fig. 3. a–d Even though graft function is good at 86 months' follow-up, it reveals new cyst formation in the native kidney in a small number of transplant recipients. A small cyst appeared in the native kidney of this 35-year-old male and became larger during subsequent follow-up. [From 13, with permission.]

graft function, i.e. one that differs from uremia-associated cyst formation [13].

Pathological Findings in the Native Kidneys after Renal Transplantation

Native kidneys following successful renal transplantation tend to be scarred and shrunken with either no, few or several cysts [7]. Thus, the number of cysts is considerably less than that observed in dialysis patients with acquired cystic disease [3]. However, despite their reduction in number, cysts may be not only of the single-layered epithelia type but also be seen to have multilayered epithelia, or be atypical cysts or papillary adenomas [7]. Little [16] or no [17] oxalate deposition has been observed in the renal parenchyma, and, in 2 transplant recipients, the cyst fluid composition was found to be closer to that of simple cysts than acquired cysts

[13]. However, on the whole, the pathological changes occurring in the native kidneys after transplantation may be quantitatively rather than qualitatively different from those found in acquired cystic disease associated with hemodialysis.

Clinical Implications of the Regression of Acquired Cysts after Successful Renal Transplantation

It is probable that few of the complications of acquired cystic disease develop after successful renal transplantation, because there is a marked subsequent regression of acquired cysts [3]. The incidence of de novo renal cell carcinoma after renal transplantation may be less than in hemodialysis patients if there was no renal cancer already present when this was carried out [3, 17].

In addition, retroperitoneal bleeding due to cyst rupture [18] and/or cyst infection [2] is likely to be very rare after renal transplantation. Actual data in support of these posits is not abundant; however, the reported incidence of renal cell carcinoma diagnosed after renal transplantation has been shown to be less than that detected during hemodialysis [17]. So far there have been no reports of retroperitoneal bleeding after successful renal transplantation. A possible link between posttransplant erythrocytosis and the regression of acquired cysts after transplantation requires further investigation [19].

Changes in the Transplanted Kidney

If graft function remains poor for a long period, acquired cysts develop in the graft [14] in a manner similar to the development of cysts in the native kidneys of patients with end-stage renal disease. The number of cysts seems to be lower in the graft than in the native kidneys of a given patient [14], probably because the total length of uremia is shorter in the transplanted kidney. Vaziri et al. [7] and Thompson et al. [9] found no cysts in grafts examined after renal transplantation; however, the possibility that simple cysts may be present in the donor's kidney prior to nephrectomy should be taken into consideration. In our study [13], 6 out of 20 donor kidneys had simple cysts; subsequent CT scans showed that 2 of these 6 had no cysts after transplantation while the other 4 retained them. Nineteen donated kidneys without simple cysts and with prolonged good function never developed acquired cysts, and in fact, there have been no reports of new cysts developing in the graft of patients with good graft function.

Renal Cell Carcinoma in Transplant Recipients

Renal Cell Carcinoma Diagnosed after Renal Transplantation (table 2)

Taking 17 cases of renal cell carcinoma collected in the literature [7, 20–31] by Ishikawa et al. [17], together with 9 more recently reported cases [32–37], it would appear that the average renal transplantation patients with renal cancer are in their 5th or 6th decade of life (48 ± 10 years; mean ± SD; range 28–67 years). All 26 cases involved males, usually with good graft function, who had been on hemodialysis for 3.1 ± 2.4 years (0.3–10.0 years) with a mean elapsed time since transplantation of 4.6 ± 3.1 years (0.3–13.0 years). Immunosuppression was achieved using prednisolone and azathioprine in all reported cases except for 25 described by Penn and Brunson [38] and Vogt et al. [39]. Metastasis was observed in 13 out of 23 patients, and renal cell carcinoma was complicated with acquired renal cysts in 16 out of 18. Histological examination showed 8 out of 15 patients to have clear cell carcinoma or mixed clear and granular cell carcinoma. In another study, Dlugosz et al. [27] have reported that 3 out of 222 deaths after renal transplantation were due to metastatic renal cell carcinoma.

Many researchers [3, 23, 26] believe that renal transplantation has an inhibitory effect on the development of renal cell carcinoma. This protective effect may be reversed by the use of immunosuppressive drugs [3, 17, 23, 26, 33]; however, reports of renal cell carcinoma are less common for transplant recipients than for dialysis patients. Penn and Brunson [38], in a registry survey, collected 101 cases of host kidney renal cell carcinoma (3.0%) among 3,320 malignancies that developed after transplantation. With the use of ciclosporin, the incidence of renal cell carcinoma after renal transplantation rose to 23 out of 412 malignancies (5.6%) [38]. Birkeland [40] has described 112 malignancies occurring among 3,936 transplant recipients. Four of these were renal cell carcinoma, 3 in the native kidney (0.1%) and 1 in the graft. This compares with a report by Vogt et al. [39] of 2 renal cell carcinomas (0.3%) among 598 renal transplant recipients treated with ciclosporin. Though the prevalence of renal cancer in dialysis patients is 1.5% [2], the incidence of renal cell carcinoma after renal transplantation is low [17]. Follow-up study has showed that the incidence of renal cell carcinoma in dialysis patients is 1 per 178 patient-years, while in transplant recipients it is 1 per 541 patient-years [17], supporting the previous statement.

Table 2. Renal cell carcinoma of the native kidney diagnosed after renal transplantation

	Author/reference	Age	Sex	Duration of hemodialysis years	Elapsed time since transplantation years	Renal cyst	Metastasis
1	Isiadinso et al., 1976 [20]	45	M	4	–	/	–
2	Vaziri et al., 1984 [7]	/	/	/	/	+	–
3	Struve et al., 1984 [21]	37	M	5	1	/	–
4	Griffin and Williams, 1985 [22]	67	M	0.5	3.5	/	+
5	Arias et al., 1986 [23]	46	M	4	5	+	–
6	Gunji et al., 1986 [28]	45	M	4.5	0.3	+	–
7	Kunikata et al., 1987 [29]	28	M	3	1.3	/	–
8	Faber and Kupin, 1987 [24]	45	M	3	2	+	+
9	Ludmerer and Kissane, 1988 [25]	49	M	0.8	8	+	+
10	O'Donnell et al., 1988 [26]	64	M	0.5	3	+	+
11	O'Donnell et al., 1988 [26]	57	M	/	13	–	+
12	Yoshimura et al., 1988 [30]	43	M	0.3	4.3	/	+
13	Basile et al., 1988 [32]	56	M	10	2	+	–
14	Basile et al., 1988 [32]	52	M	1.5	5	–	–
15	Dlugosz et al., 1989 [27]	/	/	/	/	/	+
16	Dlugosz et al., 1989 [27]	/	/	/	/	/	+
17	Dlugosz et al., 1989 [27]	/	/	/	/	/	+
18	Oda et al., 1989 [31]	34	M	1	5.3	+	–
19	Peabody et al., 1989 [37]	/	/	3.5	2	+	+
20	Almirall et al., 1989 [33]	57	M	5	7	+	+
21	Matson and Cohen, 1990 [34]	53	M	1.5	6	+	/
22	Matson and Cohen, 1990 [34]	59	M	3.2	5	+	/
23	Matson and Cohen, 1990 [34]	41	M	4	7	+	/
24	Garcia de la Oliva and Molina, 1990 [35]	47	M	6	1	+	+
25	Ishikawa et al., 1991 [13]	39	M	0.7	8.3	+	–
26	Lien et al., 1991 [36]	41	M	4	7	+	+
		47.9 ± 9.9	21/21	3.1 ± 2.4	4.6 ± 3.1	16/18	13/23

Penn and Brunson (1988) [38] collected 101 cases and Vogt et al. (1990) [39] collected 2 cases; however, the details are not known.

There is some possibility that renal cell carcinomas that have already developed before transplantation may be included among renal cell carcinomas diagnosed after surgery. In 1 case, renal cell carcinoma was diagnosed 8 years 4 months after transplantation; however, it was probable that the cancer had developed before the operation on the basis of serial CT scans [17]. Furthermore, Penn [41] has detected 12 incidental renal cancers before or at the time of renal transplantation, while Spees et al. [42] have found 3 renal cell carcinomas among 60 potential recipients and Bretan et al. [43] 3 renal cancers among 100 potential recipients. These reports [42, 43] suggest that the incidence of renal cell carcinoma in dialysis patients, who are usually young as potential recipients, is high and careful work-up for the screening of renal cell carcinoma is necessary. Screening should then continue biennially after transplantation [11].

Small numbers of cysts with single-layered epithelia, atypical cysts and also solid adenomas have often been observed in conjunction with renal cell carcinomas [7, 17, 23–26, 28, 31–37]. However, no cyst was observed in imaging studies of 1 patient with renal cell carcinoma [26, 32]. Renal cell carcinoma after renal transplantation may not be intimately related to cyst development, and may be similar to a subgroup of renal cancers described by Ishikawa and Shinoda [44] in hemodialysis patients.

The Association of Acquired Cysts and Renal Cell Carcinoma in Transplant Recipients

There are two hypotheses for the association of acquired cysts and renal cell carcinoma in transplant recipients. One is that the relationship between acquired cysts and renal cell carcinoma is the same as that seen in dialysis patients [7, 25, 28, 31, 34–37]; there is a continuum from cyst to atypical cyst to adenoma to adenocarcinoma. This would seem plausible if graft function were poor; however, it remains unclear whether the same mechanism is operating when the graft is functioning well. Another theory is that, firstly, the proliferative activity of tubular epithelium in transplant recipients is less than that in dialysis patients and, secondly, there is poor association between cyst and adenocarcinoma development [13, 23, 26, 32]. This hypothesis is supported by reports of a renal cell carcinoma forming after transplantation without concomitant cysts [26, 32]. Also, the native kidney in patients with good graft function tends to be atrophic and scarred, with few, if any, cysts. Since renal cell carcinoma after renal transplantation is reported rarely, native kidney cells after transplantation may be less biologically active than before [3, 26].

Mechanisms of Renal Cell Carcinoma Development and Regression of Acquired Cystic Disease after Renal Transplantation

Mechanism of Regression of Acquired Cysts. After successful renal transplantation, the generalized uremic state disappears, the osmotic load to the residual glomeruli of communicating cysts decreases and the cysts shrink [45]. Since the proliferative activity and amount of growth factor in the native kidneys both decrease because of good graft function, the excess of epithelial cells disappears. Residual cysts after transplantation may be of the noncommunicating or atypical types [3].

Mechanisms of New Cyst Development. A minority of transplant recipients (18% of cases) develop new cysts [13]; however, the number of these cysts is small; one such patient also developed a renal cell carcinoma [13]. It is not known whether these newly acquired cysts are simple cysts or an as yet undescribed new type, but they may develop due to a local repair response in the scarred kidneys [46].

Mechanisms of Renal Cell Carcinoma Development in Transplant Recipients. Oncogenesis after transplantation is different from that which occurs during hemodialysis [47]. Uremia-related cysts develop during hemodialysis [1] and end-stage renal disease [4] including long-term chronic rejection after transplantation. In such conditions, proliferative activity of the tubular epithelial cells is high and many cysts develop, with a continuum existing from cyst to atypical cyst to solid adenoma to renal cell carcinoma. There is, therefore, a high incidence of renal cell carcinoma, but, fortunately, with a probably low metastatic rate [11].

On the other hand, in transplant recipients with good graft function [13], factors unrelated to uremia are operating and the proliferative activity of the tubular epithelium may be low, hence the low rate of new cyst development. Many of the cysts which are seen after renal transplantation are residual ones which have not undergone regression. In this situation, the incidence of renal cell carcinoma is low if compared with that in hemodialysis and metastasis is probably due mainly to the effects of immunosuppressive drugs [17]. Further studies are necessary to confirm which of these hypotheses applies.

In summary: (1) After successful renal transplantation almost all acquired cysts regress and the native kidneys become scarred and shrunken. Therefore, acquired renal cystic disease in dialysis patients may be a good indication for renal transplantation. Very long-term follow-up reveals that new cysts may develop, but only to a limited degree and in a small number of graft recipients. (2) If the presence of renal cell carcinoma is completely excluded at the time of renal transplantation, then subsequent development of renal cell carcinoma may be rare. After renal transplantation the proliferative activity of renal tubular epithelium is probably low; however, once renal cell carcinoma develops, the rate of metastasis is high. (3) Therefore, careful screening for renal cell carcinoma before renal transplantation is important and regular screening after transplantation is recommended.

References

1 Ishikawa I, Saito Y, Onouchi Z, Kitada H, Suzuki S, Kurihara S, Yuri T, Shinoda A: Development of acquired cystic disease and adenocarcinoma of the kidney in glomerulonephritic chronic hemodialysis patients. Clin Nephrol 1980;14:1–6.
2 Ishikawa I: Uremic acquired cystic disease of kidney. Urology 1985;26:101–108.
3 Ishikawa I, Yuri T, Kitada H, Shinoda A: Regression of acquired cystic disease of the kidney after successful renal transplantation. Am J Nephrol 1983;3:310–314.
4 Bommer J, Waldherr R, van Kaick G, Strauss L, Ritz E: Acquired renal cysts in uremic patients – in vivo demonstration by computed tomography. Clin Nephrol 1980;14:299–303.
5 Krempien B, Ritz E: Acquired cystic transformation of the kidneys of haemodialysed patients. Virchows Arch [A] 1980;386:189–200.
6 Levine E, Grantham JJ, Slusher SL, Greathouse JL, Krohn BP: CT of acquired cystic kidney disease and renal tumors in long-term dialysis patients. AJR 1984;142: 125–131.
7 Vaziri ND, Darwish R, Martin DC, Hostetler J: Acquired renal cystic disease in renal transplant recipients. Nephron 1984;37:203–205.
8 Minar E, Tscholakoff D, Zazgornik J, Schmidt P, Marosi L, Czembirek H: Acquired cystic disease of the kidneys in chronic hemodialyzed and renal transplant patients. Eur Urol 1984;10:245–248.
9 Thompson BJ, Jenkins DAS, Allan PL, Winney RJ, Dick JCB, Wild SR, Anderton JL, Chisholm GD: Acquired cystic disease of the kidney: An indication for renal transplantation? Br Med J 1986;293:1209–1210.
10 Thaysen JH, Thomsen HS, Sass A, Kristensen JK: Volume changes in polycystic kidneys during chronic dialysis and after renal transplantation. Acta Med Scand 1985;217:197–204.
11 Ishikawa I: Uremic acquired renal cystic disease. Natural history and complications. Nephron 1991;58:257–267.
12 Manns RA, Burrows FGO, Adu D, Michael J: Acquired cystic disease of the kidney: Ultrasound as the primary screening procedure. Clin Radiol 1990;41:248–249.

13 Ishikawa I, Shikura N, Shinoda A: Cystic transformation in native kidneys in renal allograft recipients with long-standing good function. Am J Nephrol 1991;11:217–223.

14 Ishikawa I, Shikura N, Kitada H, Yuri T, Shinoda A, Nakazawa T: Severity of acquired renal cysts in native kidneys and renal allograft with long-standing poor function. Am J Kidney Dis 1989;14:18–24.

15 Ishikawa I: Malignancy and cystic disease; in Hatano M (ed): Nephrology. Tokyo, Springer, 1991, pp 1544–1554.

16 Cohen AH: Renal pathology forum. Am J Nephrol 1989;9:44–46.

17 Ishikawa I, Ishii H, Shinoda A, Tateishi K, Ben A, Suzuki K, Tsugawa R: Renal cell carcinoma of the native kidney after renal transplantation – A case report and review of the literature. Nephron 1991;58:354–358.

18 Ishikawa I, Saito Y, Shikura N, Kitada H, Shinoda A, Suzuki S: Ten-year prospective study on the development of renal cell carcinoma in dialysis patients. Am J Kidney Dis 1990;16:452–458.

19 Bommer J, Ritz E, Feinstein EI, Massry SG: Acquired renal cysts; in Massry SG, Glassock RJ (eds): Textbook of Nephrology. Baltimore, Williams & Wilkins, 1989, pp 792–795.

20 Isiadinso OA, Stubenbord WT, Rubin AL: Renal cell carcinoma postrenal transplant. Discovered accidentally. NY State J Med 1976;76:2024–2025.

21 Struve C, Zabel P, Roggensack H-O: Hypernephroides Nierenkarzinom einer glomerulonephritischen Schrumpfniere nach Nierentransplantation. ROFO (Stuttgart) 1984;141:587–588.

22 Griffin SC, Williams G: Metastatic hypernephroma from a native kidney after cadaver renal transplantation. Postgrad Med J 1985;61:471–472.

23 Arias M, de Francisco ALM, Ruiz L, Val F, Gonzalez M, Zubimendi A: Acquired renal cystic disease and renal adenocarcinoma in a long term renal transplant recipient. Int J Artif Organs 1986;9:271–272.

24 Faber M, Kupin W: Renal cell carcinoma and acquired cystic kidney disease after renal transplantation. Lancet 1987;i:1030–1031.

25 Ludmerer KM, Kissane JM: Clinicopathologic conference. A new chest mass in a 49-year-old man with a transplanted kidney. Am J Med 1988;84:121–128.

26 O'Donnell C, Cutner A, Williams G: Metastatic native renal carcinoma in renal transplant recipients. Nephrol Dial Transplant 1988;3:690–693.

27 Dlugosz BA, Bretan PN Jr, Novick AC, Steinmuller DR, Streem SB, Badwar K, Woodard RM: Causes of death in kidney transplant recipients: 1970 to present. Transplant Proc 1989;21:2168–2170.

28 Gunji Y, Ochiai T, Asano T, Suzuki T, Nagata M, Enomoto K, Uematsu T, Nakajima K, Goto T, Isono K, Matsuzaki O, Tsuchida H, Hamaguchi K: Atypical cysts and renal cell tumors in native kidneys in renal transplant recipients (in Japanese). Jpn J Transplant 1986;21:159–163.

29 Kunikata S, Nishioka T, Ishii T, Uemura T, Sugiyama T, Kanda H, Nagai N, Matsuura T, Akiyama T, Kurita T: Malignant tumors following renal transplantation. Clinical study of four cases (in Japanese). Jpn J Transplant 1987;22:527–533.

30 Yoshimura N, Oka T, Ohmori Y, Aikawa I, Ohmori K, Matsumura T: A kidney transplant recipient with renal cell carcinoma derived from a native kidney. Jpn J Surg 1988;18:208–212.

31 Oda H, Yanagisawa T, Kuzuhara K, Ohtsubo O, Inou T: A case with renal cell carcinoma of the native kidney found 5 years after renal transplantation (Japanese abstract). Jpn J Transplant 1989;25:649.
32 Basile JJ, McCullough DL, Harrison LH, Dyer RB: End stage renal disease associated with acquired cystic disease and neoplasia. J Urol 1988;140:938–943.
33 Almirall J, Mallofre C, Campistol JM, Montoliu J, Ribalta T, Revert L: Metastatic renal cell carcinoma in a hemodialysis patient with acquired renal cystic disease. Nephron 1989;52:96–97.
34 Matson MA, Cohen EP: Acquired cystic kidney disease: Occurrence, prevalence, and renal cancers. Medicine 1990;69:217–226.
35 Garcia de la Oliva T, Molina MG: Metastatic malignant tumor in native kidney with acquired cystic disease after renal transplantation. Eur J Radiol 1990;11:154–155.
36 Lien Y-HH, Kam I, Shanley PF, Schröter GPJ: Metastatic renal cell carcinoma associated with acquired cystic kidney disease 15 years after successful renal transplantation. Am J Kidney Dis 1991;18:711–715.
37 Peabody J, Cerny JC: Renal cell carcinoma in acquired renal cystic disease. J Urol 1989;141:466A.
38 Penn I, Brunson ME: Cancers after cyclosporine therapy. Transplant Proc 1988;20: 885–892.
39 Vogt P, Frei U, Repp H, Bunzendahl H, Oldhafer K, Pichlmayr R: Malignant tumors in renal transplant recipients receiving cyclosporin: Survey of 598 first-kidney transplantations. Nephrol Dial Transplant 1990;5:282–288.
40 Birkeland SA: Cancer in transplanted patients – The Scandia-transplant material. Transplant Proc 1983;15:1071–1078.
41 Penn I: Transplantation in patients with primary renal malignancies. Transplantation 1977;24:424–434.
42 Spees EK, Light JA, Smith EJ, Mostofi FK, Oakes DD: Transplantation in patients with a history of renal cell carcinoma: Long-term results and clinical considerations. Surgery 1982;91:282–287.
43 Bretan PN Jr, Busch MP, Hricak H, Williams RD: Chronic renal failure: A significant risk factor in the development of acquired renal cysts and renal cell carcinoma. Case reports and review of the literature. Cancer 1986;57:1871–1879.
44 Ishikawa I, Shinoda A: Renal adenocarcinoma with or without acquired cysts in chronic hemodialysis patients. Clin Nephrol 1983;20:321–322.
45 Ishikawa I: Acquired cystic disease: Mechanisms and manifestations. Semin Nephrol 1991;11:671–684.
46 Gardner KD Jr: Cystic kidneys. Kidney Int 1988;33:610–621.
47 Kjellstrand CM: Are malignancies increased in uremia? Nephron 1979;23:159–161.

Isao Ishikawa, MD, Division of Nephrology, Department of Internal Medicine, Kanazawa Medical University, Uchinada, Kahoku, Ishikawa 920-02 (Japan)

Berlyne GM (ed): The Kidney Today. Selected Topics in Renal Science.
Contrib Nephrol. Basel, Karger, 1992, vol 100, pp 269–285

Calcium-Dependent Neutrophil Activation

Marianne Haag-Weber, Walter H. Hörl

Department of Medicine, Division of Nephrology, University of Saarland, Homburg/Saar, FRG

Bacterial infections are the most common cause for hospitalization and second common cause of death in patients undergoing regular hemodialysis (HD) [1]. There is increased incidence of urinary tract infections, pulmonary infections and septicemia in patients with chronic renal failure [2]. Causes of recurrent infections in uremic patients are: (1) direct immunosuppressive effect of uremic toxins; (2) indirect effects of metabolic disturbances, e.g. protein deficiency, lack of trace elements and vitamins or generally increased catabolism, and (3) entry gate for microbes, e.g. shunt or peritoneal catheter.

It is obvious that host defence of the uremic patients is impaired. One of the main effector cells for cellular host defence are the polymorphonuclear leukocytes (PMNs). There is multiple evidence for granulocyte abnormalities in uremia. Hammerschmidt et al. [3] described functional inhibition of migration, chemotaxis, degranulation and phagocytosis. Briggs et al. [4] found metabolic disturbances, e.g. decreased oxygen consumption, decreased glucose uptake, decreased lactate production and inhibited activity of hexose monophosphate shunt. A normal metabolic situation enables granulocytes to function adequately. Intra- and extracellular calcium concentrations play an important role for neutrophil function and metabolism. This article summarizes our present knowledge of calcium-dependent neutrophil functions with focus on viral infections, hypertension and uremia.

Calcium Regulation of Neutrophil Function

Stimulated PMNs undergo conformational changes, chemotactic activation, and major changes in metabolic and secretory activity leading to phagocytosis and killing of ingested microorganisms essential for the host against bacterial infection. Interaction of neutrophils with receptor-specific or other surface-binding agents evokes an array of responses including chemotaxis, endocytosis, phagocytosis, exocytosis, and the generation of reactive oxygen species and arachidonate metabolites [5–8].

Binding of receptor-specific or plasma-membrane-perturbing agents of human neutrophils is not necessarily accompanied by the formation of reactive oxygen metabolites [9–11]. Calcium-mobilizing agents (ionophores, lectins, chemotactic peptides) induce a rise of intracellular free calcium and initiate a neutrophil oxygen burst [12]. On the other hand, binding of murine monoclonal antibodies [13, 14] induced rapid transient calcium mobilization in neutrophils but not activation of the oxidative burst. A release of azurophil granule contents (elastase, myeloperoxidase, β-glucuronidase) was also not observed under these in vitro conditions [14].

Preincubation of neutrophils with lipopolysaccharide (LPS) resulted in the enhanced release of O_2^-. Using formyl-methionyl-leucyl-phenylalanine (FMLP) as stimulus, LPS-primed neutrophils released 10.6 ± 4 more O_2^- than did similarly prepared control cells. When preincubated with LPS, phorbol myristate acetate (PMA)-stimulated cells had a greater lag time than control cells, and both FMLP- and PMA-stimulated cells had a greater route of O_2^- release [15]. LPS pretreatment significantly elevated basal $[Ca^{2+}]_i$ in unstimulated cells. The $[Ca^{2+}]_i$ response of LPS-primed cells to FMLP was significantly greater than that of control cells. The difference in $[Ca^{2+}]_i$ between control and LPS-primed cells remained significant even in the absence of extracellular Ca^{2+}. However, the enhanced O_2^- release in response to stimulation with FMLP or PMA noted in LPS-primed neutrophils was no longer evident when LPS-primed cells were first preloaded with bis(2-amino-5-methylphenoxyl)-ethane-N,N,N′,N′tetraacetic acid tetraacetoxymethyl ester (MAP-TAM) an intracellular Ca^{2+} chelator [15].

Studies by Della Bianca et al. [16] demonstrated that transmembrane signaling pathways and the rise in $[Ca^{2+}]_i$ are not essential processes for triggering the ingestion of yeast particles opsonized with IgG and C3b/bi and the activation of NADPH oxidase. In Ca^{2+}-depleted neutrophils the phagocytosis of yeast IgG was similar to that in control neutrophils; the respiratory burst was slightly depressed, while the increase in $[Ca^{2+}]_i$ and production of inositol phosphates, diacylglycerol (DAG), and phosphatidic and arachidonic acid did not occur [16]. In normal neutrophils the phagocytosis is associated with a respiratory burst, a stimulation in the formation of inositol phosphates and phosphatidic acid, the release of arachidonic acid and a rise in $[Ca^{2+}]_i$. Ca^{2+}-depleted neutrophils are able to perform the phagocytosis of yeast cells mediated by concanavalin A and to activate the respiratory burst without stimulation of inositol phosphates and phosphatidic acid formation, arachidonic acid release and rise in $[Ca^{2+}]_i$ [17]. Naccache et al. [18] could demonstrate that the decrease in cytosolic pH induced by chemoattractants was inhibited by a calcium chelator. A specific inhibitor of the NADPH oxidase had no significant effect on the cytosolic acidification by chemotactic factors. It was concluded that the initiation of the cytosolic acidification induced by chemotactic factors is a calcium-dependent event that is not directly linked to the activation of the NADPH oxidase [18].

Superoxide generation was studied in PMA- and 1-O-octadecenyl, 2-O-acetyl-glycerol (OAG)-activated leukocytes, varying intracellular calcium with the calcium ionophores ionomycin and A23187. PMA and OAG were found to induce equal activation of purified protein kinase C and translocation of protein kinase C to the membrane fraction, but differed significantly in their ability to induce superoxide generation. Using calcium ionophores and increasing $[Ca^{2+}]_i$ to more than 1 μM was found to induce increased superoxide generation in maximally OAG-stimulated but not in maximally PMA-stimulated cells [19].

In the study of Wysong et al. [20], intracellular chelation, which ablated early $[Ca^{2+}]_i$ rises, eliminated the FMLP-induced respiratory burst and profoundly reduced the response to opsonized zymosan (by 96.7%). Simultaneous $[Ca^{2+}]_i$ and $[Ca^{2+}]_e$ chelation decreased superoxide responses to opsonized zymosan and hyphae (99.4 and 90.4%, respectively) but not to unopsonized hyphae (26.7% inhibition). Marked changes in $[Ca^{2+}]_i$ are not required for FMLP-induced PMN chemotaxis [21]. The ingestion step of phagocytosis is also a Ca^{2+}-independent event [22]. A rise in $[Ca^{2+}]_i$ is not required for phagocyte motility [for review, see ref. 23].

A rise of the cytosolic free Ca^{2+} concentrations is an important intracellular messenger in neutrophils [24, 25]. It is thought that the rise consists of two components: a release of Ca^{2+} from intracellular stores and an influx of Ca^{2+} across the plasma membrane [12, 26]. To maintain low levels of $[Ca^{2+}]_i$ in resting cells (~100 n*M*) against the $[Ca^{2+}]_e$ of 2 m*M*, the intracellular compartments actively take up calcium ions, whereas the plasma membrane Ca^{2+}/Mg^{2+}-ATPase pumps out calcium ions. During cellular activation, $[Ca^{2+}]_i$ will rise to micromolar levels, due to release of calcium ions from intracellular Ca^{2+} stores and increased Ca^{2+} permeability of the plasma membrane. In homogenates, two different Ca^{2+} pools could be observed: a mitochondrial Ca^{2+} pool and a nonmitochondrial, ATP-dependent inositol-1,4,5-trisphosphate (IP_3)-responsive Ca^{2+} pool. It was shown that microsomal fractions of human neutrophils contain organelles that lower the ambient free Ca^{2+} concentration and respond to IP_3, that granules are not involved in intracellular Ca^{2+} regulation in neutrophils, but that non-endoplasmic reticulum organelles, such as endosomes or Golgi elements, play a major role [27].

The following pathways of receptor-operated Ca^{2+} influx in neutrophils are under consideration. Ca^{2+} might enter neutrophils through Ca^{2+}-selective channels, nonselective channels or a transport pathway coupled to the chemoattractant receptor directly by a guanine-nucleotide-regulatory protein (pathway 1). Ca^{2+} influx might also be coupled to the receptor via phospholipase C activation, IP_3 and eventually inositol-1,3,4,5-tetrakisphosphate production (pathway 2). Alternatively, it has been proposed that the emptying of intracellular Ca^{2+} stores by IP_3 is the signal for Ca^{2+} influx (pathway 3) [for review, see ref. 28]. Von Tscharner et al. [29] observed a time lag of 1–3 s between stimulation of neutrophils and rise in $[Ca^{2+}]_i$. This lag depended on the agonist concentration but was independent of extracellular Ca^{2+}. It was concluded that Ca^{2+} release from intracellular stores is rate limiting for the rise in $[Ca^{2+}]_i$. After this, cation channels in the plasma membrane are opened [29]. Temporal association of calcium mobilization, IP_3 generation and superoxide release was demonstrated by human neutrophils activated by serum opsonized and nonopsonized particulate stimuli [30].

Leukotriene B_4 (LTB_4), platelet-activating factor (PAF) and 5-hydroxyeicosatetraenoate stimulate rise in PMN cytosolic $[Ca^{2+}]_i$ [29, 31–34].

Data on calcium-depleted and calcium-pretreated PMNs indicate that each agonist acted principally by release of storage pool calcium through mechanisms inhibited by pertussis toxin, cholera toxin and PMA [35]. Studies by Koenderman et al. [36] show that a transient rise in $[Ca^{2+}]_i$ does not trigger the respiratory burst in granulocytes, but is sufficient to prime these cells for subsequent activation by FMLP. The ionomycin-induced priming was completely abolished by buffering $[Ca^{2+}]_i$ during exposure to the ionophore. The PAF-induced priming, however, was only partially inhibited when changes in $[Ca^{2+}]_i$ are prevented during priming [36]. Sullivan et al. [37] found that the induction of a rise in free cytosolic Ca^{2+}, which can augment but not initiate the oxidative burst, becomes capable of stimulating the generation of substantial quantities of O_2^- in granulocytes primed with the recombinant human hemopoietic growth factor granulocyte/macrophage-colony-simulating factor.

LTB_4 is synthesized by neutrophils exposed to a variety of stimuli and is also a potent chemotactic agent for these same cells. Furthermore, LTB_4 can stimulate aggregation, exocytosis and, to a lower extent, O_2^- production in neutrophils [38–41]. Lew et al. [42, 43] and others [30, 44–46] have demonstrated that an increase in $[Ca^{2+}]_i$ is an immediate response of neutrophils to LTB_4. The concentrations of LTB_4 necessary to induce a 50% maximal rise in $[Ca^{2+}]_i$ and 5.2 n*M* secondary (vitamin-B_{12}-binding protein) and primary (β-glucuronidase) content release were 0.02 and 0.1 n*M*, respectively [32].

When FMLP binds to its receptor the initial events are the hydrolysis of phosphatidylinositol bisphosphate leading to the formation of IP_3 and DAG. IP_3 then causes the release of Ca^{2+} from intracelluar stores, while DAG activates protein kinase C [47–52]. Activation of Ca^{2+} influx by FMLP depends on the elevation of $[Ca^{2+}]_i$. Of the different inositol phosphates determined in parallel and under the same conditions, accumulation of inositol-1,3,4,5-tetraphosphate correlated best with Ca^{2+} influx. IP_3 was also correlated, but to a lesser extent [53]. The distribution of protein kinase C between cytosol and membrane is a dynamic equilibrium controlled by levels of free Ca^{2+} [54]. When neutrophils were stimulated with LTB_4 O_2^- production and elevation of intracellular Ca^{2+} were initiated more rapidly than after stimulation with FMLP [55]. PMA binds directly to protein kinase C and activates O_2^- formation in neutrophils also at low $[Ca^{2+}]_i$ [56].

In a Ca^{2+}-free medium and resting $[Ca^{2+}]_i$ of 30 n*M*, LTB_4 did not induce any increase of $[Ca^{2+}]_i$ or granule content release, whereas PMA still potently stimulates the release of secondary granule content. Under these conditions, a small amount of secondary granule content release, induced by FMLP, could be detected [43]. Under physiological conditions the contribution of Ca^{2+} influx to the peak of the $[Ca^{2+}]_i$ transient is small whether LTB_4 or FMLP are added in media with or without Ca^{2+}. Maximal stimulatory doses of LTB_4 produce about 50% the amount of IP_3 generated by equimolar concentrations of FMLP, resulting in a different degree of activation of phospholipase C [43].

LTB_4 affects neutrophil function in vitro, such as aggregation, degranulation and chemokinesis [57–60]. Interaction of LTB_4 with its specific receptor is followed by a rise in cytosolic free calcium with [43, 47] and without [44] the breakdown of polyphosphoinositide hydrolysis. LTB_5 elic-

Table 1. Effect of different stimuli on intracellular and extracellular calcium and oxidative burst of PMNs

Stimulus	Extra-cellular Ca	Resting Ca_i by stimulus	Stimu-lated Ca_i	Oxidative burst	Refer-ence
FMLP	n/↑	n	↑	+++	70
	n	↓	↑	+	70
	↓	n	↑ (↑)	–	29, 70
	n	↑	↑↑	++++	14, 15
PMA	n/↓/↑	n	–	++	19, 70
	n/↓/↑	↓	–	–	19, 70
	n/↑	↑	–	+++	15
OAG	n	n	–	+	19
	↑	n	–	++	19
	n	↑	–	++	19
Substance P	n	n	↑	++	71
Lectin	n	n	↑	+++	12
Ionophore	n	n	↑	+++	12
Hexachlorocyclohexane	n	n	↑	+++	72
Opsonized *Candida*	n	n	↑	++	30
	↓	↓		+ (–)	20, 30
Opsonized zymosan	n	n	↑	++	20, 30
	n	↓	–	(+)	20
	↓	↓	–	–	20
Unopsonized *Candida*	n	n	↑	++	20, 30
	n	↓	–	++	20
	↓	↓	–	++	20
LTB_4	n	n	↑	++++	35, 55

n = Normal; ↑ = increased; ↓ = decreased; + = stimulation; – = no effect.

its weaker calcium movements than LTB_4 in fura-2-loaded human neutrophils. This might relate to the weaker biological acitivy of LTB_5 [61].

Neutrophil stimulation by chemotactic factors is accompanied by activation of the phospholipases C and A_2. The pool of calcium mobilized by arachidonic acid, the principal fatty acid liberated by phospholipase A_2, includes sensitivity to IP_3 [62]. These results also demonstrate a significant intracellular role for arachidonic acid at the level of the internal mobilization of calcium in human neutrophils.

Human neutrophils respond also to ATP. Externally applied ATP leads to increases in cytosol Ca^{2+}, inositol phosphates and exocytotic secretion of neutrophils. The responses to ATP are blocked by pertussis toxin pretreatment, indicating the involvement of a guanine-nucleotide-regulatory protein [63]. Ward et al. [64–66] and Kuhns et al [67] have shown that exogenous ATP causes enhanced O_2^- production in stimulated neutrophils. Extracellular ATP at 10 μM increased the concentrations of cytoplasmic free Ca^{2+} 3-fold in human neutrophils. The rise in $[Ca^{2+}]_i$ elicited by ATP was unaffected by the removal of extracellular Ca^{2+}, indicating that the primary source of Ca^{2+} is from intracellular stores. PMA completely inhibited the ATP-induced increases in $[Ca^{2+}]_i$ without affecting the basal $[Ca^{2+}]_i$ level [68]. The expression of putative P_2-purinergic receptors for ATP appears to precede the expression of other receptor types known to activate the inositol phospholipid signaling cascades in terminally differentiated phagocytes [69].

Table 1 demonstrates the relationship between intra- and extracellular calcium and oxidative burst considering different stimuli.

Neutrophils and Viral Infections

Influenza A virus causes inhibitory effects on human neutrophil cell function in vitro and in vivo. Chemotaxis, adherence, generation of toxic oxidative species, intracellular killing of bacteria and secretion of degradative granule enzymes are each impaired by viral infection. Bacterial superinfection in influenza-A-virus-related illness may in part be explained by virus-induced neutrohil dysfunction [73]. A rise in intracellular Ca^{2+} and efflux of Ca^{2+} from the cell occurred in direct response to influenza A virus exposure. When neutrophils were subsequently treated with the chemotactic peptide FMLP, O_2^- production and arachidonic acid release were depressed. There was also a depressed intracellular Ca^{2+} rise and lowered efflux of Ca^{2+} in response to FMLP. It was hypothesized that partial activation of neutrophils by influenza A virus leads to impaired availability of intracellular Ca^{2+} stores when subsequent stimuli are applied [73]. The virus also elicits neutrophil membrane depolarization, which is independently mediated from the Ca^{2+} increase and respiratory burst [74].

Neutrophils and Hypertension

Cytosolic free calcium and intracellular calcium stores were investigated in neutrophils from normo- and hypertensive subjects. There was no difference between patients with untreated essential hypertension compared with age- and sex-matched normotensive subjects. No significant relationship was found between blood pressure and the calcium parameters in all 20 subjects studied [75]. On the other hand, a fraction of hypertensive plasma obtained by gel filtration with a molecular range of 1,000–1,500 D causes an acceleration of Ca^{2+} uptake into organelles of permeabilized human neutrophils [76, 77].

The effect of verapamil and diltiazem on the oxidative burst of PMNs of healthy subjects is shown in table 2. Both calcium channel blockers inhibit the FMLP-stimulated oxidative burst in vitro.

As opposed to excitable tissues, Ca^{2+} influx in neutrophils is not triggered by plasma membrane depolarization and is not inhibited by blockers of voltage-dependent Ca^{2+} channels, such as nifedipine [78]. It was, there-

Table 2. In vitro effects of diltiazem and verapamil on O_2^- production by PMNs obtained from healthy subjects

	PMN O_2^- production, nmol/min/10^6 PMNs		
	without FMLP	with FMLP	difference
PMNs in PBS	1.40 ± 0.19	2.90 ± 0.30	1.50 ± 0.23
PMNs in diltiazem, 0.5 μg/ml	1.52 ± 0.26	2.40 ± 0.43	0.90 ± 0.32*
PMNs in verapamil, 0.25 μg/ml	1.35 ± 0.17	2.04 ± 0.33*	0.60 ± 0.37*

Mean values ± SEM from 12 experiments are given. PBS = Phosphate-buffered saline.
* $p < 0.05$: PBS versus calcium channel blockers.

fore, concluded that neutrophils do not possess voltage-dependent Ca^{2+} channels [28]. However, inhibition of calcium transport channels with verapamil produced a dose-related inhibition of FMLP- and PMA-stimulated O_2^- production. Verapamil inhibition of FMLP-stimulated but not PMA-stimulated O_2^- release was reversed simply by increasing $[Ca^{2+}]_e$. Verapamil had an inhibitory effect on fura-2-monitored cytosolic calcium rise induced by FMLP stimulation [70]. Verapamil-treated human neutrophils displayed 40–50% reduction in directed migration at different pharmacological concentrations (10^{-8}–10^{-6} mol/l). Activated random migration (chemokinesis) was also impaired by verapamil treatment, but random locomotion was not affected except at a high concentration (10^{-6} mol/l). Calcium antagonist binding site assays using radiolabeled dihydropyridines provided no evidence for the presence of calcium channels in neutrophil membranes. There was no $^{45}Ca^{2+}$ uptake in neutrophils by verapamil exposure. Finally, verapamil did not affect resting $[Ca^{2+}]_i$ or $[Ca^{2+}]_i$ transients after FMLP stimulation. It was suspected that the pharmacological effect of verapamil is mediated by mechanisms other than calcium antagonism [79]. Kokot et al. [80] have shown that the calcium ionophore A23187 induced release of elastase from PMNs in a dose-dependent manner and that this effect was abolished by EDTA. Verapamil inhibited this elastase release. Greenberg et al. [81] reported that verapamil exerted an inhibitory effect on NaF-stimulated neutrophils without reducing NaF-induced rise in cytosolic calcium. Massry et al. [82] studied the effects of various concentrations of verapamil on the spontaneous and 1–84 parathyroid-hormone-induced elastase release from PMNs. Verapamil produced significant decrements in the elastase levels after incubation with both vehicle and 1–84 parathyroid hormone.

Neutrophils and Uremia

The effect of oral verapamil treatment on intracellular calcium and glucose uptake in PMNs of HD patients is shown in table 3. HD patients displayed higher PMN $[Ca^{2+}]_i$ as well as lower PMN glucose uptake before and 2 months after verapamil therapy compared to healthy subjects. Both PMN parameters were normalized under oral verapamil treatment.

The effect of the calcium channel blockers nifedipine (9 and 16 μg kg^{-1} h^{-1}), diltiazem (100 and 200 μg kg^{-1} h^{-1}) and verapamil (19 μg kg^{-1} h^{-1}) continuously infused during HD on granulocyte and complement acti-

Table 3. Effect of oral verapamil treatment (120 mg daily for 2 months) on intracellular calcium (fura-2) and glucose uptake in PMNs of HD patients

	Intracellular calcium nmol/10^6 PMNs	Glucose uptake cpm/2 × 10^5 PMNs
Controls (n = 25)	58.6 ± 3.6	13,573 ± 1,720
HD patients (n = 15)		
Before verapamil therapy	75.8 ± 5.8*	6,876 ± 1,524*
At the end of verapamil therapy	49.8 ± 3.8	10,404 ± 1,044
Two months after verapamil therapy	66.7 ± 5.5*	5,146 ± 534*

* $p < 0.05$: HD patients versus controls.

vation was investigated [83]. Plasma levels of lactoferrin, elastase in complex with α_1-proteinase inhibitor (E-α_1PI) and C3a were measured in patients dialyzed with dialyzers made of cuprophane, polymethylmethacrylate (PMMA) and polyacrylonitrile (PAN). Calcium channel blockers caused no change in blood pressure during HD in all patients. There was no effect of nifedipine, diltiazem or verapamil on plasma lactoferrin, E-α_1PI or C3a levels in patients dialyzed with cuprophane. However, plasma lactoferrin and E-α_1PI values were significantly reduced by all calcium channel blockers in patients dialyzed with PMMA, and also by nifedipine and verapamil in patients dialyzed with PAN. Our data indicate that calcium channel blockers inhibit granulocyte activation occurring in dialyzers with very little anaphylatoxin formation. These drugs, however, are ineffective in patients dialyzed with cuprophane where complement activation takes place. Therefore, granulocyte activation during HD in the absence of complement activation seems to be mediated by calcium ions [83].

Plasma levels of granulocyte lactoferrin, myeloperoxidase and E-α_1PI were compared with C3a and C5a levels in 10 patients undergoing cardiopulmonary bypass [84]. At the end of the operation, plasma levels of lactoferrin increased from 97.0 ± 22.8 to 1,257 ± 139.8 ng/ml, myeloperoxidase increased from 37.1 ± 4.3 to 170.9 ± 34.9 ng/ml and E-α_1 PI rose from 89.4 ± 7.4 to 437.8 ± 97.3 ng/ml. There was also a significant anaphylatoxin formation. To investigate the relationship between complement and granulocyte activation, patients undergoing cardiopulmonary bypass received the calcium channel blocker nifedipine (orally and intravenously) and the antiplatelet drug dipyridamole. The continuous infusion of nifedipine (5.91 to 0.53 μg/kg body weight per hour) caused significantly lower levels of E-α_1PI and lactoferrin but not anaphylatoxin. Dipyridamole was without effect on complement and granulocyte activation during cardiopulmonary bypass. Our data demonstrate inhibition of granulocyte activation during cardiopulmonary bypass by continuous infusion of nifedipine, even in the presence of complement activation [84].

An indicator of PMN activation may also be the glycogen-degrading enzyme phosphorylase. Therefore, PMNs were isolated from healthy control subjects, patients undergoing continuous ambulatory peritoneal dialysis (CAPD), and patients undergoing regular HD therapy before, during and at the end of HD treatment using dialyzers made of polysulfone or PMMA. Nifedipine was continuously infused during HD with PMMA in 5 patients at a dose of 18 μg/kg body weight per hour. Glycogen, activity of glycogen synthetase and phosphorylase (active and inactive forms of both enzymes), and glucose uptake with and without stimulation with the chemotactic peptide FMLP were determined in these PMNs. During HD with PMMA, there was a significant increase in PMN phosphorylase *a* activity 15 and 30 min after the start of HD. HD with polysulfone did not stimulate the active *a* form of the glycogen-degrading enzyme in PMNs. HD with PMMA significantly inhibited the active I form of glyocgen synthetase, whereas polysulfone activated glycogen synthetase I. NIF inhibited phosphorylase *a* activation during HD with PMMA. PMN glycogen content and glucose uptake were improved during HD with polysulfone, but not with PMMA. PMN glycogen content, activity of glycogen synthetase and glucose uptake were significantly lower also in CAPD patients compared with healthy controls. These data show that HD with PMMA activates PMN glycogenolysis. This effect can be inhibited by calcium channel blockers. PMN glycogen content of patients on regular HD therapy and CAPD is significantly lower compared with healthy controls due to inhibition of glycogen synthesis. Elimination of dialyzable factor(s) improves, but does not restore, PMN glycogen synthesis and glucose uptake [85, 86].

Conclusion

$[Ca^{2+}]_i$ plays an important role in neutrophil function and metabolism. Different stimuli bind to neutrophil receptors, increase DAG and IP_3. This leads to an increase in intracellular calcium both from intracellular compartments and from extracellular medium. This rise of intracellular calcium correlates with neutrophil function.

Influenza A virus deactivates neutrophils via an increased resting intracellular calcium. It is suggested that high resting intracellular calcium has a negative influence on neutrophil function with focus on bacterial killing.

Uremia is also a state where increased resting intracellular calcium in neutrophils and abnormalities of neutrophil function were found.

Up to now the existence of voltage-sensitive calcium channels could not be certainly proved. On the other hand, calcium channel blockers could improve neutrophil function and calcium content especially in uremia and during extracorporeal circulation. These data suggest that there are indirect evidences for calcium channels or the effect is mediated by mechanisms other than calcium antagonism.

References

1 Nsouli KA, Lazarus M, Schoenbaum SC, Gottlieb MN, Lowrie EG, Shocair M: Bacteremic infection in hemodialysis. Arch Intern Med 1979;139:1255–1258.

2 Montgomerie JZ, Kulmanson GM, Guze LB: Renal failure and infection. Medicine 1968;47:1–32.

3 Hammerschmidt DE, Goldberg R, Raij L, Kay NE: Leukocyte abnormalities in renal failure and hemodialysis. Semin Nephrol 1985;5:91–103.

4 Briggs WA, Sillix DH, Mahajan S, McDonald FD: Leukocyte metabolism and function in uremia. Kidney Int 1983;24(suppl 16):93–96.

5 Sklar LA; Jesaitis AJ, Painter RG: The neutrophil N-formyl peptide receptor: The dynamics of ligand/receptor interactions and their relationship to cellular responses. Contemp Top Immunol 1984;14:29–82.

6 Sklar LA; Finney DA, Oades ZG, Jesaitis AJ, Painter RG, Cochrane CG: The dynamics of ligand-receptor interaction: Real-time analysis of association, dissociation and internalization of a N-formyl peptide and its receptors on human neutrophils. J Biol Chem 1984;259:5661–5669.

7 Painter RG, Sklar LA, Jesaitis AJ, Schmitt M, Cochrane CG: Activation of neutrophils by N-formyl chemotactic peptides. Fed Proc 1984;43:2737–2742.

8 Snyderman R, Goetzl EJ: Molecular and cellular mechanisms of leukocyte chemotaxis. Science 1981;213:830–837.

9 McPhail LC, Clayton C, Snyderman R: The NADPH oxidase of human polymorphonuclear leukocytes: Evidence for regulation by multiple signals. J Biol Chem 1984;259:5768–5775.

10 Fletcher MP, Seligman BE, Gallin JI: Correlation of human neutrophil secretion, chemoattractant receptor mobilization, and enhanced functional capacity. J Immunol 1982;128:941–948.

11 Kitagawa S, Takaku F, Sakamoto S: A comparison of the superoxide-releasing response in human polymorphonuclear leukocytes and monocytes. J Immunol 1980; 125:359–364.

12 Korchak HM, Vienne K, Rutherford LE; Wilkenfeld C, Finkelstein MC, Weissmann G: Stimulus-response coupling in the neutrophil. II. Temporal analysis of changes in cytosolic calcium and calcium efflux. J Biol Chem 1984;259:4076–4082.

13 Apfeldorf WJ, Melnick DA, Mehulam T, Rasmussen H, Malech HL: A transient rise in intracellular free calcium is not a sufficient stimulus for respiratory burst activation in human polymorphonuclear leukocytes. Biochem Biophys Res Commun 1985;132:674–680.

14 Schmitt M, Painter RG, Von Tscharner V, Sklar LA, Jesaitis AJ, Fayle DH, Cochrane CG; Monoclonal antibody NMS-1 increases N-formyl chemotactic peptide-mediated oxidative burst generation in human neutrophils. J Immunol 1987;139: 4178–4185.

15 Forehand JR, Pabst MJ, Philips WA, Johnston RB Jr: Lipopolysaccharide priming of human neutrophils for an enhanced respiratory burst: Role of intracellular free calcium. J Clin Invest 1989;83:74–83.

16 Della Bianca V, Grzeskowiak M, Rossi F: Studies on molecular regulation of phagocytosis and activation of the NADPH oxidase in neutrophils. IgG-and C3b-mediated ingestion and associated respiratory burst independent of phospholipid turnover and Ca^{2+} transients. J Immunol 1990;144:1411–1417.

17 Rossi F, Della Bianca V, Grzeskowiak M, Bazzoni F: Studies on molecular regulation of phagocytosis in neutrophils: Con A-mediated ingestion and associated respiratory burst independent of phosphoinositide turnover, rise in $[Ca^{2+}]_i$ and arachidonic acid release. J Immunol 1989;142:1652–1660.

18 Naccache PH, Therrien S, Caon AC, Liao N, Gilbert C, McColl SR: Chemoattractant-induced cytoplasmic pH changes and cytoskeletal reorganization in human neutrophils: Relationship to the stimulated calcium transients and oxidative burst. J Immunol 1989;142:2438–2444.

19 Christiansen NO, Larsen CS, Esmann V: A study on the role of protein kinase C and intracellular calcium in the activation of superoxide generation. Biochim Biophys Acta 1988;971:317–324.

20 Wysong DR, Lyman CA, Diamond RD: Independence of neutrophil respiratory burst oxidant generation from the early cytosolic calcium response after stimulation with unopsonized *Candida albicans* hyphae. Infect Immun 1989;57:1499–1505.

21 Perez HD, Elfman F, Marder S, Lobo E, Ives HE: Formyl peptide-induced chemotaxis of human polymorphonuclear leukocytes does not require either marked changes in cytosolic calcium or specific granule discharge: Role of formyl peptide receptor reexpression (or recycling). J Clin Invest 1989;83:1963–1970.

22 Jaconi ME, Lew DP, Carpentier JL, Magnusson KE, Sjögren M, Stendahl O: Cytosolic free calcium elevation mediates the phagosome-lysosome fusion during phagocytosis in human neutrophils. J Cell Biol 1990;110:1555–1564.

23 Di Virgilio F, Pizzo P, Picello E: Mechanisms of neutrophil and macrophage motility: in Hörl WH, Schollmeyer PJ (eds): New Aspects of Human Polymorphonuclear Leukocytes. New York, Plenum Press 1991, pp 13–22.

24 Pozzan T, Lew PD, Wollheim CB, Tsien R: Is cytosolic free Ca^{2+} concentration regulating neutrophil activation? Science 1983;221:1413–1415.

25 Gennaro R, Pozzan T, Romeo D: Monitoring of cytosolic free Ca^{2+} in C5a-stimulated neutrophils: Loss of receptor-modulated Ca^{2+} stores and Ca^{2+} uptake in granule-free cytoplasts. Proc Natl Acad Sci USA 1984;81:1416–1420.
26 Korchak HM, Rutherford LE, Weissmann G: Stimulus response coupling in the human neutrophil. I. Kinetic analysis of changes in calcium permeability. J Biol Chem 1984;259:4070–4075.
27 Krause KH, Lew PD: Subcellular distribution of Ca^{2+} pumping sites in human neutrophils. J Clin Invest 1987;80:107–116.
28 Krause KH, Lew DP, Welsh MJ: Electrophysiological properties of human neutrophils; in Hörl WH, Schollmeyer PJ (eds): New Aspects of Polymorphonuclear Leukocytes. New York, Plenum Press, 1991, pp 1–11.
29 Von Tscharner V, Prod'hom B, Baggiolini M, Reuter H: Ion channels in human neutrophils activated by a rise in free cytosolic calcium concentration. Nature 1986; 324:369–372.
30 Meshulam T, Diamond RD, Lyman CA, Wysong DR, Melnick DA: Temporal association of calcium mobilization, inositol trisphosphate generation, and superoxide anion release by human neutrophils activated by serum opsonized and nonopsonized particulate stimuli. Biochem Biophys Res Commun 1988;150:532–539.
31 White JR, Naccache PH, Molski TF, Borgeat P, Sha'afi RI: Direct demonstration of increased intracellular concentration of free calcium in rabbit and human neutrophils following stimulation by chemotactic factor. Biochem Biophys Res commun 1983;113:44–50.
32 Lad PM, Olson CV, Grewal IS, Scott SJ: A pertussis toxin-sensitive GTP-binding protein in the human neutrophil regulates multiple receptors, calcium mobilization, and lectin-induced capping. Proc Natl Acad Sci USA 1985;82:8643–8647.
33 Kikuchi A, Kozawa O, Kalibuchi K, Katada T, Ui M, Takai Y: Direct evidence for involvement of a guanine nucleotide-binding protein in chemotactic peptide-stimulated formation of inositol bisphosphate and triphosphate in differentiated human leukemic (HL-60) cells. Reconstitution with G_i or G_o of the plasma membranes ADP-ribosylated by pertussis toxin. J Biol Chem 1986;261:11558–11562.
34 O'Flaherty JT, Nishihira J: 5-Hydroxyeicosatetraenoate promotes Ca^{2+} and protein kinase C mobilization in neutrophils. Biochem Biophys Res Commun 1987;148: 575–581.
35 O'Flaherty JT, Jacobson D, Redman J: Mechanism involved in the mobilization of neutrophil calcium by 5-hydroxyeicosatetraenoate. J Immunol 1988;140:4323–4328.
36 Koenderman L, Yazdanbakhsh M, Roos D, Verhoeven AJ: Dual mechanisms in priming of the chemoattractant-induced respiratory burst in human granulocytes. J Immunol 1989;142:623–628.
37 Sullivan R, Fredette JP, Griffin JD, Leavitt JL, Simons ER, Melnick DA: An elevation in the concentration of free cytosolic calcium is sufficient to activate the oxidative burst of granulocytes primed with recombinant human granulocyte-macrophage colony-stimulating factor. J Biol Chem 1989;264:6302–6309.
38 Samuelsson G: Leukotrienes: Mediators of immediate hypersensitivity reactions and inflammation. Science 1983;220:568–575.
39 Goetzl EJ: Leukocyte recognition and metabolism of leukotrienes. Fed Proc 1983; 42:3128–3131.

40 Gimbrone MA, Brock AF, Schafer AI: Leukotriene B_4 stimulates polymorphonuclear leukocyte adhesion to cultured vascular endothelial cells. J Clin Invest 1984;74:1552–1555.
41 Goldman DA, Goetzl EJ: Heterogeneity of human polymorphonuclear leukocyte receptors for leukotriene B_4: Identification of a subset of high affinity receptors that transduce the chemotactic response. J Exp Med 1984;159:1027–1041.
42 Lew PD, Dayer JM, Wollheim CB, Pozzan T: Effect of leukotriene B_4, prostaglandin E_2 and arachidonic acid on cytosolic-free calcium in human neutrophils. FEBS Lett 1984;166:44–48.
43 Lew PD, Monod A, Waldvogel FA, Pozzan T: Role of cytosolic free calcium and phospholipase C in leukotriene-B_4-stimulated secretion in human neutrophils. Comparison with the chemotactic peptide formyl-methionyl-leucyl-phenylalanine. Eur J Biochem 1987;162:161–168.
44 Verghese MW, Charles L, Jakoi L, Dillon SB, Snyderman R: Role of a guanine nucleotide regulatory protein in the activation of phospholipase C by different chemoattractants. J Immunol 1987;138:4374–4380.
45 Volpi M, Yassin R, Tao W, Molski TF, Naccache PH, Sha'afi RI: Leukotriene B_4 mobilizes calcium without the breakdown of polyphosphoinositides and the production of phosphatidic acid in rabbit neutrophils. Proc Natl Acad Sci USA 1984;81:5966–5969.
46 Bradford PG, Rubin RP: Characterization of formylmethionyl-leucyl-phenylalanine stimulation of inositol trisphosphate accumulation in rabbit neutrophils. Mol Pharmacol 1985;27:74–78.
47 Berridge MJ, Irvine RF: Inositol trisphosphate, a novel second messenger in cellular signal transduction. Nature (Lond) 1984;312:315–321.
48 Cockcroft S, Gomperts BD: Role of guanine nucleotide binding protein in the activation of polyphosphoinositide phosphodiesterase. Nature (Lond) 1985;314:534–536.
49 Prentki M, Wollheim CB, Lew PD: Ca^{2+} homeostasis in permeabilized human neutrophils. Characterization of Ca^{2+}-sequestering pools and the action of inositol 1,4,5-trisphosphate. J Biol Chem 1984;259:13777–13782.
50 Dougherty RW, Godfrey PP, Hoyle PC, Putney JW, Freer RJ: Secretagogue-induced phosphoinositide metabolism in human leukocytes. Biochem J 1984;222:307–314.
51 Lew DP: Receptor signalling and intracellular calcium in neutrophil activation. Eur J Clin Invest 1989;19:338–346.
52 Lew PD: Receptors and intracellular signalling in human neutrophils. Am Rev Respir Dis 1990;141:S127–S131.
53 Pittet D, Lew PD, Mayr GW, Monod A, Schlegel W: Chemoattractant receptor promotion of Ca^{2+} influx across the plasma membrane. A role for cytosolic free calcium elevations and inositol 1,3,4,5-tetrakisphosphate production. J Biol Chem 1989;264:7251–7261.
54 Phillips WA, Fujiki T, Rossi MW, Korchak HM, Johnston RB Jr: Influence of calcium on the subcellular distribution of protein kinase C in human neutrophils. Extraction conditions determine partitioning of histone-phosphorylating activity and immunoreactivity between cytosol and particulate fractions. J Biol Chem 1989;264:8361–8365.
55 Gyllenhammar H: Correlation between neutrophil superoxide formation, luminol-augmented chemiluminescence and intracellular Ca^{2+} levels upon stimulation with

leukotriene B_4, formylpeptide and phorbolester. Scand J Clin Lab Invest 1989;49: 317–322.

56 Di Virgilio F, Lew DP, Pozzan T: Protein kinase C activation of physiological processes in human neutrophils at vanishingly small cytosolic Ca^{2+} levels. Nature 1984; 310:691–693.

57 Goldman DW, Pickett WC, Goetzl EJ: Human neutrophil chemotactic and degranulation activities of leukotriene B_5 (LTB_5) derived from eicosapentaenoic acid. Biochem Biophys Res Commun 1983;117:282–288.

58 Prescott SM: The effect of eicosapentaenoic acid on leukotriene B production by human neutrophils. J Biol Chem 1984;259:7615–7621.

59 Lee TH, Mencia-Huerta JM, Shih C, Corey EJ, Lewis RA, Austen KF: Characterization and biologic properties of 5,12-dihydroxy derivates of eicosapentaenoic acid, including leukotriene B_5 and the double lipoxygenase product. J Biol Chem 1984; 259:2383–2389.

60 Terano T, Salmon JA, Moncada S: Biosynthesis and biological activity of leukotriene B_5. Prostaglandins 1984;27:217–232.

61 Seya A, Terano T, Tamura Y, Yoshida S: Comparative effect of leukotriene B_4 and leukotriene B_5 on calcium mobilization in human neutrophils. Prostaglandins Leukot Essent Fatty Acids 1988;34:47–50.

62 Beaumier L, Faucher N, Naccache PH: Arachidonic acid-induced release of calcium in permeabilized human neutrophils. FEBS Lett 1987;221:289–292.

63 Cockcroft S, Stutchfield J: ATP sitmulates secretion in human neutrophils and HL 60 cells via a pertussis toxin-sensitive guanine nucleotide-binding protein coupled to phospholipase C. FEBS Lett 1989;245:25–29.

64 Ward PA, Cunningham TW, McCulloch KK, Phan SH, Powell J, Johnson KJ: Platelet enhancement of O_2^- responses in stimulated human neutrophils: Identification of platelet factor as adenine nucleotide. Lab Invest 1988;58:37–47.

65 Ward PA, Cunningham TW, McCulloch KK, Johnson KJ: Regulatory effects of adenosine and adenine nucleotides on oxygen radical responses of neutrophils. Lab Invest 1988;58:438–447.

66 Ward PA, Cunningham TW, Walker BAM, Johnson KJ: Differing calcium requirements for regulatory effects of ATP, ATP gamma S and adenosine on O_2^- responses of human neutrophils. Biochem Biophys Res Commun 1988;154:746–751.

67 Kuhns DB, Wright DG, Nath J, Kaplan SS, Basford RE: ATP induces transient elevations of $[Ca^{2+}]_i$ in human neutrophils and primes these cells for enhanced O_2^- generation. J Lab Invest 1988;58:448–453.

68 Kuroki M, Takeshige K, Minakami S: ATP-induced calcium mobilization in human neutrophils. Biochim Biophys Acta 1989;1012:103–106.

69 Cowen DS, Lazarus HM, Shurin SB, Stoll SE, Dubyak GR: Extracellular adenosine trisphosphate activates calcium mobilization in human phagocytic leukocytes and neutrophil/monocyte progenitor cells. J Clin Invest 1989;83:1651–1660.

70 Weisdorf DJ, Thayer MS: Occult intracellular calcium pools: Relevance to neutrophil oxidant production. J Lab Clin Med 1989;114:260–265.

71 Serra MC, Bazzoni F, Della Bianca V, Grescowiak M, Rossi F: Activation of human neutrophils by substance P. Effect on oxidative metabolism, exocytosis, cytosolic Ca^{2+} concentration and inositol phosphate formation. J Immunol 1988;141:2118–2124.

72 Kaplan SS, Zdziarski UE, Kuhns DB, Basford RE: Inhibition of cell movement and associated changes by hexachlorocyclohexanes due to unregulated intracellular calcium increases. Blood 1988;71:677–683.

73 Hartshorn KL, Collamer M, Auerbach M, Myers JB, Pavlotsky N, Tauber AI: Effects of influenza A virus on human neutrophil calcium metabolism. J Immunol 1988;141:1295–1301.

74 Hartshorn KL, Collamer M, With MR, Schwartz JH, Tauber AI: Characterization of influenza A virus of the human neutrophil. Blood 1990;75:218–226.

75 Lew PD, Favre L, Waldvogel FA, Vallotton MB: Cytosolic free calcium and intracellular calcium stores in neutrophils from hypertensive subjects. Clin Sci 1985;69: 227–230.

76 Zidek W, Sachinidis A, Spieker C, Storkebaum W: Effect of hypertensive plasma on Ca^{2+} transport in human neutrophils. Atherosclerosis 1988;69:185–190.

77 Zidek W, Sachinidis A, Spieker C, Storkebaum W: Effect of plasma from hypertensive subjects on Ca^{2+} transport in permeabilized human neutrophils. Clin Sci 1988; 74:53–56.

78 Andersson T, Dahlgren C, Pozzan T, Stendahl O, Lew PD: Characterization of fmet-leu-phe receptor-mediated Ca^{2+} influx across the plasma membrane of human neutrophils. Mol Pharmacol 1986;30:437–443.

79 Pennington JE, Kemmerich B, Kazanjian PH, Marsh JD, Boerth LW: Verapamil impairs human neutrophil chemotaxis by a non-calcium-mediated mechanism. J Lab Clin Med 1986;108:44–52.

80 Kokot K, Teschner M, Schaefer RM, Heidland A: Stimulation and inhibition of elastase release from human neutrophils – dependent on the calcium messenger system. Miner Electrolyte Metab 1987;13:133–140.

81 Greenberg J, Feinberg H, Wencel JD, Anderson BR: Stimulation of canine neutrophils: The role of calcium binding. Proc Soc Exp Biol Med 1982;170:333–340.

82 Massry SG, Schaefer RM, Teschner M, Roeder M, Zull JF, Heidland A: Effect of parathyroid hormone on elastase release from human polymorphonuclear leucocytes. Kidney Int 1989;36:883–890.

83 Haag-Weber M, Schollmeyer P, Hörl WH: Granulocyte activation during haemodialysis in the absence of complement activation: Inhibition by calcium channel blockers. Eur J Clin Invest 1988;18:380–385.

84 Riegel W, Spillner G, Schlosser V, Hörl WH: Plasma levels of main granulocyte components during cardiopulmonary bypass. J Thorac Cardiovasc Surg 1988;95: 1014–1019.

85 Haag-Weber M, Hable M, Fiedler G, Blum I, Schollmeyer P, Kreusser W, Hörl WH: Alterations of polymorphonuclear leukocyte glycogen metabolism and glucose uptake in dialysis patients. Am J Kidney Dis 1991;17:562–568.

86 Hörl WH, Haag-Weber M, Georgopoulos A, Block LH: Physicochemical characterization of a polypeptide present in uremic serum that inhibits the biological activity of polymorphonuclear cells. Proc Natl Acad Sci USA 1990;87:6353–6357.

Prof. Dr. Dr. W.H. Hörl, Medizinische Universitäts-Klinik IV,
D-W-6650 Homburg/Saar (FRG)

Subject Index